Total Pressure Measurements in Vacuum Technology

Total Pressure Measurements in Vacuum Technology

A. BERMAN
Vacuum Calibration Laboratory
Soreq Nuclear Research Centre
Yavne, Israel

1985

ACADEMIC PRESS, INC.
(Harcourt Brace Jovanovich, Publishers)
Orlando San Diego New York London
Toronto Montreal Sydney Tokyo

ACADEMIC PRESS, INC.
Orlando, Florida 32887

United Kingdom Edition published by
ACADEMIC PRESS INC. (LONDON) LTD.
24–28 Oval Road, London NW1 7DX

LIBRARY OF CONGRESS CATALOGING IN PUBLICATION DATA

Berman, A. (Armand)
Total pressure measurements in vacuum technology.

Bibliography: p.
Includes index.
1. Pressure—Measurement. 2. Pressure-gauges.
3. Vacuum technology. I. Title.
QC165.B45 1985 533'.5 84-28454
ISBN 0-12-092440-4 (alk. paper)

PRINTED IN THE UNITED STATES OF AMERICA

85 86 87 88 9 8 7 6 5 4 3 2 1

To
Nadya-Ilane and Eric

Contents

Chapter 4. Gauges for Low-Pressure Measurement

Chapter 5. Methods for Vacuum Gauge Calibration

Chapter 6. Calibration of Vacuum Gauges for Different Gases

Chapter 7. Performance Determination of Hot-Cathode Ionization Gauges

Chapter 8. Pressure Measurement in Confined Environments

Preface

> When you can measure what you are speaking about and express it in numbers you know something about it; but when you cannot express it in numbers, your knowledge is of a meagre and unsatisfactory kind.
>
> Lord Kelvin

Vacuum technology is an integral part of a whole range of modern industries, for example, those involved with integrated circuits, semi-conductors, plasma research and metallurgy. In addition, it is of vital importance to research and development in science and engineering. In industry, the quality of the end product is directly related to the quality of the vacuum. This depends on the accuracy of the most adequate method of measuring the rarefied environment and on the correct interpretation of the results yielded by measurements.

Vacuum measurement has, however, been a rather neglected area and only in the last two decades has the question of measurement accuracy and diagnosis assumed proper significance and attention. Interest in the subject is therefore increasing and likely to keep increasing with the development of new technologies requiring very low residual pressures.

The quantitative determination of the rarefied gas environment as well as the assessment of performance typical of vacuum equipment (e.g., flow rate, throughput, conductance of ducts) involve measuring pressure. Pressure measurement is far more complicated than the measurement of any of the fundamental quantities—length, mass and time—in that the parameters characterizing both the rarefied environment and the measuring process are neither unique nor invariable. Indeed, as the degree of rarefaction reaches lower values, both density and chemical composition of the gas keep changing. The measuring process itself interferes with the gas measured, modifying its chemical and physical properties.

Unfortunately, pressure measuring techniques are used before all the necessary factors contributing to the rarefied environment are known, and the interpretation of the results can be misleading. This situation is further complicated by the fact that there is no single well-established standard measuring technique for the quantities involved and that the choice of units is still a problem. On the one hand there is the self-consistent system of SI units which unfortunately is not widespread even in countries that have adopted it; on the other hand there is a natural desire to preserve the use of "Torr" a term familiar to vacuum workers. We decided to use the highly logical system of SI units, and to express pressure in both Pa and Torr.

The technical literature concerned with pressure measurement in vacuum technology amounts to hundreds of papers, a few dozen chapters in specialized books on vacuum technology, and only one excellent textbook published twenty years ago. In all this literature little has been included on problems concerning particular aspects of low total pressure measurement, such as in hostile environments of corrosive or radioactive gases or in the presence of magnetic fields.

In writing this book we have attempted to select and organize an immense store of information so as to bridge the existing gap in the literature on the measurement of *low total pressure*. Emphasis is placed on the general processes and problems involved in measurement techniques as well as on the physical principles on which vacuum gauges operate, rather than on the detailed description of the gauges. However, where special instruments are necessary for the determination of "pressure" or gas density, such as pressure converters or radioactive gauges, both the design and techniques involved in their use are fully described.

The text is mainly intended for both graduate students and research scientists who have a good general background in physics and engineering.

Acknowledgments

I should like to thank Professor A. Roth for his invaluable help in the in-depth reading and criticizing of substantial sections of the text. I am grateful to Mrs. Hilda Krumbein and Miss Rhea Plottel who kindly undertook the editing of the manuscript and made many suggestions which were useful in improving the text. For her skillful and patient preparation of figures, I express my grateful appreciation to Mrs. Sara Saphier. The burden of typing has been carried by Mrs. Linda Wolff to whom many thanks are due. I am grateful to the authors, journals, publishers and industrial establishments cited in the text for permission to reproduce figures and tables. The excellent cooperation of Academic Press in the preparation of this book is gratefully acknowledged.

Finally, I owe more than I can say to my wife for allowing me to work long hours at home without demanding much of my attention and for continuous encouragement.

List of Symbols

A	Area
a	Cross-sectional area or thickness
B	Magnetic field
C	Conductance
C_p	Electrical capacitance
$\bar{c}$	Average speed of gas molecules
c_p	Heat capacity (constant pressure)
c_v	Heat capacity (constant volume)
D	Diffusion coefficient of gas
E	Electric field
E_m	Elasticity module
E_T	Rate of energy transfer per unit area
F	Force
g	Acceleration due to gravity
h	Height
I	Current or current density
I^+	Positive ion current in the gas phase
I^-	Electron (emission) current
I_c	Collected ion current
I_r	Residual current
I_x	Photo current
K	Thermal conductivity
K_n	Knudsen's number
k	Boltzmann's constant
L	Leak rate
l	Distance
M	Molecular weight
m	Mass (of molecule)
N	Total number (of molecules)
N_r	Number of revolutions
N_s	Number of scattered molecules
n	Refraction index (light)
n_m	Number density of molecules
P	Total pressure
p	Partial pressure
p_v	Vapor pressure
Q	Throughput
Q_h	Heat transfer
Q_m	Rate of transfer of momentum of mass motion
R_0	Universal gas constant
R_e	Electrical resistance
S	Volume rate of flow of gas
s	Absolute sensitivity
s_R	Relative sensitivity
T	Thermodynamic temperature
t	Time in general
$t_{0.5}$	Half-time
t_m	Time to build a monolayer
u	Velocity
V	Volume
W	Constriction of the gas flow
α	Modulation factor for I^+
α_t	Thermal accomodation coefficient
α_g	Polarizability of gas molecules
α_a	Attenuation coefficient
β	Volume expansion coefficient
γ	Surface tension
δ_m	Drift from the average value
δ	Molecular diameter
ε	Positive charges produced per electron, per cm path, per Pa at 0°C
η	Viscosity
η_{EID}	Efficiency of electron impact desorption
θ	Contact angle
λ	Mean free path
λ_0	Wave length
μ_a	Actual mean value
ν	Modulation factor for I_r
ξ	Condensing probability
ρ	Density
σ_m	Macroscopic coefficient of tangential momentum transfer
σ	Standard Deviation
τ_c	Time constant of electric cables
τ_k	Gas-kinetic time constant of a vessel
τ_p	Time constant of a measuring system
τ_t	Time constant of a pressure transducer
ϕ_t	Light yield of fluorescent light
ϕ	Molecular incidence rate
ω	Angular velocity

CHAPTER 1

Units and Terminology in Vacuum Technology

1.1 Pressure Units in Different Systems

1.1.1 Introduction

The root of the word *vacuum* is the Latin word *vacuus* (pl. *vacua*), which means a space devoid of matter. Such a state, however, can never be practically attained in a laboratory, nor even in outer space, where there are a few hydrogen atoms per cubic centimeter at 10^{-14} Pa (10^{-16} Torr). In modern usage vacuum is considered to exist in an enclosed space when the pressure of the gaseous environment is lower than atmospheric pressure or has been reduced as much as necessary to prevent the influence of some gas on a process being carried out in that space.

Vacuum developed from the state of an art to a precise science in the past four decades when scientists throughout the world began the systematic study of the physics and chemistry of vacuum techniques. As a result of these studies, the physical phenomena which occur in vacuum could be explained and quantified by the kinetic theory of gases, the theory of gas flow through impedances, and the theory of elementary gas transport. These theories are treated in detail in many excellent reference books, and it is assumed that the reader is familiar with them.

As a result of the evolution of vacuum science and technology, an increasing number of physical quantities had to be considered and measured, and a correct terminology had to be established. Pressure, used to measure the degree of rarefaction (even though not correct for all situations) and to calculate gas throughput and conductance of ducts and orifices to gas flow, was being reported in no fewer than 15 units in 1945. The decision taken to use the pascal (newtons per square meter) belonging to the International System of Units (SI) as a unit for pressure seemed to make order in the chaos created by the use of so many units, but it was not until the past decade that countries all over the world started using the pascal or the bar (also belonging to SI).

This section is restricted primarily to a short presentation of units for pressure, throughput, and conductance to gas flow. Nomenclature in vacuum technique and in standardization of measuring methods and interchangeable parts is also presented.

1.1.2 Expressions for Pressure

The pressure units presently in use in vacuum science and technology fall into two categories. In the first, pressure units are grouped within coherent unit systems; the second contains units that do not belong to such systems.

Pressure P is defined as the force F which a gas or vapor exerts perpendicularly on an area A and is expressed by:

$$P = F/A \tag{1.1}$$

Pressure can also be written

$$P = h\rho g \tag{1.2}$$

This expression is obtained by applying the general theorem of the states of perfect liquids to the equilibrium of a liquid column under the conditions of Torricelli's experiment. Here h is the height of the liquid column, ρ the density of the liquid in the column, and g the acceleration due to gravity at the location of the measurement.

The derived unit of pressure, expressed by Eq. (1.1), is in either of the forms

$$P = [m][l]^{-1}[x]^{-2} \quad (1.3)$$

$$P = [F][l]^{-2} \quad (1.4)$$

depending on which of the base units length l, mass m, time t, or length l, force F of a coherent system of units have been used to express P.

The derived unit of pressure in Eq. (1.2) expressed in the same units as Eq. (1.3) does not belong to a coherent system.

Figure 1.1 illustrates the derived pressure units in coherent as well as other systems. Some of the pressure units shown, such as pascal, Torr, millimeters of mercury, and bar, are frequently used, others rarely, and some (vac, gaede) never. The *vac* was proposed to supersede the millimeter of mercury, which does not belong to a coherent system of units (Florescu, 1960, 1961). The unit, having a value 1 vac = 10^3 dyn/cm^2, was not accepted, since it represents another name for the millibar which was in use at the time (Volet, 1960; Bigg, 1960, 1961). The *gaede* was intended to provide a smaller unit than the picotorr, supposed not to suffice for the measurement of pressures less than 10^{-12} Torr (Thomas *et al.*, 1959).

Coherent systems

$P = FA^{-1}$

$[P] = [\ell]^{-1}[m][t]^{-2}$

Base units of the system length (ℓ) mass (m) time (t)

Unit system	Derived pressure unit
SI (MKS)	Pa (N/m^2)
CGS	dyn/cm^2
MTS	pz
British	pdℓ/ft^2

Decimal (sub) multiples of pressure units

The system:
bar = 10^{-5} Pa (N/m^2)
millibar
microbar (barye)* = fdyn/cm^2

Other systems:
Gd (gaede) = 10^6 Pa (N/m^2)
vac = millibar

$[P] = [F][\ell]^{-2}$

Base units of the system length (ℓ) force (f) time (t)

Unit system	Derived pressure unit
Technical	kg f/m^2
British	ℓb f/ft^2

kg f/cm^2 (at) (technical atmosphere)

cm H_2O (guericke)**

Other systems

$P = h\rho g$

Normal atmosphere (atm)
mm Hg
Torr
in Hg
mm H_2O
in H_2O
ft H_2O

mTorr = μHg = μmHg

Fig. 1.1 Various units of pressure P. * French literature; ** German literature.

The use of units other than pressure in order to characterize the degree of gas rarefaction in the ranges of high and ultrahigh vacuum has been proposed. Worell (1963), Chutbert (1964), and Chuan (1965) proposed the measurement of the molecular density of a gas, and Bobenrieth (1959) suggested the measurement of the "main" spacing between gas molecules. This spacing varies as the third root of the inverse of the number density of gas molecules.

1.1.3 Pressure in Coherent Systems of Units

A coherent system of units is one in which any quantities other than base units are derived by simple multiplication or division, without the use of a numerical factor. The base units of such a system, presumably suited to practical requirements, have to be adaptable to all scientific disciplines. Thus, for instance, in the CGS system arbitrary constants of proportionality between derived and base units have been eliminated, except for electromagnetic and electrostatic quantities. The magnitudes of the units are not practical for all cases. This drawback of the CGS system is eliminated in the Systeme Internationale (SI) units adopted in 1954 by the Conference Generale des Poids et Mesures (CGPM). The SI system, based on the MKSA system, is obviously coherent, but the pressure unit pascal is not matched to practical requirements. Therefore the use of a unit outside the SI, the bar, with submultiples millibar and microbar, is permitted. The bar is recognized by the Comite International des Poids et Mesures (CIPM) as having to be retained either because of its importance in metrology or because of its use in other specialized fields (BS 5555, 1976). Different expressions for pressure in different coherent systems of units are given in Table 1.1. The graph in Fig. 1.2 permits the conversion of torr and millibar to pascals in the range 1.013×10^5 Pa (1 atm) to 10^{-16} Pa.

For more details regarding coherent systems of units, the reader is referred to Bailey (1982), Mosbacher (1974), Nicholson (1965), Thomas (1963a,b, 1968), and Thomas and Leiniers (1962a,b, 1967).

1.1.4 Pressure Units Not Belonging to Coherent Systems

Early pressure measurements were made with manometers of the liquid column type. The liquids were either mercury ($\rho = 13.5951$ g/cm^3 at 0°C) or water ($\rho = 0.999972$ g/cm^3 at 4°C). The acceleration due to gravity was taken to be $g = 980.665$ cm/s^2. Use of these values to calculate pressure P according to Eq. (1.2) results in the following values.

Table 1.1

PRESSURE IN COHERENT SYSTEMS OF UNITS

	Base units			Force		Derived units	Pressure		
System	Name	Symbol	Unit of area	Name	Symbol	Relation to SI	Name	Symbol	Relation to SI
SI (MKS)	meter kilogram (mass) second	m kg s	m^2	newton	N	1 N = 1 kg m/s^2	pascal newton per square meter	Pa N/m^2	1 N/m^2 = 1 Pa
CGS	centimeter gram (mass) second	cm g s	cm^2	dyne	dyn	1 dyn = 1 g cm/s^2 = 10^{-5} N	dyne per square centimeter	dyn/cm^2	1 dyn/cm^2 = 10^{-1} N/m^2
Technical	meter kilogram (force) second	m kgf s	m^2	kilogram force	kgf	1 kgf = 9.81 N	kilogram force per square meter	kgf/m^2	1 kgf/m^2 = 9.81 N/m^2
MTS	meter ton (mass) second	m t s	m^2	stène	sn	1 sn = 1 tm/s^2 = 10^3 N	pieze	pz	1 pz = 10^3 N/m^2
British	foot pound (mass) second	ft lb s	ft^2	poundal	pdl	1 pdl = 1 lb ft/s^2 ≈ 0.138 N	poundal per square foot	pdl/ft^2	1 pdl/ft^2 ≈ 1.488 N/m^2
	foot pound (force) second	ft lbf s		pound-force	lbf	1 lbf ≈ 4.448 N	pound-force per square foot	lbf/ft^2	1 lbf/ft^2 ≈ 47.880 N/m^2

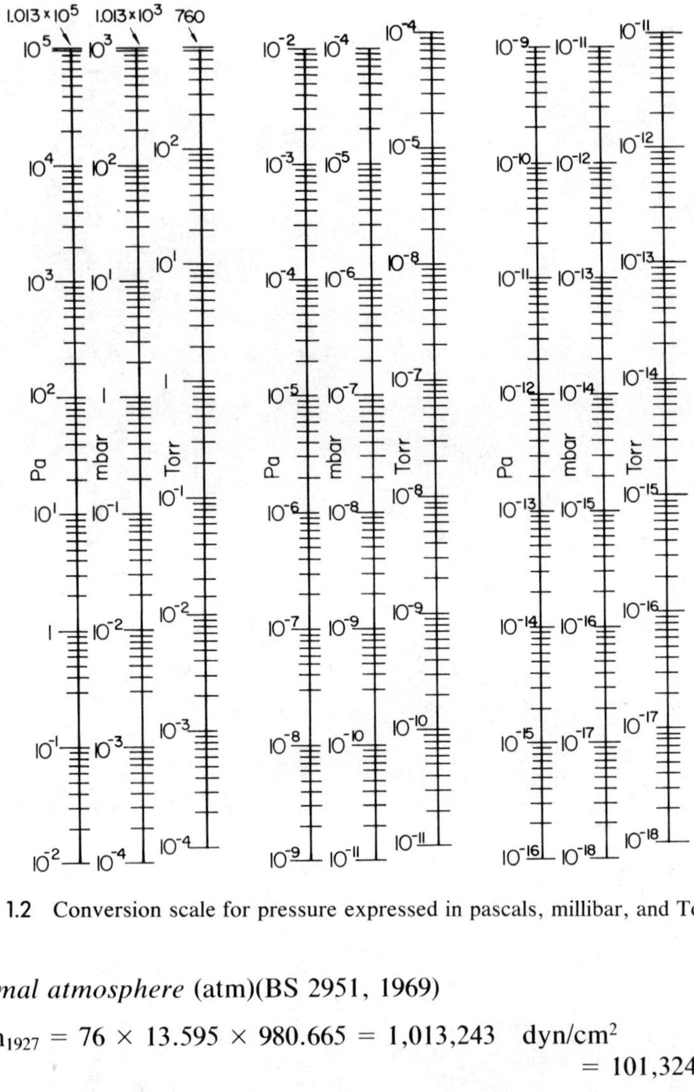

Fig. 1.2 Conversion scale for pressure expressed in pascals, millibar, and Torr.

Normal atmosphere (atm)(BS 2951, 1969)

$$1 \quad \text{atm}_{1927} = 76 \times 13.595 \times 980.665 = 1{,}013{,}243 \quad \text{dyn/cm}^2$$
$$= 101{,}324.3 \quad \text{Pa}$$

In 1954 the value of the atm_{1927} was rounded by CIMP to

$$1 \quad \text{atm}_{1954} = 1{,}013{,}250 \quad \text{dyn/cm}^2$$

Conventional millimeter of mercury (mm Hg)(ISO/DIS 3529/I, 1981)

$$1 \quad \text{mm Hg} = 0.1 \times 13.595 \times 980.665 \cong 133.22 \quad \text{dyn/cm}^2$$
$$\cong 133.322 \quad \text{Pa}$$

torr (Torr)(ISO/DIS 3529/I, 1981)

$$1 \quad \text{Torr} = 1 \quad \text{atm}_{1927}/760 \cong 1333.32 \quad \text{dyn/cm}^2 \cong 133.322 \quad \text{Pa}$$

Due to the rounding off of the normal atmosphere in 1954, 1 Torr differs from 1 mm Hg by 1.4 parts in 10^7. For practical reasons, the difference between these units is disregarded. The torr has submultiples expressed by negative powers of 10.

Conventional inch of mercury (in. Hg)(ISO/DIS 3529/I, 1981)

$$1 \quad \text{in. Hg} \cong 33863.9 \quad \text{dyn/cm}^2 \cong 3386.39 \quad \text{Pa}$$

Conventional millimeter of water (mm H_2O)(ISO/DIS 3529/I, 1981)

$$1 \quad \text{mm H}_2\text{O} \cong 98.0665 \quad \text{dyn/cm}^2 \cong 9.80665 \quad \text{Pa}$$ (value rounded to 6 significant digits).

Conventional inch of water (in. H_2O)(ISO/DIS 3529/I, 1981)

$$1 \quad \text{in. H}_2\text{O} \cong 2490.89 \quad \text{dyn/cm}^2 \cong 249.089 \quad \text{Pa}$$

Conventional foot of water (ft H_2O)(ISO/DIS 3529/I, 1981)

$$1 \quad \text{ft H}_2\text{O} \cong 29890.7 \quad \text{dyn/cm}^2 \cong 2989.07 \quad \text{Pa}$$

The conversion of various noncoherent systems to coherent systems of units is given in Table 1.2.

1.2 The Logarithmic Representation of Pressure

1.2.1 Pressure Scales

Pressures lower than a certain value are expressed in negative powers of 10, depending on the system of units selected. This has led to the paradox of expressing degrees of higher gas rarefaction in the high and very high vacuum ranges by very small numerical values (see Tables 2.1 and 2.2).

To eliminate negative exponents, proposals were advanced to represent pressure on a logarithmic scale selected such that the lowest value of the scale has a positive value. Three systems have been proposed; two are analogous to the *bel* and *decibel* systems used mainly in electronics, the third is analogous to the *pH* concept used by chemists.

Table 1.2

CONVERSION FACTORS FOR PRESSURE UNITS ($1X = nY$)

X \ Y		Coherent systems					Noncoherent systems			
			CGS							
X		SI (MKS) Pa (N/m^2)	(dyn/cm^2)	(bar)	Technical (kgf/m^2)	British (lbf/in^2)	Millitorr (mTorr)	Inch of mercury (in. Hg)	Torr (Torr)	Normal atmosphere (atm)
SI (MKS)	Pa N/m^2	1	10	10^{-5}	1.02×10^{-1}	1.45×10^{-4}	7.50×10^{-6}	2.95×10^{-4}	7.50×10^{-3}	9.87×10^{-6}
CGS	dyn/cm^2	10^{-1}	1	10^{-6}	1.02×10^{-2}	1.45×10^{-5}	7.50×10^{-1}	2.95×10^{-5}	7.50×10^{-4}	9.87×10^{-7}
	bar	10^{5}	10^{6}	1	1.02×10^{4}	1.45×10	7.50×10^{5}	2.95×10	7.50×10^{2}	9.87×10^{-1}
Technical	kgf/m^2	9.81	9.81×10	9.81×10^{-5}	1	1.42×10^{-3}	7.35×10	2.89×10^{-3}	7.35×10^{-2}	9.68×10^{-5}
British	lbf/in^2	9.89×10^{3}	9.89×10^{4}	6.89×10^{-2}	7.03×10^{2}	1	5.17×10^{4}	2.03	5.17×10	6.80×10^{-2}
Millitorr	mTorr	1.33×10^{-1}	1.33	1.33×10^{-6}	1.35×10^{-2}	1.93×10^{-5}	1	3.94×10^{-5}	10^{-3}	1.31×10^{-6}
Inch of mercury	in. Hg	3.39×10^{3}	3.39×10^{4}	3.39×10^{-2}	3.45×10^{2}	4.91×10^{-1}	2.54×10^{3}	1	2.54	3.34×10^{-2}
Torr	Torr	1.33×10^{2}	1.33×10^{3}	1.33×10^{-3}	1.36×10	1.93×10^{-2}	10^{3}	3.94×10^{-2}	1	1.31×10^{-3}
Normal atmosphere	atm	1.01×10^{5}	1.01×10^{6}	1.01	1.03×10^{4}	1.47×10	7.6×10^{5}	2.99×10	7.60	1

1.2.2 Pressure Scale Analogous to the Bel Concept

A unit pressure on a scale analogous to the bel concept equals the negative value of the logarithm of the ratio P/P_0. The pressures P and P_0 are the actual and the reference pressures, respectively. Upon assigning a value to P_0 such that $\log P_0 = 0$, the resulting scale is

$$\text{``unit of pressure''} = -\log P \tag{1.5}$$

The following pressure scales have been proposed:

$$\text{McLeod} = -\log P_{\text{mm Hg}} \quad \text{(Feinberg, 1945)} \tag{1.6}$$

$$\text{pA} = -\log P_{\text{atm}} \quad \text{(Chutbert, 1964)} \tag{1.7}$$

$$\text{VAC} = -\log P_{\text{Torr}} \quad \text{(Male, 1965)} \tag{1.8}$$

where the reference level P_0 = 1 mm Hg, 1 atm, and 1 Torr, respectively. Table 1.3 lists values of pressure lower than atmospheric in Chutbert's scale.

1.2.3 Pressure Scale Analogous to the Decibel Concept

On the decibel scale, a unit of pressure is equal to 10 times the negative value of the logarithm of the ratio of the actual pressure to a reference pressure. Thus,

$$\text{``unit of pressure''} = -10 \log(P/P_0) \tag{1.9}$$

Table 1.3

LOGARITHMIC SCALE OF PRESSURE ACCORDING TO THE RELATIONSHIP $\text{pA} = -\log P_{\text{atm}}$

Pressure(Torr)	pA
760 (atm)	0
1	2.9
10^{-3}	5.9
10^{-6}	8.9
10^{-9}	11.9
10^{-12}	14.9

On this scale, proposed by Townsend (1954), the reference pressure is $P_0 = 1$ mm Hg, which he considers to be the distinct border between high and low vacuum. The scale bears no special name. A similar scale called *decilog* (dL), where pressure is measured in bar and $P_0 = 1$ bar, was proposed by Rose (1945). Forrest (1961) proposed a scale called *pVac* where

$$\text{pVac} = -\log(P_{\text{vac}}/10^3) = -\log P_{\text{vac}} + 3 \tag{1.10}$$

Here $P_{\text{vac}} = 1$ vac [see Eq. (1.1)] and $P_0 = 10^3$ vac, a pressure a little lower than normal atmosphere. Baldwin and Tonks (1964, 1966) and Tonks and Baldwin (1965) proposed the scale *deciboyle* expressed as

$$\text{dB} = 10 \log(P/P_0) + C \tag{1.11}$$

essentially based on the logarithm of the bar. In Eq. (1.11) P is the actual pressure in whatever units one considers and C a constant having a value 10 times the logarithm of the pressure P in bars. Values of C for some systems of units are shown in Table 1.4. On the *decibar* scale proposed by Gundersen (1965), *bar* was substituted for *Boyle* in Eq. (1.11). Thus,

$$\text{dbar} = 10 \log(P_{\text{Torr}}) - 28.7509 \tag{1.12}$$

Nieweld's (1965) scale, a modified version of the Baldwin and Tonks scale is expressed as

$$\text{Bo} = \log(P/P_0) \tag{1.13}$$

where P is the actual pressure in any system of units converted to pascals, and $P_0 = 1$ Pa.

1.2.4 Pressure Scale Analogous to the pH Concept

The pressure scale analogous to the pH concept is based on the use of the logarithmic representation of particle density to measure pressure.

Table 1.4
SOME CONVERSION CONSTANTS FOR PRESSURE TO DECIBOYLE SCALE

Unit for P	Conversion of the unit for P in bar	C
bar	1	0
Torr	1.3332×10^{-3}	−28.7510
in. Hg	3.386×10^{-2}	−14.7031
in. H_2O (20°C)	2.486×10^{-3}	−26.0449
normal atmosphere	1.01325	0.00572

Table 1.5
SOME VALUES OF n_L FOR DIFFERENT PRESSURES P

Density			P		
n_L	n_m (particles/m³)	T(K)	Torr	bar	Pa
25.48	3.0×10^{25}	273.15 (0°C)	760	1.01325	101.325
25.44	2.8×10^{25}	300 (27°C)	760	1.01325	101.325
22.51	3.2×10^{22}	300	1	1.333×10^{-3}	133.3
10.51	3.2×10^{10}	300	10^{-12}	1.333×10^{-15}	1.333×10^{-10}
7.51	3.2×10^{7}	300	10^{-15}	1.333×10^{-18}	1.333×10^{-13}

The particle density (PD) system of Herskowitz (1965) is given by

$$\text{PD value} = \log n = 22.518 + \log p \tag{1.14}$$

where n is the number of particles per cubic meter and p the pressure expressed in Torr. Equation (1.14) is deduced from

$$P = nkT \tag{1.15}$$

where k is Boltzmann's constant and T the thermodynamic temperature. For convenience, T is considered constant and has the value $T = 293$ K. The beginning of the scale PD = 0 corresponds to one particle per cubic meter (3.035×10^{-23} Torr). The scale ends at a value PD = 43.84 obtained by considering a substance composed solely of closest packed hydrogen nuclei. Such a substance would contain 7×10^{43} particles per cubic meter. The PD scale spans 44 decades and generates only positive values.

Borson (1965) expressed Eq. (1.15) as

$$\log P = n_L + \log K + \log T \tag{1.16}$$

where n_L, the *logarithmic number density,* represents the logarithm of the ratio n_m/n_o. Here n_m and n_o are the actual particle density and a reference density (taken as one particle per cubic meter), respectively. Some values of n_L for different pressures are listed in Table 1.5.

1.3 Throughput and Conductance in the SI System

The throughput is the quantity of gas, in pressure–volume units, at a specified temperature flowing per unit time across a specified open cross section (AVS 1980 Dictionary of terms). Since the quantity of gas mea-

Table 1.6

CONVERSION FACTORS FOR THROUGHPUT UNITS

	Pa m^3/s	Pa L/s	Torr L/s	atm cm^3/y	atm cm^3/h	atm cm^3/s	atm ft^3/y	atm ft^3/h	atm ft^3/min
Pa m^3/s	1	10^3	7.50	3.11×10^8	3.55×10^4	9.87	1.09×10^4	1.25	2.08×10^{-2}
Pa L/s	10^{-3}	1	7.50×10^{-3}	3.11×10^5	35.53	9.87×10^{-3}	1.09×10	1.25×10^{-3}	2.08×10^{-5}
Torr L/s	1.33×10^{-1}	1.33×10^2	1	4.15×10^7	4.73×10^3	1.32	1.46×10^3	1.67×10^{-1}	2.78×10^{-3}
atm cm^3/y	3.21×10^{-9}	3.21×10^{-6}	2.40×10^{-8}	1	1.14×10^{-4}	3.17×10^{-8}	3.53×10^{-5}	4.03×10^{-9}	6.72×10^{-11}
atm cm^3/h	2.81×10^{-5}	2.81×10^{-2}	2.11×10^{-4}	8.76×10^3	1	2.78×10^{-4}	3.1×10^{-1}	3.53×10^{-5}	5.88×10^{-7}
atm cm^3/s	1.01×10^{-1}	1.01×10^2	7.60×10^{-1}	3.15×10^7	3.60×10^3	1	1.11×10^3	1.27×10^{-1}	2.12×10^{-3}
atm ft^3/y	3.09×10^{-5}	9.09×10^{-2}	6.82×10^{-4}	2.83×10^4	3.23	8.98×10^{-4}	1	1.27×10^{-4}	2.12×10^{-6}
atm ft^3/h	7.97×10^{-1}	7.97×10^2	5.97	2.48×10^8	2.83×10^4	7.87	8.76×10^3	1	2.78×10^{-4}
atm ft^3/min	4.80×10	4.80×10^4	3.59×10^2	1.49×10^{10}	1.7×10^6	4.72×10^2	5.26×10^5	60	1

sured in pressure–volume units is q_v, the expression of the throughput Q is

$$Q = dq_v/dt = d(PV)/dt \tag{1.17}$$

If S is the pumping speed at the cross section where throughput is considered, then for steady-state conditions and constant pressure P,

$$S = dV/dt \tag{1.18}$$

Hence the throughput can be written

$$Q = SP \tag{1.19}$$

Since in the SI system the unit for pressure is the pascal and that for pumping speed is cubic meters per second, the derived unit for pumping speed in this system is joules per second or watts. Although perfectly coherent, the SI units is misleading since it represents throughput as a power quantity. The significance of Q as a mass per unit time and also the analogy with power could be maintained by rewriting Eq. (1.15) as

$$Q = Sn_m \tag{1.20}$$

For practical considerations, however, this is unrealistic (Lisberger and Parker, 1970). Since the derived unit for throughput is not matched to practical requirements, vacuum users measure throughput in pascal liters per second. The conversion of throughput units in different systems is given in Table 1.6.

Conductance is defined as the ratio of throughput, under steadystate conditions, to the pressure differential between two specified cross sections inside a pumping system (AVS 1980 Dictionary of terms). The derived SI unit for conductance is cubic meters per second. Other more useful units are liters per second and liters per minute. Table 1.7 shows the conversion factors for different pumping-speed units.

1.4 Terminology and Standardization in Vacuum Technology

1.4.1 Terms Related to Specific Vacuum Applications and Graphic Symbols

To avoid the confusion caused by the multiplicity of terms used to specify the same fundamental concept, the Committee on Vacuum Techniques decided in 1953 to set down a nomenclature for vacuum technology. This nomenclature was presented to vacuum users all over the world for their consent. The nomenclature is based on Dayton's proposals (Day-

Table 1.7

CONVERSION FOR PUMPING SPEED UNITS (1X = nY)

X \ Y	m^3/s	m^3/h	L/s	L/min	cm^3/s	ft^3/min	ft^3/h
m^3/s	1	3.6×10^3	10^3	6.0×10^4	10^6	2.119×10^3	1.271×10^5
m^3/h	2.778×10^4	1	2.778×10^{-1}	1.667×10	2.778×10^2	5.888×10^{-1}	3.531×10
L/s	10^{-3}	3.6	1	6×10	10^3	2.119	1.271×10^2
L/min	1.667×10^{-5}	6×10^2	1.667×10^{-2}	1	1.667×10	3.531×10^{-2}	2.119
cm^3/s	10^{-6}	3.6×10^{-3}	10^{-3}	6×10^{-2}	1	2.119×10^{-3}	1.271×10^{-1}
ft^3/min	4.719×10^{-4}	1.699	4.719×10^{-1}	2.831×10	4.719×10^2	1	6×10
ft^3/h	7.865×10^{-6}	2.831×10^{-2}	7.865×10^{-3}	4.719×10^{-1}	7.865	1.667×10^{-2}	1

ton, 1955) supplemented with terms related to specific applications (Dayton, 1955, 1956) and was published on the condition that it be tentative until approved or amended by the American Standard Association (Glossary of Terms Used in Vacuum Technology, 1958). The same year, the British Standard Institution published a standard for vacuum terminology (BS 2951, 1958), which was revised in 1969 (BS 2951, Part 1, 1969) and supplemented in 1975 (BS 2951, Part 2, 1975) with terms of scientific applications. Vocabularies of vacuum technology were also published in Germany (DIN 28400, 1972), France (NFX 10-500, 1975) and the United States (ISO/DIS 1975, 1976, 1982).

Less critical for the terminology, but worth mentioning, are proposals to use the Greek root *ken* instead of the Latin *vacuus* (Langmuir, 1953; Schram, 1964). The extension of the latin root to activities in vacuum technology, e.g., *vacuology* as the *science du vide* and *vacuotechnique* as the *technique du vide* were also proposed (Thomas and Leyniers, 1962a,b, 1964).

Graphic symbols for equipment components and their assembly have been standardized in many countries (AVS 7.1, 1966; ISO/DIS, 1975; JIS, 1964; PN-04-201, 1962). Although slight dissimilarities differentiate between them, their use proved to be of real help to vacuum users.

1.4.2 Standardization of Interchangeable Parts, Measuring Methods, and Equipment Performance

The development of industries and research fields depending on vacuum technology steadily increased the demand for standardizing the different parts commonly used in vacuum systems as well as the measuring methods and the performance characteristics of the equipment. Interchangeable parts, like flanges, and the methods of testing them are standardized. To give a survey of the international situation to date would surpass the aim of this book. The related standards represent the best thinking on the subject at the time and fully serve their purpose.

Measuring methods for some of the parameters of vacuum systems as well as vacuum equipment performance are standardized in many countries and will be quoted in the following sections. One of the fundamental problems of vacuum technology, pressure measurement standardization, however, is not yet solved. Attempts to arrive at a standard procedure for measurements by means of the McLeod gauge, for instance, have shown the complexity of the problem; about 50 pages of conditions for procedures did not cover the whole area of interest.

The fact that the measurement of the same parameter according to different national standards does not lead to consistent results shows that in the standardization of measuring methods much remains to be done. As an example, the discrepancy in the results obtained when measuring the pumping speed of pumps according to AVS and IUVST standards (Denison, 1974 and Denison and McKee, 1974; respectively) can be cited.

For more details concerning the standardization problem, the reader is referred to Günter (1968).

References

AVS 7.1 (1969). Graphic Symbols in Vacuum Technology. *J. Vac. Sci. Technol* **4,** 139.

AVS (1980). Dictionary of Terms for Vacuum Science and Technology, Surface Science, Thin Film Technology, Vacuum Metalurgy, Electronic Materials. (M. S. Kaminsky and J. Lafferty, eds.). Amer. Inst. Phys., New York.

Bailey, A. E. (1982). *J. Phys. E* **15,** 849.

Baldwin, G. C., and Tonks, L. (1964). *Nature* **203,** 633.

Baldwin, G. C., and Tonks, L. (1966). *Res./Dev.* **17,** 66.

Bigg, P. H. (1960). *Nature* **188,** 1017.

Bigg, P. H. (1961). *Nature* **189,** 523.

Bobenrieth, A. (1959). *Vide* **14,** 271.

Borson, E. N. (1965). *Res./Dev.* **16,** 94.

BS 2951 (1958). Glossary of Terms Used in High Vacuum Technology. British Standards Institution.

BS 2951 (1969). Glossary of Terms Used in Vacuum Technology. Part 1. Terms of General Application. British Standards Institution.

BS 2951 (1975). Glossary of Terms Used in Vacuum Technology. Part 2. Terms of Specific Application. British Standards Institution.

BS 5555 (1976). SI Units and Recommendations for the Use of Their Multiples and of Certain Other Units.

Chuan, R. L. (1965). *Res./Dev.* **16,** 94.

Chutbert, J. (1964). *Nature* **201,** 61.

Dayton, B. B. (1955). *Trans Vac. Symp., U.S.A. 1954,* p. 132. Committee on Vacuum Technology, New York.

Dayton, B. B. (1956). *Trans. Vac. Symp., U.S.A. 1955,* p. 99. Committee on Vacuum Technology, New York.

Denison, D. R. (1974). *Proc. 6th Int. Vac. Congr., Kyoto 1974,* p. 155. *Jpn. Appl. Phys. Suppl. 2 Part 1.*

Denison, D. R., and McKee, E. S. (1974). *J. Vac. Sci. Technol.* **11,** 337.

DIN 28400 (1972). Vakuumtecnik Bennenungen und Definitionen, pp. 1–3.

Feinberg, R. (1954). *Nature* **126,** 85.

Florescu, N. A. (1960). *Nature* **188,** 303.

Florescu, N. A. (1961). *Nature* **190,** 522.

Forrest, W. W. (1961). *Nature* **189,** 476.

Glossary of Terms Used in Vacuum Technology (1958). Pergamon Press, New York.

Gundersen, N. A. (1965). *Res./Dev.* **16,** 94.

Günter, K. G. (1968). *Proc. 4th Int. Vac. Congr., Manchester 1968,* p. 48. *Inst. Phys. Phys. Soc. Conf. Ser. No. 5.,* England.

Herskowitz, Sh. B. (1965). *Res./Dev.* **16,** 94.
ISO/DIS (1975). 3529/I Vacuum Technology Vocabulary Part I. General Terms.
ISO/DIS (1976). 3529/II Vacuum Technology Vocabulary Part II. Vacuum Pumps and Related Terms.
ISO/DIS (1981). 3529/I Vacuum Technology Vocabulary Part I. General Terms.
JIS (1964). Z 8027 Graphical Symbols for Vacuum Apparatus. Dictionary of Terms for Vacuum Science and Technology, Surface Science, Thin Film Technology, Vacuum Metalurgy, Electronic Materials (Kaminsky and Lafferty, eds.) American Institute of Physics 1980, U.S.A.
Langmuir, I. (1953). *Vacuum* **3,** 113.
Liseberger, P. H., and Parker, M. R. (1970). *Vacuum* **20,** 75.
Male, D. W. (1965). *Res./Dev.* **16,** 94.
Mosbacher, C. J. (1974). *Res./Dev.* **25,** 41.
NF X 10-500 (1975). Technique de Vide. Vocabulaire. Terms Généraux. Societé Francaise des Ingenieurs et Techniciens Du Vide.
Nicholson, B. J. (1965). *J. Vac. Sci. Tecnol.* **2,** 161.
Nieweld, A. (1965). *Res./Dev.* **16,** 94.
PN E 04-201 (1962). Représentation Symbolique des Appareils de Vide. Societé Francaise des Ingenieurs et Techniciens Du Vide.
Rose, F. C. (1945). *Nature* **124,** 268.
Schram, A. (1964). *Vide* **110,** 79.
Thomas, E. (1963a). *Proc. Symp. Electron Vac. Phys.* p. 29. Publishing House of the Hungarian Academy of Sciences, Balatonfoldvar, Hungary.
Thomas, E. (1963b). *Vacuum* **13,** 376.
Thomas, E. (1968). *Proc. 4th Int. Vac. Congr., Manchester 1968,* p. 43. *Inst. Phys. Phys. Soc. Conf. Ser. No 5,* England.
Thomas, E., and Leyniers, R. (1962a). *Vacuum* **12,** 180.
Thomas, E., and Leyniers, R. (1962b). *Trans. 8th Natl. Vac. Symp., U.S.A. 1961,* p. 499. Pergamon Press, New York.
Thomas, E., and Leyniers, R. (1964). *Vide* **19,** 112.
Thomas, E., and Leyniers, R. (1967). *3rd Czech. Conf. Electron. Vac. Phys., Prague 1967,* p. 457. Academic Publishing House, Prague.
Thomas, E., Servanckx, R., and Leyniers, R. (1959). *Vacuum* **19,** 207.
Tonks, L., and Baldwin, G. C. (1965). *Res./Dev.* **16,** 57.
Townsend, F. H. (1954). *Nature* **127,** 545.
Volet, Ch. (1960). *Nature* **188,** 1017.
Worell, F. T. (1963). *Nature* **199,** 477.

CHAPTER 2

Basic Concepts of Pressure Measurement

2.1 The Rarefied Environment

A gaseous environment where pressure or density is below that of standard atmosphere is termed a vacuum. The physical quantity used to characterize such an environment, though not often the most appropriate, is the gas pressure.

Usually vacuum is obtained by removing gas from an airtight vessel by some means of pumping. In the absence of a gas load, the degree of vacuum in the vessel increases as the pressure of the residual gases decreases. Lower and lower pressures obtained by continuously pumping down the vessel are conventionally classified within the ranges listed in Table 2.1. The bordering values of the SI and Torr systems in the four different ranges do not exactly overlap. For convenience, rounded num-

Table 2.1
CLASSIFICATION OF VACUUM RANGES[a]

Vacuum ranges	Systems of units: SI, Pa	SI, bar	Torr
Low (rough)	10^5–10^2	1–10^{-3}	760–1
Medium	10^2–10^{-1}	10^{-3}–10^{-6}	1–10^{-3}
High	10^{-1}–10^{-6}	10^{-6}–10^{-11}	10^{-3}–10^{-8}
Ultrahigh	below 10^{-6}	below 10^{-11}	below 10^{-8}

[a] From BS (1969) and DIN (1972).

bers are used in both systems. This drawback has been overcome in the new classification of the AVS, as shown in Table 2.2.

The practice of classifying the rarefied environment as high and ultrahigh vacuum despite the low values measured is universally accepted. This practice could be changed, but the resulting confusion would probably be worse than the relatively minor inconvenience caused by the present convention.

Generally speaking, pressure is a convenient parameter to characterize vacuum. However, below certain limits of gas rarefaction, pressure is not adequate to explain phenomena that take place in the vacuum, and other physical quantities, such as the number density of molecules and the mean free path, can characterize it more exactly. The meaning and magnitude of these quantities is given by the kinetic theory of gases. According to this theory, a gas is a collection of an enormous number of particles

Table 2.2
CLASSIFICATION OF VACUUM RANGES[a]

Vacuum range	Pressure units: Pa	Torr
Low	$3.3 \times 10^3 < P < 10^5$	$25 < P < \sim 750$
Medium	$10^{-1} < P \leq 3.3 \times 10^3$	$7.5 \times 10^{-4} < P < 25$
High	$10^{-4} < P \leq 10^{-1}$	$7.5 \times 10^{-7} < P < 7.5 \times 10^{-4}$
Very high	$10^{-7} < P \leq 10^{-4}$	$7.5 \times 10^{-10} < P < 7.5 \times 10^{-7}$
Ultrahigh	$10^{-10} < P \leq 10^{-7}$	$7.5 \times 10^{-13} < P < 7.5 \times 10^{-10}$
Extreme ultrahigh	$P < 10^{-10}$	$P < 7.5 \times 10^{-13}$

[a] From AVS (1980).

(molecules) moving along random directions, colliding with each other, and changing their direction of motion with each collision. For a given gas, having a known temperature T and molecule diameter δ_m, the number density of molecules n_m and their mean free path λ are a function of the pressure P:

$$n_m = 7.24 \times 10^{10} P_{Pa} T^{-1} \qquad \text{molec/m}^3 \tag{2.1}$$

$$\lambda = 3.107 \times 10^{-16} T^{-2}\, \delta_m P_{Pa}^{-1} \quad \text{m} \tag{2.2}$$

By reducing pressure in an airtight vessel under constant temperature conditions, some of the molecules are eliminated, thus creating conditions of less close packing of the remaining molecules. At 10^2 Pa (1 Torr) the number of molecules in the volume of the vessel is large, and intermolecular collisions are by far more frequent than collisions of molecules with the walls of the vessel. Upon continuing pumping, the number of molecules continuously diminishes, and the mean free paths increase. Collisions of molecules with the walls of the vessel occur more frequently, and soon an instant is reached (depending also upon the dimensions of the vessel) when the ratio of the number of intermolecular collisions to that of molecules striking the walls equals unity. This situation occurs in most industrial and laboratory equipment at room temperature at pressures of about 10^{-1} Pa (10^{-3} Torr) or below.

Upon further pumping, gas molecules start colliding with the walls more frequently, and the composition of the gas differs from that at the beginning of pumping. The equilibrium conditions occurring at the solid–gas interface are upset by the deliberate removal of adsorbed or absorbed gas. About 80 to 90% of the gas is composed of water molecules freed from the walls. Generally, this situation occurs within the pressure range 10^{-1}–10^{-6} Pa (10^{-3}–10^{-8} Torr). Beside n_m and λ, two other physical quantities are significant in characterizing the rarefied gas environment. These are the *molecular incidence rate* ϕ (the number of molecules perpendicularly striking a unit area per unit time) and the *time* t_m *to build a monolayer* expressed as

$$\phi = 2.635 \times 10^{16} P_{Pa}/(MT)^{1/2} \qquad \text{molecules/m}^2\text{s} \tag{2.3}$$

$$t_m = 4.38 \times 10^{-21} (MT)^{1/2}/\delta_m P_{Pa} \quad \text{s} \tag{2.4}$$

where M and T are the molecular weight and the thermodynamic temperature of the gas, respectively.

Further pumping with simultaneous heating of the vessel walls results once again in a change in the composition of the gas. Gases released either from the walls or permeating through them, together with contaminants coming from the pumping system, are now the main components of the residual gas. This is the ultrahigh vacuum range below 10^{-6} Pa (10^{-8} Torr).

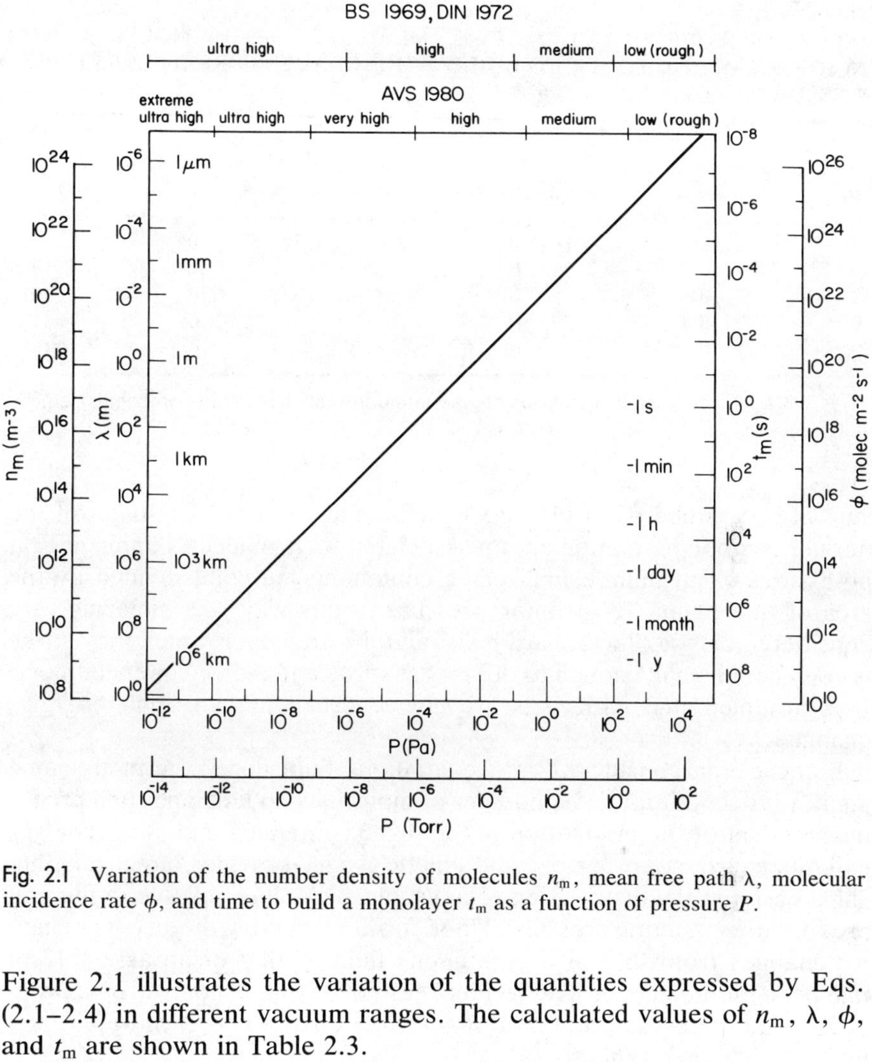

Fig. 2.1 Variation of the number density of molecules n_m, mean free path λ, molecular incidence rate ϕ, and time to build a monolayer t_m as a function of pressure P.

Figure 2.1 illustrates the variation of the quantities expressed by Eqs. (2.1–2.4) in different vacuum ranges. The calculated values of n_m, λ, ϕ, and t_m are shown in Table 2.3.

Within the framework of physical processes taking place when evacuating an airtight vessel, the main question requiring clarification is how pressure can characterize the rarefied environment.

2.2 The Pressure Concept

The pressure exerted by gas molecules on a boundary surface is expressed as the rate of transfer of the normal component of their momentum with respect to a unit area divided by that area. Analogously, pres-

Table 2.3

VALUES OF MOLECULAR DENSITY n_m, MEAN FREE PATH λ, MOLECULAR INCIDENCE ϕ, AND TIME t_m TO BUILD A MONOLAYER FOR DIFFERENT AIR PRESSURES[a]

P		n	λ	ϕ	t_m
Pa	Torr	(m^{-3})	(m)	($m^{-2}s^{-1}$)	(s)
10^5	7.6×10^2	2.43×10^{13}	6.62×10^{-4}	2.83×10^{19}	2.91×10^{-9}
10^0	7.5×10^{-3}	$\times 10^8$	$\times 10^1$	$\times 10^{14}$	$\times 10^{-4}$
10^{-4}	$\times 10^{-7}$	$\times 10^3$	$\times 10^6$	$\times 10^9$	$\times 10^0$
10^{-8}	$\times 10^{-11}$	$\times 10^{-2}$	$\times 10^{11}$	$\times 10^4$	$\times 10^4$
10^{-12}	$\times 10^{-15}$	$\times 10^{-7}$	$\times 10^{16}$	$\times 10^{-1}$	$\times 10^8$

[a] $T = 298.16$ K; t_m was calculated on the assumption that the sticking probability equals unity.

sure at a specified point in a gas is defined as the rate of transfer of the normal component of momentum associated with molecules impinging on both sides of an infinitesimal plane containing the point divided by the area of that plane. By defining pressure in this way, gas molecules are considered elastic, like billiard balls, and the area upon which they strike is considered small enough to offer a flat surface to molecular encounters. It is assumed that molecules are not adsorbed nor are their energies changed.

If the gas is considered as a continuous fluid under thermodynamic equilibrium conditions, the number of molecules colliding per unit area is independent of the orientation of the plane containing that area. The gas exhibits isotropic properties, and kinetic theory provides direct relationships between the distribution of molecular velocities relative to the unit area and the resulting pressure. Upon further rarefying the gas, its behavior changes from that of a continuous fluid to that of an assembly of independent molecules. The residual gas now exhibits nonisotropic properties, and the assumption of a Maxwellian distribution of molecular velocities no longer applies.

Industrial vacuum equipment subject to dynamic working conditions does not permit isotropic gas conditions at reduced degrees of rarefaction. For instance, let us consider a simple vacuum system consisting of an airtight bell jar evacuated by a diffusion pump. For a certain regime of gas flow towards the pump, the gas isotropy at the pumping port fundamentally differs from that in the bell jar. Thus, the pressure exerted on a unit area in a plane parallel to the pumping port would be different on the two faces of the plane. Venema (1954) showed that the pressure on the side facing the vessel is about 30% higher than that exerted on the side facing the pumping port. Further, the pressure P_i exerted on an area of a plane immersed in the gas can vary between P_i (isotropic) and $5P_i$ depending upon the orientation of the plane with respect to the nonisotropic component (Giorgi and Pisani, 1966).

Some of the main reasons for nonisotropic properties of the gaseous environment at high degrees of rarefaction are different flow regimes of gas during the pumping process, steady or transient evolution of gas sometimes characterized by directional effects, gas molecules sticking or evaporating on or off the surfaces of the system, respectively, environmental nonuniformities due to temperature gradients, plasma concentrations, etc.

For nonisotropic conditions, pressure as defined at the beginning of this section fails to characterize the rarefied gaseous environment. Other physical quantities, such as those expressed by Eqs. (2.1–2.4), or other relationships between degassing rates, capture probability, and temperature are more significant for defining the rarefied environment than pressure, defined as force per unit area (Beynon, 1970; Close and Yarwood, 1970; Denis and Heppel, 1968; Chuan, 1964).

2.3 Influence of the Rarefied Environment on the Measuring Process

2.3.1 Measurement Problems

The conventional measurement of pressure is based on the concept that in a *stationary* mass of gas, uniform in composition and in both thermal and mechanical equilibrium with the containing vessel, the Maxwell–Boltzmann velocity distribution of gas molecules applies. (A gas mass within a volume is *stationary* when each element of the surface confining the mass emits gas at the same rate at which it receives gas.) It is assumed that gas molecules neither interfere with each other nor interact with the surface at which they collide. After collision with the surface, molecules leave it in a diffuse way according to the cosine distribution of directions and reenter the gas phase with an average kinetic energy corresponding to the temperature of the surface. The rate of incidence of molecules on a plane within the volume of the vessel is independent of the orientation of the normal to the plane. This is the case, when the gas has isotropic properties, for which the kinetic theory of gases provides a straightforward relationship between the number density of molecules and pressure.

Within the usual uncertainty in pressure measurement of $\pm 10\%$ in industrial determination or less than ±5% in careful research work, the gas in a pumped-down vessel approximates isotropic conditions when

(i) the vessel is at room temperature, or
(ii) the removal rate of gas from the vessel is low or moderate (<700 L/s for air).

The isotropic state discussed is often disturbed by the dynamic molecular flow of gas resulting from localized or distributed (usually both) sinks and/or sources (e.g., pumps of all kinds, adsorption, gas deliberately admitted via leaks, outgassing, specular reflection of molecules from surfaces, temperature nonuniformities). Depending on their contribution to the process, these sources and/or sinks cause nonuniform and nonisothermal gas flow distribution in the pumped vessel. The properties of the rarefied environment vary with direction. Thus, the ratio of the gas density to gas pressure varies throughout the volume of the vessel. Under such circumstances knowledge of pressure, as defined physically, becomes of reduced concern in expressing the properties of the gas phase. The quality of the gas environment is more meaningfully specified by the mean free paths of the different particles or the impingement rate of molecules (Beynon, 1970).

2.3.2 Sinks of Gas

The gas molecules confined in a vessel being evacuated are removed by (a) the main pumps, (b) ionization gauges, residual gas analyzers, etc., and (c) the internal surfaces of the vacuum system.

(a) The main pumps produce a preferential direction of gas flow along a path of decreasing molecular density due to the distance between the pumps and the gas source. (Some pumping systems, such as cryopanels uniformly distributed around the gas sources, e.g., space simulator chambers or plasma machines, are exceptions.) Thus, in the molecular range of flow, the gas leaving a vessel is beamed at the inlet of the tubing connecting to the pump according to a certain distribution (Dayton, 1957). The pattern of the gas beam distribution is a function of the ratio l/r, where l is the length of the connecting duct (cylindrical with circular cross section) and r the radius of the duct. Figure 2.2 illustrates such a pattern for different l/r ratios. [For details concerning the patterns of gas beam distribution through slots see Reynolds and Richley (1964).]

(b) Ionization gauge heads and/or other instruments used to ionize the gas for measuring purposes, such as residual gas analyzers (RGA), pump-off gas by a combination of physical and chemical processes. The rate of pumping depends on several parameters: the vacuum history of the in-

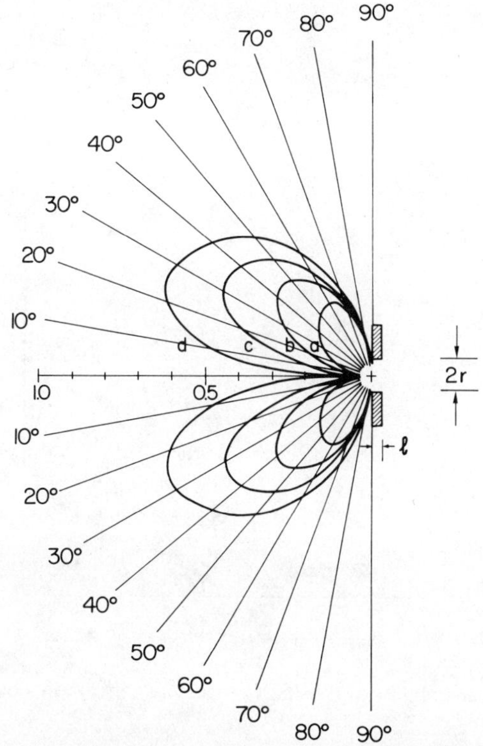

Fig. 2.2 Flow pattern of a gas beam distribution at the inlet of a thin aperture short tube. $l/r = 1$ (curve a), 2 (curve b), 4 (curve c), 10 (curve d). Adapted with permission from *Trans. Natl. Symp. Vac. Technol. 1956*, pp. 9–10. B. B. Dayton. Copyright 1957 Pergamon Press.

strument, the species of gas present in the vacuum vessel, the electrical characteristics of the measuring transducer, and its temperature (Alpert, 1962; Lee *et al.*, 1962) (see also Section 7.4). Figure 2.3 illustrates the pumping speed of a Bayard–Alpert gauge head for carbon monoxide and nitrogen.

(c) Generally, in a vacuum system, residual gases having molecular mean free paths in excess of the dimensions of the confining vessel may be sorbed at the internal surfaces. Sorption depends on parameters such as the physical nature and temperature of the surface, the chemical composition of colliding species, as well as the kinds of colliding species (Hobson, 1963). Re-emission and desorption processes and other phenomena which are difficult to follow and interpret, particularly when several gas species are involved, contribute to further complicate the situation (Moore, 1960, 1961).

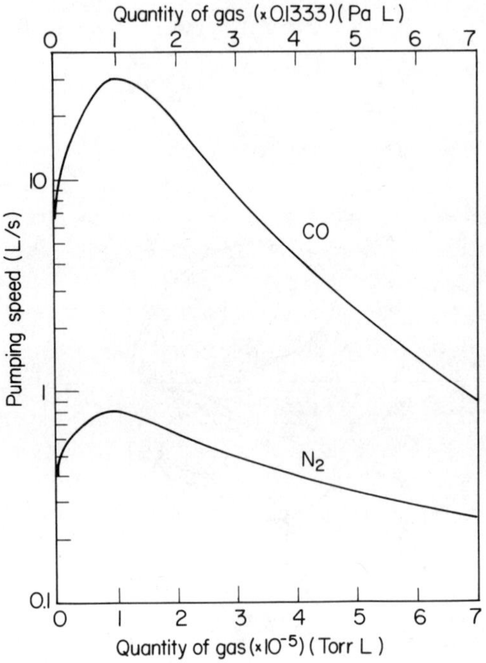

Fig. 2.3 Pumping speed of a Bayard–Alpert gauge head for carbon monoxide and nitrogen. (After Alpert, 1962.)

An obvious example of sorption phenomena is the effect occurring when pressure is measured with tubulated gauge heads. Under certain circumstances, the readings of a tubulated gauge head may be one order of magnitude lower than those of a gauge head immersed in the vacuum system (nude gauge head). Blears (1946) attributed this discrepancy in the readings to the adsorption of oil vapors (present in the rarefied environment due to the diffusion pump) at the walls of the gauge head tubulation. The inconsistency in the readings, referred to in literature as the *Blears effect,* was proved by Reich (1961) to be due to oil vapor. Other explanations of this effect have been formulated by Haefer and Hengevos (1961) and Apgar (1963).

2.3.3 Sources of Gas

Due to operating conditions in an evacuated system, different populations of gas, fragments of molecules, multimolecular agglomerations, and electrically charged particles are always present. The contributing sources are distributed throughout the system and can be attributed to one or a combination of the following mechanisms: (a) molecular gas leaks, (b) gas desorption, (c) transport of matter through the system, (d) decomposition of residual gases, and (e) permeation of gas.

(a) *Molecular gas leaks* through the ducts are used to introduce a certain amount of gas into the vacuum vessel (like in systems used for calibrating gauge heads). Such leaks generate molecular beams at the outlet of the duct. The molecular flux density in the beam is dependent on the ratio of the duct length *l* to the radius *r* of the duct (Dayton, 1965). Figure 2.4 illustrates such a pattern for different *l*/*r* ratios.

(b) *Gas desorption* occurs either spontaneously (outgassing) or deliberately (degassing) from the bulk of the surfaces of the components of the system, as a result of the energy transferred by

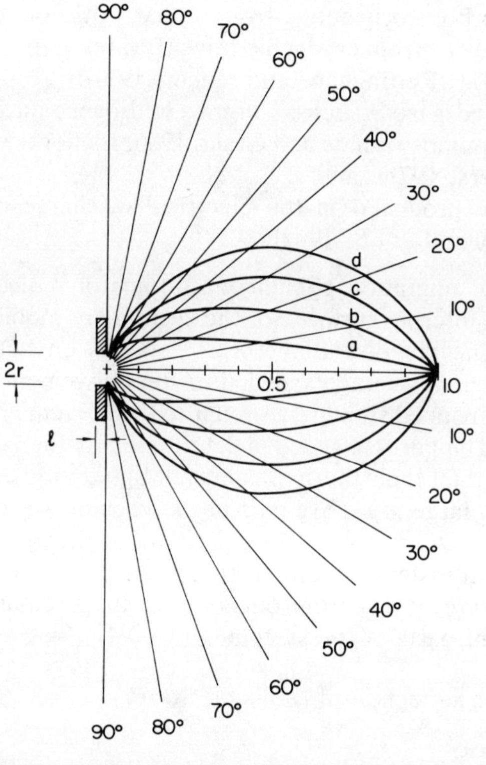

Fig. 2.4 Flow pattern of a gas beam distribution at the outlet of a thin aperture short tube. $l/r = 1$ (curve a), 2 (curve b), 4 (curve c), 10 (curve d). Adapted with permission from *Trans. Natl. Symp. Vac. Technol. 1956,* pp. 9–10. B. B. Dayton. Copyright 1957 Pergamon Press.

(i) heat, electrons (Leck and Stimpson, 1972; Madey and Yates, 1971),
(ii) ions (Clampit, 1972; Lichtman, 1965),
(iii) protons (Erents and McCracken, 1973),
(iv) chemical changes in the adsorbed layers (Martinez and Hudson, 1973),
(v) thermal radiation (Benvenuti and Calder, 1971),
(vi) photons (Kronauer and Menzel, 1972; Lange, 1965; Menzel, 1977), or
(vii) laser pulses (Gauthier *et al.*, 1977).

(c) *Transport of matter through the system* from the pumping system or the associated equipment towards the evacuated enclosure is caused by backstreaming and/or migration (creeping) phenomena. The transport due to backstreaming includes

(i) hydrocarbons originating from rotary, diffusion, or turbomolecular pumps, as well as from cryogenic traps (Backer *et al.*, 1972; Hablanian and Herrick, 1971; Rettinghaus and Huber, 1974);
(ii) re-emitted gases, such as argon, hydrogen, nitrogen previously pumped by ion pumps (Bance and Craig, 1966; Lamont, 1973; Williams *et al.*, 1972; Winters, 1971); and
(iii) methane produced in the electrical discharge of getter pumps (Bills, 1973; Holland *et al.*, 1962).

In the case of migration phenomena, atoms or molecules previously adsorbed at the internal surfaces of the system are mobile in two-dimensional space. The distance they travel depends on both their sticking probability and their average speed. For free movement conditions over the surface, their speed should be similar to the movement of molecules in the gas phase. The particle speed is determined by the activation energy, which is larger than their thermal energy. Migration phenomena become important when large areas are part of the vacuum system.

(d) *Decomposition of residual gases* within the system results from thermal energy transferred from electrical discharges, hot bodies, etc.

(e) *Permeation of gas* from outside into the vacuum system occurs through different parts of the system.

2.3.4 Specular Reflection of Molecules from Surfaces

Under certain conditions, specular reflection of molecules may occur, which affects the spatial distribution of the gas in the volume considered. Specular reflection develops during molecular flow of gas through ducts (Edmonds and Hobson, 1964; Gottwald, 1973; Teubner, 1967) and as a result of the interaction of atomic beams with cleaned crystal surfaces (Smith Jr., 1964) or with epitaxially grown films. Smith and Saltsburg (1964) assign the specular reflection of atomic beams to both the temperature and the gas adsorbed at the surface (or film). Light gases, such as hydrogen and helium, as well as heavier ones, such as neon, argon and xenon, can be scattered in the processes mentioned above.

2.3.5 Temperature Nonuniformities

Temperature nonuniformities within a vacuum system are produced by sources at temperatures different from that of the vessel (e.g., hot plasma discharges, cryo-panels, measuring instruments at liquid nitrogen temperature). Since in most cases the walls of the vacuum vessel are at room temperature, the gas flux, density, and pressure will vary according to the location considered. Localized gas densities will therefore affect the output of ionization gauge heads, which are unable to distinguish between isotropic and nonisotropic conditions (Moore *et al.*, 1966).

Moore (1964) gives a detailed theoretical treatment of nonuniformity of residual gas pressure and density due to temperature gradients in a chamber at high vacuum. He considers the cases of parallel planes and concentric spheres under the following experimental conditions: molecular flow, nonequilibrium temperature, and stationary gas. By using the term *stationary gas* it is meant that the number of molecules in the chamber is fixed, i.e., no sinks or sources of gas are present. All molecules are diffusively reflected, but their distribution of speeds is not Maxwellian. He concludes that when estimating pressure from density measurement in a chamber where the temperature of two opposite walls is T and a^2T, respectively, the maximum uncertainty in the observations is the temperature ratio a^2. The maximum uncertainty in pressure observations from the gaseous flux is the temperature ratio a. Both uncertainties may be reduced if the geometry and accommodation coefficients are known.

Equations expressing pressure under conditions of nonuniform temperature distribution must be corrected even if measurements are made in spherical geometry (Moore, 1965). In complex systems where nonuniform gas distributions prevail, Moore (1969) suggests the use of topological methods to determine the gaseous flux patterns. (For pressure measurement in confined environments under nonisotropic conditions, see Section 8.2.2.)

2.3.6 Transient Pressures

The properties of a rarefied gas are the same everywhere (isotropic) if temperature is constant throughout the vacuum system and no sources or sinks of gas exist (see Section 2.3.2, 2.3.3). An instrument which measures pressure by either direct or indirect methods (see Section 2.4) will indicate the same value regardless of its position and location relative to the gas mass, provided that

(a) the energy (if any) injected to operate the instrument does not disturb the prevailing thermal equilibrium,
(b) the instrument and the gas are at the same temperature, and
(c) the instrument is not a source or sink of gas.

The physical quantity the instrument measures (see Section 4.1), as long as such *steady-state* conditions exist, represents the property of the gas mass *at the instant* of determination. The steady state changes to a *transient regime* when the gas load (dq_v/dt) throughout the vacuum system equals the difference between the ultimate (Q_u) and instantaneous (Q_i) gas load, divided by a time constant t_c. Thus,

$$dq_v/dt = (Q_u - Q_i)/t_c \tag{2.5}$$

The value of t_c depends on the parameters of the measuring system. Expressed in pressure–volume units, Eq. (2.5) becomes

$$dP/dt = (P_u - P_i)/t_c \tag{2.6}$$

Here the subscripts u and i have the same meaning as in Eq. (2.5). Transient regimes either occur as a result of situations that cannot be controlled (e.g., pressure bursts in diffusion pumps under some circumstances) or are deliberately generated as part of an experiment (e.g., injection of gas into a vacuum vessel for the operation of arc-jets, plasma wind tunnels, and plasma generators or reduction of pressure according to a certain schedule in order to scale the ascent conditions in atmosphere of a sounding rocket).

Usually because of technological restrictions, low pressure measurement under a transient regime is made by locating the gauge head at a point different from that where pressure determination is required. It is therefore necessary to correct the readings of the measuring system to allow for the time t elapsed until the pressure at the output meter becomes the same as that at the point required.

The time constant (referred to as time response, dynamic response, rise time, lag coefficient) is the time required for the current (or voltage) in the gauge head power-supply system to rise to 63% ($1 - 1/e$) of its steady, ultimate value or fall to 37% ($1/e$) of its initial value, as a result of a step pressure input.

The value of t_c depends on (a) the gas flow regime, (b) both the type of the instrument used as pressure transducer and its construction in a gauge head (nude or tubulated), (c) the components interposed between the gauge head and the point where the measurement is required, (d) the electronic (electrical) circuitry of the power supply to energize the gauge head, and (e) the cables which connect the gauge head to the power supply.

(a) The gas flow within the vessel of a vacuum system can be viscous (turbulent or laminar), intermediate, or molecular. In a viscous flow regime, where gas can be considered as a continuous fluid (see Section 2.2), the time lag depends on the velocity of the gas pressure front and the distance between the gauge head and the point where pressure is required to be measured. Pressure measurement under transient regime in the viscous range of gas flow is of no interest to vacuum users.

In the intermediate and molecular ranges of flow the lag in the gauge head response is dependent on the time for the gas molecules to reach equilibrium conditions with the neighborhood of the pressure transducer (see Section 4.1). Gas desorption and/or adsorption at the components of the measuring assembly (e.g., the box which houses the pressure transducer, the tubing connecting the gauge head to the location where pressure is determined, apertures in the measuring system) contribute to this equilibrium.

(b) In the intermediate and molecular ranges of gas flow, pressures are determined by direct and indirect methods of measurement, respectively (see Section 2.4). Tubulated or nude-type gauge heads are used (see Chapter 4) depending on whether or not the pressure transducer is protected.

The time constant of a tubulated gauge head is made up of the gas kinetic time constant of the transducer holder τ_k and the time constant of the transducer itself τ_t.

The time required to reduce pressure to a given amount within a volume V, evacuated through an aperture of conductance C, is $\tau_k = V/C$. Small values of τ_k can be obtained by reducing V and increasing C. For special measuring purposes requiring a small time constant, such as in magnetically confined plasma research, miniature tubulated hot-cathode-ionization gauge heads have been designed (see Section 8.2.3). The box which accommodates the ionization transducer has either a rectangular (Lewin and Martin, 1962) or a circular (Masaaki and Kyoichi, 1972) cross section. Lewin and Martin (1963) designed a miniature hot-cathode-ionization transducer accommodated in a box 1.8 × 3.1 × 1.2 cm. The box was set flush (1.8 × 3.1 cm) with the wall where transient pressures were measured. The calculated time constant of their gauge head for deuterium was $\tau_k = 130$ s.

Since only thermal and ionization gauge heads are routinely used for the measurement of transient pressures, only these types of instruments will be considered (see Chapter 4).

The time constant for *thermal gauge heads* is specified in either of two ways.

(i) The time for the gauge head, held at reduced pressure (about 10^{-3} Pa; 10^{-1} Torr) to respond to a sudden inrush of air at atmospheric pressure. Specified in this way the value obtained for the time constant is misleading in that it is too short, as a result of both the change in thermal conductivity of the gas inside the gauge head and the forced cooling effect caused by the gas motion over the heated elements of the transducer. The response characterized by such a short value of τ_t is, however, used when the gauge head is required to protect the vacuum system against failures (e.g., electrical cutoff of the filament of ionization gauge heads in the vacuum system in case of its rupture to atmospheric pressure).

(ii) The time taken for the gauge head, stabilized at some pressure within the range where it is particularly useful, to respond to a sudden, small change in that pressure. A fiducial arbitrary value for the determination of τ_t is that pressure at which the gauge head reads midscale on the output meter of its power supply (Benson, 1962). Thus, for a gauge head designed for pressure determination within the range 10–10^5 Pa (10^{-1}–760 Torr), the midscale (nonlinear) value is 2×10 Pa (2×10^{-1} Torr).

The same response to a variation of pressure, small compared with the midscale value at which the gauge head is stabilized, can be obtained if the power supplied to heat the transducer is changed by a small step function [a function defined by two states $f(t) = 0$ for $t \leq t_0$ and $f(t) = Ct$ for $t > t_0$; this function can be expressed by a Fourier integral). This kind of simulation of the response to a sudden pressure change eliminates the errors in the values of τ_t resulting from gas adsorption and/or desorption at the transducer box.

Benson (1962) used this technique to determine the time constant of thermocouple and thermopile gauge heads. The transducers of the gauge heads were heated with ac, rather than dc, current in order to avoid the stray electromotive forces due to an *IR* drop in the heating circuit at the junction produced by dc current. The time constant of a thermopile gauge head, having a useful measuring range 0–2666.4 Pa (0–20 Torr), was determined by both methods. In the experiment where the heating input was varied, the gauge head was stabilized at 200 Pa (1.5 Torr) midscale value. The resulting τ_t was 0.10 s (Fig. 2.5). Then the gauge head, previously evacuated to low pressure, was connected to a gas reservoir (Fig. 2.6). The gas in the reservoir, at 200 Pa (1.5 Torr) pressure, could suddenly be released into the instrument box by means of a quick-opening

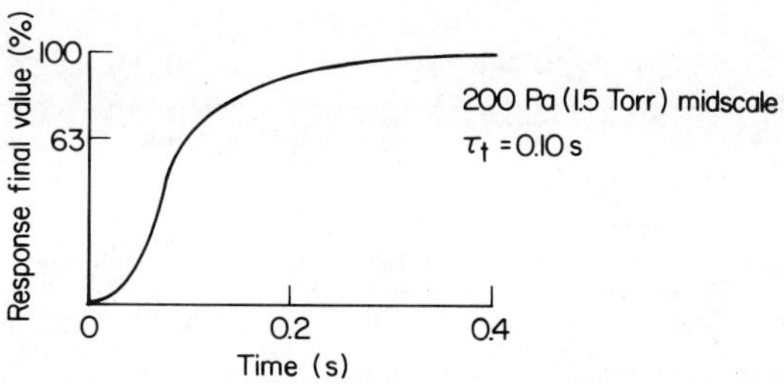

Fig. 2.5 Response of a thermopile gauge head to a variation in heating input. Reprinted with permission from *Tran. 8th Nat. Vac. Symp. 1961*. J. M. Benson. p. 492. Copyright 1962 Pergamon Press.

valve. The time constant ($\tau_t = 0.12$ s) found was in good agreement with that determined by the former method.

Some values of the time constant of different types of thermal gauge heads are shown in Table 2.4.

The time constant for *ionization gauge heads* depends on the mechanism yielding the ionized gas particles (thermionic emission, cold discharge) and is limited by the transit time of charged particles within the electrode system. For nude hot-cathode gauge heads, this time is extremely short (less than 10^{-6} s). For gauge heads of the cold-cathode type, the time constant τ_t is proportional to the transit times of electrons τ_t^- and ions τ_t^+ involved in electron–atom collision. Redhead (1958) calculated the transit times for electrons and ions in an ionization gauge head with crossed electric and magnetic fields (see Section 4.4.8) assuming that electrons move in a constant electric field between collisions and that the transit time of ions is not influenced by either the space charges or the magnetic field.

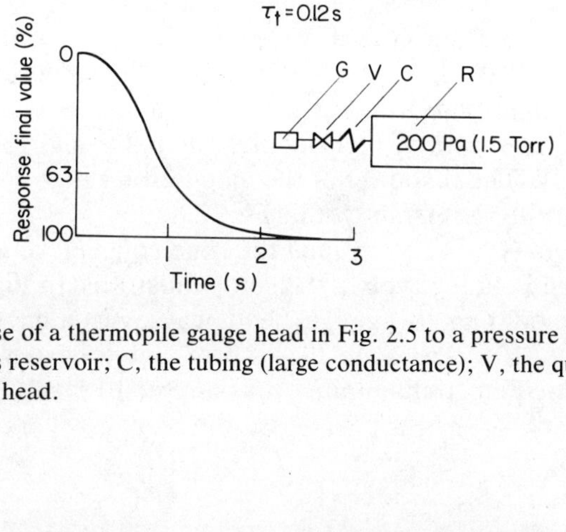

Fig. 2.6 Response of a thermopile gauge head in Fig. 2.5 to a pressure burst of 200 Pa (1.5 Torr). R is the gas reservoir; C, the tubing (large conductance); V, the quick opening valve; and G, the gauge head.

Table 2.4

TIME CONSTANTS OF DIFFERENT THERMAL GAUGE HEADS[a]

Gauge head type	Construction of the box housing the transducer	Temperature range (°C)		Response time within the range 10^{-1}–10 Pa (10^{-3}–10^{-1} Torr)
		Operating	Maximum	
Pirani	Al alloy	0–50	0–50	<0.2
	Hard glass, bakeable			2×10^{-2}
Thermistor	Metal	15–35	100	$<5 \times 10^{-2}$
Thermocouple	Metal, glass	30	250	0.3
	Metal, glass	1.5	48	25
	Metal	15–35	100	$<5 \times 10^{-2}$
	Metal, pyrex	−20	100	1
	Metal, pyrex	−20	100	15
	Metal, pyrex	−20	100	25
	Metal, pyrex	—	—	3

[a] According to Holland *et al.* (1974).

The values he found for τ_t^- and τ_t^+ are very approximate because of the distortion produced by the electric fields at the end plates of the gauge head. However, the resulting value for τ_t^+ (about 10^{-6} s for pressures within the range 10^{-2}–10^{-10} Pa (10^{-4}–10^{-12} Torr) may be used to provide a rough estimate of the time constant τ_t.

Smith and Melfi (1966) and Smith (1971) also measured the time constants of ionization gauge heads with crossed electric and magnetic fields. The setup used by Smith for magnetron gauge heads within the pressure range 10^{-1}–10^{-4} Pa (20^{-3}–10^{-6} Torr) is shown in Fig. 2.7. The gauge heads under test maintained at initial pressures in the 10^2-Pa (10-Torr) range were submitted to a sudden pressure increase produced by injecting small quantities of gas into the system by means of quick-acting microvalve (Gorowitz *et al.*, 1960). Two nude, hot-cathode gauge heads, mounted at the outlet of the microvalve and near the instrument tested, were used to indicate the average time of flight of the gas pulse and to monitor pressure, respectively. The response of the magnetron gauge head to a sudden pressure rise of nitrogen is shown in Fig. 2.8.

Smith and Melfi (1966) determined the time constant of a nude cathode gauge head of the Penning type in the setup illustrated in Fig. 2.9. The bell jar (B_j) accommodating the gauge head was evacuated to its ultimate pressure (6.67×10^{-5} Pa, 5×10^{-7} Torr) and then dry nitrogen was leaked into the bell jar to maintain a pressure of 10^{-4} Pa (10^{-6} Torr). The

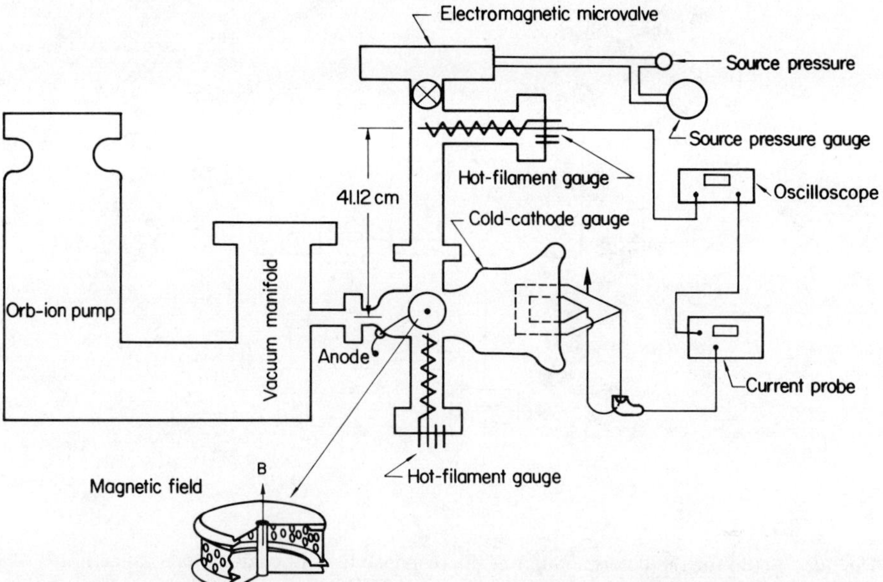

Fig. 2.7 Diagram of a dynamic response test apparatus (After Smith, 1971. Work for the U.S. government.)

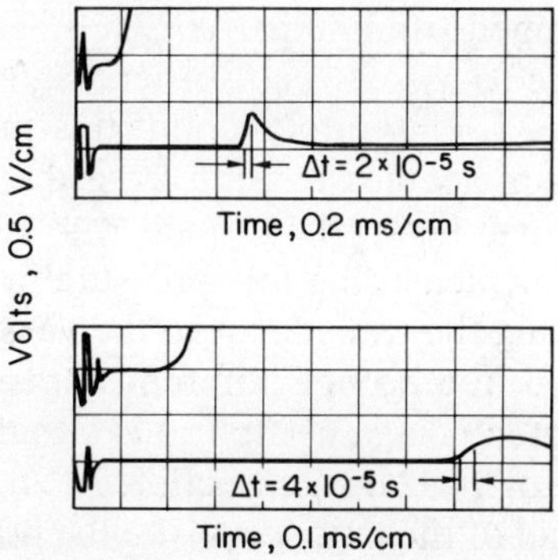

Fig. 2.8 Dynamic response measurement (nitrogen). Lower trace: response of a cold-cathode gauge to a gas pulse (microvalve source pressure 0.5 atm). Upper trace: response of the reference gauge at the gas inlet (microvalve source pressure 1 atm). (After Smith, 1971. Work for the U.S. government.)

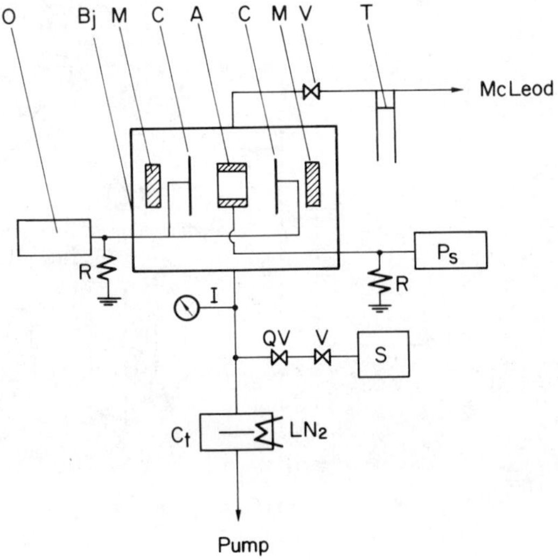

Fig. 2.9 Setup for the determination of the response time of cold-cathode gauge heads of the Penning type. A is the anode; B_j, the bell jar; C, the cathode; C_t, the cold trap; I, the ionization gauge head; M, the magnet poles; O, the oscilloscope; QV, the quickacting microvalve; P_s, the power supply; R, the resistor; S, the gas reservoir; T, the symmetrically cooled **U** trap; and V, the valve.

quick-acting microvalve (QV), designed based on a principle similar to the one used by Gorowitz *et al.* (1960), had a capacity of 1 to 400 μg of gas at a single opening. The reservoir S was pressurized at 14.053 $\times$ 10^5 Pa. The time the QV was open ranged from 40 to 100 μs.

Observations were made to assess the transient signal occurring at the start of the valve opening. Taking into account this time, as well as the time delay due to the gas molecules passing from the exit of QV to the discharge region of the gauge head, they found τ_t = 100 μs.

(c) A typical system for measuring low pressure in a vacuum system consists of a pipe connecting the transducer to the vessel (Fig. 2.10). The transducer (usually housed in a box of different dimensions than that of the pipe) is sometimes protected against spurious signals produced by the gas measured by a louvre or screen (see Section 7.5.6). An aperture may also be necessary at the end of the pipe, opposite the pressure transducer. The time lag in the response of the transducer to a pressure variation in the vacuum vessel depends upon the time taken by the gas flowing in and out of the pipe to reach thermal and mechanical equilibrium with the measuring system. For each set up of a given vacuum and measuring

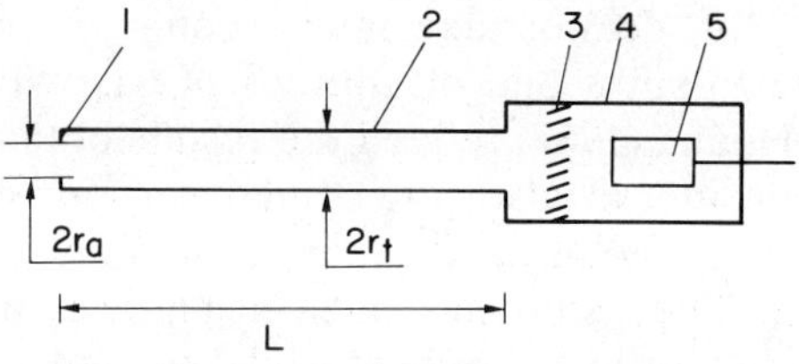

Fig. 2.10 Typical system for measuring gas at low pressure. 1, aperture; 2, tubing; 3, louvre; 4, box; 5, transducer. r_a is the aperture radius; r_t, tubing radius.

system, there exists an optimum piping diameter (Schaaf and Cyr, 1949). Too large a diameter increases the time lag because the gas must reach equilibrium conditions within a large volume; too small a diameter also increases the time lag because the small conductance reduces the gas flow. Schaaf and Cyr (1949) calculated the time constant of a measuring system like that illustrated in Fig. 2.10 (without a louvre or screen) in the molecular range of gas flow. They assumed that for steady isothermal flow in a large pipe, the instantaneous flow rate of gas is proportional only to the instantaneous pressure gradient. The time constant of the measuring system is the sum of the time constants of each of the components of the system:

$$\tau_p = \mathrm{Ct}|VL/r_t^3 + 8V/3r_a^2 + 8\pi r_t^2 L/3r_a^2 + 4L^2/\pi r_t| \tag{2.7}$$

where V is the volume of the box accommodating the transducer and Ct a constant having the value $\mathrm{Ct} = \frac{3}{4}\,|1/(2\pi RT)^{1/2}|$, R and T being the gas constant and the thermodynamic temperature, respectively.

The four time constants in Eq. (2.7) represent the flow of gas: in the box accommodating the transducer through the resistance of the tube; through the resistance of the aperture; in the tube volume through the resistance of the aperture; and in the tube through the distributed resistance of the tube itself. In the case of a measuring system with no aperture, the second and third terms in Eq. (2.7) can be neglected and the time constant τ_p expressed as

$$\tau_p = \mathrm{Ct}|VL/r_t^3 + 4L^2/\pi r_t| \tag{2.8}$$

Usually the length L of the tubing (2 in Fig. 2.10) is as short as possible, so that the connection has a small volume. For small values of L and a rather large opening of the aperture of radius r_a, the fourth term in Eq. (2.7) can be neglected. Schaff and Cyr also deduced the optimum value of r for the tubing as

$$r_{opt} = (0.18 r_a^2 V)^{1/5} \tag{2.9}$$

Equations (2.7–2.9) do not take into account the effect on pressure determination due to sinks and/or sources of gas within the measuring system. Soo and Huang (1962) refined the results obtained by Schaaf and Cyr in that they considered adsorption and desorption phenomena, which become an important factor in the molecular flow of gas through the measuring system. They studied a measuring system having the same configuration as that used by Schaaf and Cyr and assumed simplified isothermal conditions as well as a semiempirical adsorption–desorption model. Three basic situations were studied, i.e., stepwise pressure changes, linear rates of pressure change, and periodic variations of pressure. The experimental results obtained are in good agreement with the theoretical ones.

Matricon (1962) theoretically examined the problem of time constants of both a pumping and a measuring system for the case when the pumping load is continuously reduced with time. He analyzed the measuring system characterized by *uniformly distributed constants* or by *localized constants* according to whether the conductances, throughputs, and volumes are considered uniform over all the measuring system or can be separated and attributed to each part of the system. According to the analysis, the time constant of a measuring system with a tubulated gauge head having a volume that does not include that of the connecting tubing can be calculated with an error of less than 4%. The time constant in this case is obtained by summing the time constants of the tubing and gauge head housing. He concluded that for gas loads of short duration in a volume where pressure is continuously reduced, it is advisable to use the approach of uniformly distributed constants in the calculation of the time constant. In his study Matricon (1962) did not take into account either sinks and/or sources of gas within the measuring system.

Hord (1967) presented a simple method for estimating the time response of a measuring system in the case of a nonperiodic input of a step function. The measuring system is connected to either a quick-connecting valve or a bursting diaphragm to obtain the step input required. Assuming quasi-steady gas flow conditions, Hord obtained an algebraic formula called the *time response function* (TRF), which predicts the response time of a measuring system in the molecular, transition, and continuum flow regimes with an accuracy of $\pm 20\%$. The formulas which are limited to the upper boundary of the viscous regime, but not restricted for the molecular range of gas flow, also account for thermal transpiration effects.

(d) The time constant τ_e of the electrical and/or electronic circuitry of the power supply, including the output meter (usually an oscilloscope), has an extremely low value, differing in usual experiments by several orders of magnitude from the gas kinetic time constant.

(e) The time constant τ_c of the cables which connect the gauge head to the power supply is expressed by

$$\tau_c = R_e C_p \tag{2.10}$$

where R_e and C_p are the electrical resistance and capacitance, respectively, of the cables. The values usually found in practice ($R \cong 10^4 \Omega$ and $C_p \cong 1000$ pF) lead to $\tau_c = 10$ μs.

Thus, the time constant T_c of the entire measuring system is given by

$$T_c = \tau_p + \tau_e + \tau_c \tag{2.11}$$

where τ_p is given by

$$\tau_p = \tau_k + \tau_t \tag{2.12}$$

2.4 Methods of Pressure Measurement

2.4.1 Direct Method

The direct method, absolute in that the determination of pressure does not depend on the gas species, is based on the measurement of the elastic properties of the gas. Measurements can be made by balancing the force exerted by the gas on a unit area of a plane against

(a) the hydrostatic force of a liquid or a gas column,
(b) the strain yielded in an elastic element,
(c) gravitational forces, or
(d) centrifugal forces (Brombacher, 1970).

Regardless of the technique for the measurement of pressures higher than 10 mPa (10^{-4} Torr), the method offers sufficient confidence, as proved by the uncertainty analyses of extensive experimental data.

The measurement of pressures below about 10 mPa (10^{-4} Torr) becomes difficult due to the limited ability of present-day instruments to accurately measure minute mechanical forces. The amplification of the primary signal by mechanical, optical, or electrical means (vibration, hysteresis, mercury vapor flow, etc.) constitutes an additional source of uncertainty associated in most cases with the measuring instrument.

2.4.2 Indirect Method

In the indirect method a particular physical quantity of the gas phase is measured, and pressure is deduced by conventional kinetic theory relationships. These relationships hold true for the molecular flow of gas having a stationary distribution of velocities, uniform temperature and composition, and thermal and mechanical equilibrium with the confining container. This last condition requires that the area of the aperture through which the gas is removed be relatively small so that equilibrium is not disturbed by the particles entering the container from the pump. Usually quantities which are well-defined functions of the number density of gas molecules, such as viscosity, thermal conductivity, and ionizability, are measured.

In the low and medium vacuum ranges where the gas transport mechanism undergoes a distinct change, viscosity or thermal conductivity are usually determined. In the high and ultrahigh vacuum ranges the positive ion current resulting from the bombardment of the residual gas with accelerated particles, usually electrons, is measured. However, the ionization of gas molecules can also be used to measure pressures in the medium vacuum range by using a special gauge head design of the Schultz–Phelps type. Provided that the heating power of the electron emitter is not drastically limited due to the increased thermal conductivity of the gas, pressures as high as 133.3 Pa (1 Torr) can be measured. The direct and indirect methods overlap in the medium vacuum range.

Since the magnitude of each of the quantities measured by the indirect method varies according to the gas species, pressure determination depends on the chemical composition of the gas.

Conceptual errors in the principle of the indirect method arise when attempting to attribute the pressure measured at a given point in a dynamic gas to the gaseous environment as a whole. The readings of a gauge head in such an environment will change as it is moved with respect to the sinks or sources of gas as a function of both the gauge head position and orientation in the gas flow. In such cases one resorts to measuring pressure profiles that can be obtained by a reasonable distribution of an appropriate number of gauge heads at points where pressure measurements are required (Laurent *et al.*, 1977).

The response of the gauge head complicates the interpretation of pressure readings for one or a combination of the following reasons.

(a) The chemical composition, temperature, or amount of gas can be altered by either the gauge head or the tubing connecting the head to the measuring point.

(b) It is not always possible to locate the gauge head at the position where pressure is required, but only at a technologically feasible point. An unknown pressure differential will therefore exist between these two locations.

(c) The gas is not all of one species.

If ionization gauge heads are considered, additional uncertainties may be contributed by spurious currents not related to gas pressure, uneven sensitivity throughout the range of pressures measured, instability of the electronic circuitry of the power supply, and Barkhausen–Kurtz oscillations. Despite their drawbacks, however, ionization gauge heads are used almost exclusively for pressure determination in the high and ultrahigh vacuum ranges.

The following characteristics of ionization gauge heads should be borne in mind when considering their application.

(a) Nude gauge heads do not pump (see Section 7.5). When the electrode system is of symmetric design (conventional triode), the gauge heads do not differentiate between molecules entering into their ionization volume with random or preferred directions of flow.

(b) Changes in the local number density of molecules as a result of nonisotropic gas distribution alter the gauge head readings (Moore *et al.*, 1966). When the electrode system is nonsymmetric (e.g., Bayard-Alpert type), the response of the gauge head is sensitive to the direction of the gas flow (Kleber, 1975). Regardless of the symmetry of the electrode system, the readings of nude gauge heads are practically independent of molecular velocities, due to the high velocity of electrons.

(c) Tubulated gauge heads pump, dissociate, and react with the gas being measured under certain circumstances and may also generate gases. Their response depends on the molecular flow rate density and on the incident direction of flow. To a first approximation, their response is proportional to the cosine of the angle between the normal to the plane of the tubulation aperture and the direction of gas flow (Feakes *et al.*, 1965).

Regardless of the method selected for pressure measurement (direct or indirect) care must be taken in relating the gas flow direction to the gauge aperture. Thus, *static* or *dynamic pressure* will be read according to whether the normal to the aperture is oriented perpendicular or parallel to the mass velocity vector.

2.5 The Process of Pressure Measurement

2.5.1 Object of Measurement

All measurements are essentially inaccurate to some extent, so that what is called the "actual" or the "accurate" value of a physical quantity is not attainable. It is reasonable, however, to assume that the accurate value exists, and the experimentalist is concerned with estimating the limits between which this value lies. The closer the limits, the more accurate the measurement. The required accuracy of a reported value depends on the intended use of that value, and in practice, it is not always useful to require very high accuracy. Cruder results may serve the purpose of an experiment well enough, provided that the conclusions inferred from the measurements are not affected.

Measurement of a physical quantity consists of finding how many times it contains a fixed, known quantity of the same kind, called a unit. Measurement is usually a repeatable process and its accuracy can be specified by (i) the systematic error or bias of the observations, (ii) the precision obtained, and (iii) the distribution of the individual measurements about the average. It is also necessary to specify end results within *confidence limits,* corresponding to a selected *confidence level,* in order to determine credible bounds to the measurement. It therefore becomes obvious that a measurement process can yield accurate results *only* if developed under the concepts and techniques of statistical estimation and decision theory. The statistical interpretation of results of a measurement process is beyond the scope of this book. Many excellent textbooks concerned with the subject of error observation and related topics are available. The aim of the following paragraphs is to point out the characteristic parameters of the measurement process, in general, and of the instruments used for low pressure measurement, in particular.

2.5.2 Characteristic of the Measurement Process

The difference between the measured value and the true one is called the *error of observation*. It can be due to the observer (*accidental errors*) or to both the observer and the instrument (*systematic errors,* or *bias*). Accidental errors are usually random and of variable magnitude, either positive or negative. These errors can be easily eliminated. The systematic error or bias, defined as the difference between the estimated limiting mean of the output values and the true value measured at the input, is sometimes difficult to determine.

Accuracy, also referred to as *uncertainty,* designates the degree of *agreement of the value of a physical quantity with its true value*. Sometimes *inaccuracy* is used to express *disagreement* between the measured and true values. Thus, accuracy may be considered as the degree of correctness with which a measurement process yields the *true value* of the measured quantity. The units may be those of the measured quantity or the ratio (or percent) of the error to the full scale value or to the actual value. Usually, when a large number of observations is involved, the difference between the true value and the mean value of a distribution of observed values indicates the *mean accuracy*. This, however, involves stable input signals under invariable operating conditions. The effort necessary to attain high accuracy is directly related to the cost involved in the measurement and constitutes the main factor to be taken into account in planning the measuring process.

Uncertainty (sometimes called *average deviation, probable error, standard deviation*) is an estimate of the amount by which the observed or calculated value departs from the actual value of the physical quantity.

Confidence limits are defined as follows: If σ_s and S are, respectively, the standard deviation and the normal sampling distribution (sample size $N \geq 30$), then $S \pm 1.96\sigma_s$ and $S \pm 2.58\sigma_s$ are 95 and 99% confidence limits, respectively, for S. The confidence limit is often called the confidence level. The numbers 1.96, 2.58, etc., in the confidence limits are called *confidence coefficients* or *critical values*.

Precision is the degree of agreement between a number of independent observations of the same physical quantity obtained under the same procedural conditions. The precision of a measurement process can be evaluated by the standard deviation (σ) of the determinations. Precision and accuracy become synonymous when there is no bias in the experimental and estimation procedures. However, even when bias exists, accuracy and precision are seldom confused. To point out the difference, the results of two measurement processes are plotted in Fig. 2.11. Thus, if μ_a and M are, respectively, the actual mean value and the observed mean value of the physical quantity measured, curve 1 represents a measuring process of high accuracy and low precision, whereas curve 2 shows a high-precision, low-accuracy measuring process.

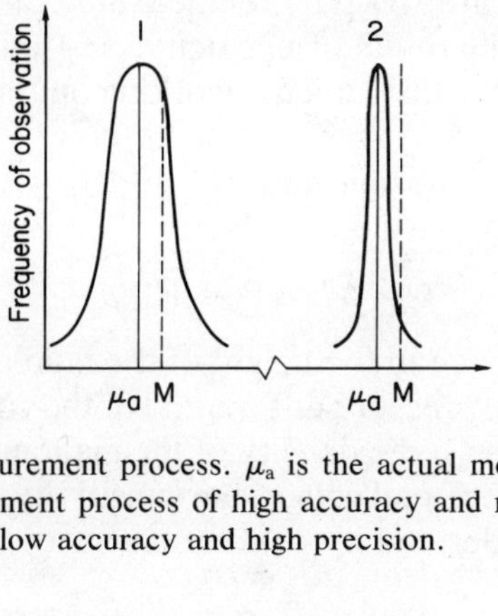

Fig. 2.11 Plot of a measurement process. μ_a is the actual mean value; M, the observed value. Curve 1: Measurement process of high accuracy and reduced precision. Curve 2: Measurement process of low accuracy and high precision.

Resolution is the ability to discriminate the smallest incremental values of the physical quantity measured.

Repeatability is the extent of agreement among a number of consecutive measurements, made under the same operating conditions. The observations are approached from the same direction and for full range traverses. This parameter does not include hysteresis.

Stability is the tendency of an instrument used for determinations to remain in a given state or condition, without spontaneous change, under complete freedom from oscillatory tendencies.

Traceability is the chain of measurements by which an instrument to be used as a secondary standard is calibrated against a primary (absolute) standard.

2.5.3 Characteristic Parameters of Gauge Heads

Sensitivity is, if not otherwise specified, the change in the output response of a gauge head to a specified change in the measured variable at the input. Expressed this way sensitivity represents a slope, usually measured on the linear region of the gauge head output plot. According to whether the instrument is primary (absolute) (see Section 3.1) or secondary, sensitivity is calculated or measured. For the common vacuum manometers or gauge heads, sensitivity is expressed as follows.

(a) For gauge heads measuring force effect by the deformation of an elastic element (see Section 4.2.1), sensitivity is

$$s = \Delta l/\Delta P = \mathrm{Ct}r^3/E_{\mathrm{m}}a^3 \tag{2.13}$$

In the deflection pressure ($\Delta l/\Delta P$) relationship, Ct is a constant, r, E_{m}, and a are the radius, modulus of elasticity, and thickness of the elastic membrane, respectively. Instruments working on this principle need calibration.

(b) For manometers measuring force effect by balancing hydrostatic forces (see 4.2.2), sensitivity is

$$s = \Delta h/\Delta P = 1/(g\rho) \tag{2.14}$$

where Δh is the difference in the height of the fluid in the two columns of the manometer, ΔP the pressure differential of the columns, g the acceleration due to gravity, and ρ the density of the manometer fluid (usually Hg or oil) Except for the McLeod gauge, the sensitivity of manometers of the liquid column type is derived by calibration.

(c) For manometers measuring the gas kinetic effect due to gas viscosity (see Section 4.3.4), we have

$$s = 40\eta r^2/(3\bar{c}g_B) \tag{2.15}$$

Equation (2.15) expresses the sensitivity of a spinning-rotor viscosity gauge head, the only rugged one in the class of viscosity manometers. Here r and g_B are the radius and the mass of the rotating ball, respectively, η is the viscosity coefficient, and $\bar{c}$ the average speed of the gas molecules. According to Reich (1982) calibration is required if the range of η is not wide enough.

(d) There are two type of gauge heads measuring gas density by ionizing the gas (see Section 4.4.1)

(i) *Hot-cathode-ionization gauge heads.* The absolute sensitivity s referred to as the gas ion constant (Schuetze and Stork, 1962) or tube calibration factor (AVS 6.4, 1969) is expressed as

$$s = I_c/I^-P \tag{2.16}$$

where I_c and I^- are the collected ion current and the emission current and P is the gas pressure. The absolute sensitivity can be regarded as the number of collected ions per electron emitted at the corresponding pressure P and has the units of reciprocal pressure (1/Pa, 1/Torr) provided that both I_c and I^- are expressed in the same units. Sensitivity is also known as the conversion constant (AVS, 1980) denoting the ratio of the collected ion current to the corresponding pressure A/Pa (A/Torr). The conversion constant is given for a specified emission current.

For the determination of ionization gauge performance (see Section 7.3), Eq. (2.16) is written

$$s = (I_c - I_r^-)/I^-(P - P_r) \tag{2.17}$$

where I_r is the residual current to the ion collector at pressure P_r.

(ii) *Cold-cathode-ionization gauge heads.* The expression for the absolute sensitivity is

$$s = I_c^+/P^n \tag{2.18}$$

where I_c is the positive ion current collected at the cathode, P the gas pressure, and n a numerical exponent. The value of n is usually greater than unity; for Penning gauge heads, $1.1 \leq n \leq 1.4$.

The absolute sensitivity of hot- and cold-cathode gauge heads is determined by calibration (see chapter 5 and Section 7.3). The ratio of the absolute sensitivity for a given gas to the absolute sensitivity for nitrogen

Table 2.5

RELATIVE SENSITIVITY OF DIFFERENT GAUGE HEADS NORMALIZED TO ARGON AND NITROGEN

Gauge head type	Gas						Pressure		V_g	I^-	Magnetic flux density		Reference
	N_2	Air	O_2	Ar	He	Ne	Pa	Torr	V	A	T	G	
Hot cathode	—	—	—	1	0.1283	0.2407	—	—	180	10^{-3}	—	—	Moesta and Penn, 1957
Triode	0.67	—	—	1	0.14	0.22	10^{-3} up	10^{-5} up	140	10^{-4}	—	—	Schultz, 1957
	0.75	0.81	0.94	1	0.13	—	10^{-3}–10^{-1}	10^{-5}–10^{-2}	127.5	10^{-4}	—	—	McGowan and Kervin 1960
	—	—	—	1	0.17	0.26	10^{-7}–10^{-2}	10^{-9}–10^{-4}	240	10^{-3}	—	—	Cobic et al. 1961
	1	—	—	—	—	—	$<5 \times 10^{-3}$	$<5 \times 10^{-5}$	145	—	—	—	Ehrlich 1963
	0.53	—	0.48	1	—	—	10^{-2}	10^{-4}	150	10^{-2}	—	—	Anderson 1963
Bayard-Alpert	—	—	—	1	0.134	0.258	10^{-3}–10^{-1}	10^{-5}–10^{-3}	180	10^{-4}	—	—	Bennewitz and Dohmann 1966
	0.56	—	—	1	0.15	—	10^{-4}–10^{-3}	10^{-5}–10^{-5}	140	10^{-4}	—	—	Shaw 1966
	—	—	—	1	—	0.22	10^{-6}–10^{-3}	10^{-8}–10^{-5}	105	8×10^{-5}	—	—	Redhead et al. 1968
	0.705	—	0.616	1	0.127	0.22	10^{-2}–3×10^{-1}	10^{-4}–3×10^{-3}	145	9.3×10^{-5}	—	—	Utherbach and Griffiths 1966
Cold cathode													
Magnetron	1	—	0.99	1.76	0.24	—	10^{-5}–10^{-3}	10^{-7}–10^{-5}	5000	—	0.1	1000	Barnes et al. 1962
Inverted megnetron	1	—	—	—	0.15	—	10^{-6}–10^{-3}	10^{-8}–10^{-5}	6000	—	0.2	2000	Hobson and Redhead 1958
Trigger discharge	1	—	0.66	—	—	—	5×10^{-8}	5×10^{-10}	2000	—	0.1	1000	Davis 1968

Table 2.6
RELATIVE SENSITIVITIES OF BAYARD–ALPERT GAUGE HEADS FOR DIFFERENT GASES[a]

	Bayard–Alpert gauge head number				
Gas	1	2	3	4	5
N_2	1.00	1.00	1.00	1.00	1.00
CO	0.947	0.954	0.942	0.956	0.951
CO_2	1.36	1.36	1.35	1.34	1.35
CH_4	1.59	1.56	1.58	1.59	1.57
C_2H_4	1.30	1.34	1.28	1.30	1.23
C_2H_6	2.60	2.54	2.61	2.54	2.60
C_3H_6	1.75	1.79	1.79	1.75	1.76
NH_3	0.660	0.654	0.615	0.625	0.668

[a] $I_e = 1$, mA, $v_f = 125$ V, $v_e = -25$ V. After Nabayama and Hojo (1974).

(or argon) is called *relative sensitivity*. Tables 2.5 and 2.6 show the relative sensitivity of different gauge heads for different gases.

Linearity expresses the proportionality between the output and the input of the gauge head. Linearity is the degree to which a set of successive readings approximates a straight line (usually in the region where sensitivity is linear).

References

Alpert, D. (1962). *Vide* **17,** 19.

Apgar, E. (1963). *2nd European Symposium "Vacuum"*, p. 223. R. A. Lang. Esch. (Taunus).

AVS 6.4 (1969). Procedure for calibrating hot filament ionization gauges against a reference manometer in the range 10^{-2}–10^{-5} Torr.

AVS (1980). Dictionary of Terms for Vacuum Science and Technology, Surface Science, Thin Film Technology, Vacuum Metalurgy, Electronic Materials (M. S. Kaminsky and J. Lafferty, eds.). Amer. Inst. Phys., New York.

Backer, M. A., Holland, L., and Stanton, D. A. G. (1972). *J. Vac. Sci. Technol.* **9,** 412.

Bance, U. R., and Craig, R. D. (1966). *Vacuum* **16,** 647.

Benson, J. M. (1962). *Trans. 8th Natl. Vac. Symp. USA 1961,* p. 489. Pergamon Press, New York.

Benvenuti, C., and Calder, R. S. (1971). *Phys. Lett.* **35A,** 291.

Beynon, J. (1970). *Vacuum* **20,** 443.

Bills, D. G. (1973). *J. Vac. Sci. Technol.* **10,** 65.

Blears, J. (1946). *Proc. Roy. Soc. A* **188,** 62.

Brombacher, W. G. (1970). *NBS Monograph 114.*

BS 2951: Part 1: 1969 Glossary Of Terms Used in Vacuum Technology, Part 1. Terms of general application. British Standard Institution.

Clampit, R. (1972). *In* "Adsorption–Desorption Phenomena" (F. Ricca, ed.), p. 203. Academic Press, New York.

Close, K., and Yarwood, J. (1970). *Vacuum* **20,** 56.

Chuan, R. L. (1964). *Res./Dev.* **15,** 144.

Dayton, B. B. (1957). *Trans. Natl. Symp. Vac. Technol. USA 1956,* p. 5. Pergamon Press, New York.

Denis, N. T. M., and Heppel, T. A. (1968). "Vacuum Systems Design," pp. 1–3. Chapan and Hall, London.

DIN 28400 Blatt 1 Ianuar (1972). Vakuumtechnik Bennenungen und Definitionen.

Edmonds, T., and Hobson, J. P. (1964). *J. Vac. Sci. Technol.* **2,** 182.

Erents, S. K., and McCracken, G. M. (1973). *J. Appl. Phys.* **44,** 3139.

Feakes, F., Torney, Jr., F. L., and Brock, F. J. (1965). *NASA CR-167.*

Gauthier, R., Babout, M. and Guittard, C. (1977). *Proc. 7th Int. Vac. Congr. Vienna 1977* (Dobrozemsky *et al.,* eds.), p. 1213. Bergher and Sohne, Horn, Austria.

Giorgy, T. A., and Pisani, C. (1966). *Vacuum* **16,** 669.

Gorowitz, B., Moses, K., and Gloersen, P. (1960). *Rev. Sci. Instrum.* **31,** 146.

Gottwald, B. A. (1973). *Vac. Tech.* **22,** 106.

Hablanian, M. H., and Herrick, R. F. (1971). *J. Vac. Sci. Technol.* **8,** 317.

Haefer, R. A., and Hengevos, J. (1961). *Trans. 7th Nat. Symp. Vac. Technol. USA 1960,* p. 67. Pergamon Press, New York.

Hobson, J. P. (1963). *Brit. J. Appl. Phys.* **14,** 544.

Holland, L., Laurenson, L., and Allen, P. G. W. (1962). *Trans. 8th Natl. Vac. Symp. USA 1961,* p. 208. Pergamon Press, New York.

Holland, L., Steckelmacher, W., and Yarwood, J. (1974). "Vacuum Manual," pp. 209–245. E. and F. N. Spon, London.

Hord, J. (1967). *Instrum. Soc. Am. Trans.* **6,** 252.

Kleber, P. (1975). *Vacuum* **25,** 191.

Kronauer, P., and Menzel, D. (1972). *In* "Adsorption–Desorption Phenomena" (F. Ricca, ed.), p. 301. Academic Press, New York.

Lange, W. J., and Eriksen, D. P. (1966). *J. Vac. Sci. Technol.* **3,** 303.

Lamont, Jr., L. T. (1973). *J. Vac. Sci. Technol.* **10,** 251.

Laurent, J. M., Benvenuti, C., and Scalambrin, F. (1977). *Proc 7th Int. Vac. Congr. Vienna 1977* (Dobrozemsky *et al.,* eds.), p. 113.

Leck, J. H., and Stimpson, B. P. (1972). *J. Vac. Sci. Technol.* **9,** 293.

Lee, D., Tomaschke, H., and Alpert, D. (1962). *Trans. 8th Natl. Vac. Symp. USA 1961,* p. 151. Pergamon Press, New York.

Lewin, G., and Martin, G. (1962). *Rev. Sci. Instrum.* **33,** 447.

Lewin, G., and Martin, G. (1963). *Rev. Sci. Instrum.* **34,** 942.

Lichtman, D. (1965). *J. Vac. Sci. Technol.* **2,** 270.

Madey, T. E., and Yates, Jr., J. T. (1971). *J. Vac. Sci. Technol.* **8,** 525.

Martinez, J. M., and Hudson, J. B. (1973). *J. Vac. Sci. Technol.* **10,** 35.

Matricon, M. M. (1962). *Vacuum* **12,** 59.

Masaaki, I., and Kyoichi, K. (1972). *Rev. Sci. Instrum.* **43,** 1670.

Menzel, D. (1977). *Proc. 7th Int. Vac. Congr. Vienna 1977* (Dobrozemsky *et al.,* eds.), p. 1003. Bergher and Sohne, Horn, Austria.

Moore, G. E. (1960). *Trans. 6th Natl. Symp. Vac. Technol. USA 1959,* p. 16. Pergamon Press, New York.

Moore, G. E. (1961). *J. Appl. Phys.* **32,** 1241.
Moore, B. C. (1964). *J. Vac. Sci. Technol.* **1,** 10
Moore, B. C. (1965). *J. Vac. Sci. Technol.* **2,** 211.
Moore, B. C. (1969). *J. Vac. Sci. Technol.* **6,** 246.
Moore, B. C., Bergquist, L. E., Camarillo, R. G., and Larson, R. (1966). *Trans. 3rd Int. Vac. Congr. Stuttgart 1965, p. 251.* Pergamon Press, New York.
Nakayama, K., and Hojo, H. (1974). *Proc. 6th Int. Vac. Congr. Kyoto 1974,* p. 113. *Jpn. J. Appl. Phys. Suppl.* **2,** (*Part 1*).
Reich, G. (1961). *Trans. 7th Natl. Symp. Vac. Technol. USA 1960,* p. 112. Pergamon Press, New York.
Reich, G. (1982). *J. Vac. Sci. Technol.* **20,** 1148.
Redhead, P. A. (1958). *Can. J. Phys.* **36,** 525.
Rettinghaus, G., and Huber, W. K. (1974). *Vacuum* **24,** 249.
Reynolds, T. W., and Richley, E. A. (1964). *NASA TND-1864.*
Schaaf, S. A., and Cyr, R. R. (1949). *J. Appl. Phys.* **20,** 860.
Schütze, H. J. and Stork. (1962). *Vak. Tech.* **11,** 133
Smith, Jr., J. N. (1964). *J. Chem. Phys.* **40,** 2520.
Smith, A. (1971). *J. Vac. Sci. Technol.* **8,** 458.
Smith, A., and Melfi, L. T. (1966). *NASA TND-3266.*
Smith, Jr., J. N., and Saltsburg, H. (1964). *J. Chem. Phys.* **40,** 3585.
Soo, S. L., and Huang, A. B. (1962). *Trans. 9th Natl. Vac. Symp. AVS. USA 1962,* p. 443. Macmillan, New York.
Teubner, W. (1967). *Vak. Tech.* **16,** 69.
Venema, A. (1954). *Vacuum* **4,** 227.
Williams, F., Boudart, M., and Frennet, A. (1972). *J. Vac. Sci. Technol.* **9,** 362.
Winters, H. F. (1971). *J. Vac. Sci. Technol.* **8,** 17.

CHAPTER 3

Pressure Standards

3.1 Classification of Pressure Standards

An *absolute pressure standard,* usually referred to as a primary standard, is an instrument whose *calibration is calculated* from the knowledge of its significant dimensions and physical constants. The calibration of such an instrument therefore needs no reference to another instrument for pressure determination and remains unchanged for all ideal gases.

Pressure resulting from a measurement M is expressed as

$$P = \text{const} \times M \tag{3.1}$$

The value of the constant can be calculated in terms of fundamental units of length and mass.

The error $\Delta P/P$ in measuring the pressure is composed of the systematic error Δ(const)/(const) resulting from errors in dimensional measurements and a random error $\Delta M/M$ arising from errors in experimental observations. The latter error contributes to the measurement uncertainty, thus limiting the accuracy of pressure determinations.

A *secondary pressure standard,* referred to as a transfer gauge, is an instrument which measures either gas pressure effects (e.g., the strain in an elastic element) or some physical quantity of gas (e.g., viscosity or ionization). In the latter the quantity measured is related to pressure by the kinetic theory of gas. Secondary pressure standards *must be calibrated* (see Section 5) and must meet the following main requirements:

(a) The physical and/or chemical properties of the gas measured must not be changed by the measuring process.

(b) The physical quantity which is measured must depend as little as possible on other environmental variables such as temperature or magnetic fields.

(c) The output of the instrument must be linear, hysteresis free, and highly reproducible.

(d) The instrument must be very stable, i.e., should conserve the relationship between the quantity at the output and the signal at the input.

Primary pressure standards can be classified according to the operating principle into two classes.

The first class includes instruments which *measure* pressure directly (see Section 2.4.1), based on the concept expressed by Eq. (1.2). Instruments in this class are the precision liquid-column manometer and the McLeod gauge.

The second class comprises instruments or systems which *generate* steady known pressures. The piston manometer and static (see Section 5.4.2) and dynamic gas expander systems (see Sections 5.4.2 and 5.5.2) belong to this class.

The capacitance manometer, the spinning-rotor viscosity gauge, and the conventional triode ionization gauge (see Section 4.4.7) are utilized as secondary standards.

Precision liquid-column manometers and pressure expanders of the static or dynamic type as well as the capacitance manometer and, most recently, the spinning-rotor viscosity gauge are intensively used in vacuum metrology. The piston manometer (Warshawsky, 1972) and the conventional triode gauge (Sutton and Poulter, 1982) also have been utilized as primary and secondary standards, respectively, in the past few years.

It is worth mentioning that *particle counting* is another means for measuring *absolutely* total low pressures. Adam (1971) advanced the idea of counting techniques which make use of the processes that take place at the walls of the gauge (molecular range) instead of those processes that occur within the gauge volume. Marton (1974) suggested the use of light scattering (Rayleigh scattering) as a basis for an absolute low-pressure standard (see Section 4.4.5). Recently, the use of *atom counting* has been proposed instead of Marton's *molecular counting* as the operating principle of a low-pressure standard as well. It consists in selectively removing electrons from the gas atoms in the volume of the gauge and subsequently counting the resulting ions for different gas species. This would be feasible if a tunable laser and proportional counting techniques permitting one-atom counting were used. However, no routine gauges based on molecule or atom counting as discussed have yet been thoroughly studied.

Unlike in other languages, where different words are used to differentiate an absolute instrument (e.g., *etalon primaire* in French, *Absolute Eichinstrument* in German) from national or international documents related to measurement specifications (e.g., *norme* in French, *Norme* in German), the English word *standard* is equivocal. Its meaning must therefore be cautiously approached since, in relation to measurements, it is used to designate an absolute instrument, the physical embodiment of the unit of some quantity, e.g., atmosphere, or a national or international document specifying conditions of some kind. In this chapter, as well as throughout the book, *pressure standard* will be used to designate an absolute instrument for low-pressure measurement.

3.2 Precision Liquid-Column Manometers

3.2.1 Principle

The U-tube liquid-column manometer (called hereafter U-tube manometer, Fig. 3.1) is the instrument commonly used for the measurement of pressures in the range 10^3–10^{-4} Pa (10–10^{-6} Torr) (Brombacher, 1970). The measuring range depends upon the manometer fluid used, the design of the instrument, and the techniques used for the measurement of the liquid-column heights.

The pressure balance at the liquid-connecting port of the two manometer tubes shown in Fig. 3.1 is

$$P_1 + h_1\rho_1 g + P_{c1} + P_{v1} = P_2 + h_2\rho_2 g + P_{c2} + P_{v2} \tag{3.2}$$

Here P is pressure, h the height of the liquid column, ρ the density of the manometer fluid, g the acceleration due to gravity at the location of the experiment, P_c the pressure due to capillarity, and P_v the pressure due to

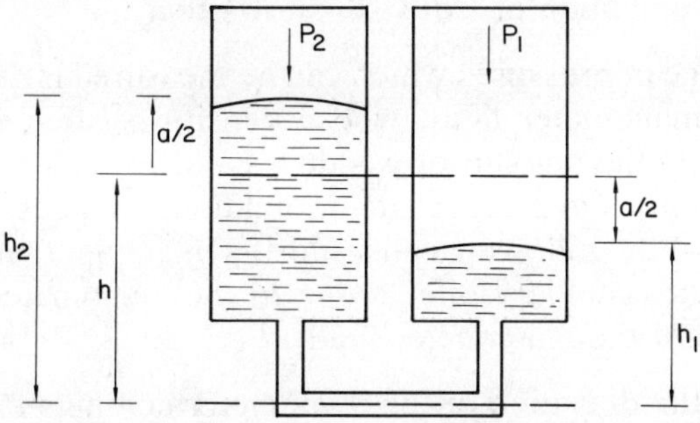

Fig. 3.1 U-tube liquid-column manometer.

the vapor of the manometer fluid. Subscripts 1 and 2 refer to the two arms of the manometer.

The density ρ of the manometer fluid varies with temperature T as

$$\rho = \rho_r/[1 + \beta_v(T - T_r)] \tag{3.3}$$

where b_v is the volume expansion coefficient and the subscripts r to the density ρ and temperature T indicate reference values.

Upon substituting the value ρ given by Eq. (3.3) in Eq. (3.2), the resulting value for the differential pressure $P_1 - P_2$ is

$$\begin{aligned}(P_1 - P_2)/\rho_r g = {} & h_2/[1 + \beta_v(T_2 - T_r)] - h_1/[1 + \beta_v(T_1 - T_r)] \\ & + (p_{c2} - p_{c1})/\rho_r g + (p_{v2} - p_{v1})/\rho_r g \end{aligned} \tag{3.4}$$

Equation (3.4) can be rewritten in a more explicit form upon referencing

(a) the column heights h_1 and h_2 to an average position h (see Fig. 3.1) so that

$$h_1 = h - a/2 \qquad \text{and} \qquad h_2 = h + a/2 \tag{3.5}$$

(b) the temperatures t_1 and t_2 to an average temperature T_m. In this case,

$$T_1 = T_m + \delta T_1 \qquad \text{and} \qquad T_2 = T_m + \delta T_2 \tag{3.6}$$

When

$$\beta_v(\delta T_1 + \delta T_2) \ll 1 \qquad \text{and} \qquad \beta_v(T_m - T_r) \ll 1 \tag{3.7}$$

Eq. (3.4) becomes

$$\begin{aligned}(P_1 - P_2)/\rho_r g = {} & a[1 + a\beta_v(T_m - T_r)] + h\beta_v(\delta T_1 - \delta T_2) \\ & + (a\beta_v/2)(\delta T_1 + \delta T_2) \\ & + (p_{c2} - p_{c1})/\rho_r g \\ & + (p_{v2} - p_{v1})/\rho_r g \end{aligned} \tag{3.8}$$

A careful examination of Eq. (3.8) shows that:

(a) The range of pressures which can be measured is dependent on the choice of the manometer fluid, whose density is affected by pressure, temperature, and the amount of gas dissolved.
(b) Uncertainties in pressure determinations may result from each or a combination of the following causes: drifts in ambient temperature from its average value, capillary effects due to the manometer fluid, and the vapor pressure of the manometer fluid.

In addition, the deviation of the manometer columns from the vertical position (referred to as *leveling*) as well as the accuracy resulting from the measurement of heights h_1 and h_2 contribute an important share to the uncertainty in pressure determination.

3.2.2 Factors Affecting the Measurement Accuracy

The range of pressures spanned by U-tube manometers and the accuracy of pressure determinations depend mainly on (a) the manometer fluid, (b) temperature, (c) the measurement of the column heights.

(a) *The manometer fluid.* Mercury and low-pressure diffusion pump oils of the DC 704 type, di-2-ethyl hexyl sebacate, and silicone oil MS 200 are most utilized. Their main physical characteristics are listed in Table 3.1.

The quantitative data which must be accurately known for either mercury or oil manometer fluid are the following.

(i) *The density and volume expansion coefficient.* Since data available for commercial products vary among batches it is advisable to determine these quantities rather than accept published values which represent mean values.

The *density of mercury* can be determined by using the absolute displacement or the content methods (Cook and Stone, 1957; Cook, 1961). Cook and Stone (1957) determined the density of mercury at 20°C with 1-ppm accuracy by means of the displacement method. Gas absorption in mercury is negligible but its density may significantly change because of its ability to dissolve alkaline and earth alkaline metals, Ag, Cd, Cu (clean), Mg, Pb, Sn, and Zn (Espe, 1966, vol. I, p. 614).

The volume expansion coefficient of mercury can be determined by measuring density at two temperatures.

The *density of DC 704 oil* was determined by means of mass and weight techniques by Orcutt (1973) and that of MS 200 oil by Poulter and Nash (1979) by magnetically controlled float methods (Hales, 1970).

Table 3.1

SOME PHYSICAL PROPERTIES OF MANOMETER LIQUIDS AT 20°C

Liquid	Density ρ (kg/m³)	Volume expansion coefficient β (1/°C)	Surface tension (mN/m)	Dynamic viscosity (Pa s)	Gas solubility	Metal solubility
Mercury	13545.8[d]	18.115×10^{-5}	485[e]	1.55×10^{-3}[e]	Very low	Some metals (see text)
DC 704 tetra phenyl–tetra methil trisilioxane[a]	1067.9[f]	72×10^{-5}[g]	38[h]	0.47[i]	About 10% of the volume of air measured at equilibrium temperature[c] (DC 704, DEHS, MS 200)	None (DC 704, DEHS, MS 200)
DEHS di-2-ethyl hexil sebacate (di-octyl sebacate)	913.6[j]	72×10^{-5}[k]	30[j]	0.245[j]		
MS 200 silicone oil	818[l]	135×10^{-5}[m]	17.5[n]	1.0×10^{-3}[b,j]		

[a] Manufactured by Dow-Corning Corp.
[b] At 25°C.
[c] Thomas *et al.* (1962).
[d] Cook (1961).
[e] Kaye and Laby (1973).
[f] Poulter and Nash (1979).
[g] Orcutt (1973).
[h] Thomas and Cross (1967).
[i] Power (1966).
[j] Peggs (1980).
[k] Esso data sheet.
[l] Smith and Murphy (1955).
[m] Boyer and Pankhurst (1971).
[n] Bradshaw (1965).

Diffusion pump oils may absorb gas to as much as 10% of their volume. Thomas and Cross (1967) experimentally determined that gas dissolved in DC 704 oil affects its density by one part in 10^4.

Changes in oil density may occur as a result of wetting agents used to reduce the contact angle of oil at the interface with glass.

The volume *expansion coefficient of oil* is determined by the same method as that used for mercury.

(ii) *The capillary depression.* The forces which act at the interface between the arm of the instrument, the manometer fluid, and gas depress the meniscus of the manometer fluid by a linear amount referred to as *capillary depression*. Theoretically, this amount is a function of the surface tension of the fluid, the meniscus height, and the bore of the manometer tube. Readings of the column height must therefore be corrected for capillary depression.

For mercury, the shape of the meniscus (convex) (Fig. 3.2) can be expressed according to Laplace (1816) as

$$-1/b = 2Z/a^2 - \sin V/u \tag{3.9}$$

where b is the radius of curvature at a point of ordinate Z and abscissa u; V is the angle between the tangent at that point and the horizontal. The term a^2, called capillary constant, has the value

$$a^2 = 2\gamma/\Delta\rho g \tag{3.10}$$

where γ is the surface tension, $\Delta\rho$ the difference in density between the mercury and the surrounding gas, and g the acceleration due to gravity at the location of determination.

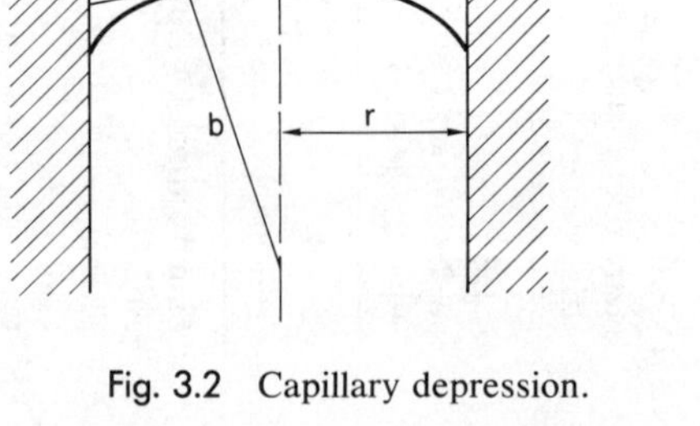

Fig. 3.2 Capillary depression.

Equation (3.9) has not been solved explicitly, and the usual approach is the use of tables based on numerical solutions (Blaisdell, 1940; Kistemaker, 1945; Gould and Vickers, 1952; Brombacher *et al.*, 1960). There is, however, no certainty that the calculated values will hold for conditions slightly different from those considered at calculations. This is mainly due to the tendency of the capillary depression to vary with the contamination of both the mercury and the walls of the glass tubing, the age of the mercury, local differences in both the bore of the tube and the glass surface, the direction of the change in pressure, etc.

Huang *et al.* (1972) developed an empirical equation for computing the capillary depression of mercury in tubes of diameter larger or equal to 0.4 cm, expressed as

$$h_c = [(0.2936r - 0.42325)\Delta h_c^2 + 0.255r^3 - 0.325r^2 + 0.009r + 0.130] \Delta h_c/(\Delta h^2 + r^2) \qquad (3.11)$$

Here h_c is the capillary depression, r the radius of the tube, and Δh_c the height of the meniscus, all in centimeters. The authors claim that h_c can be computed within about 1 μm.

Mercury does not wet most materials. In glass tubes it exhibits an erratic behavior partially attributed to the electrostatic charges generated when the mercury column shifts its position in the glass tube. The electrostatic charge yielded by the movement of mercury in a glass tube was approximated to be a maximum of 6×10^{-9} C (Miller and Jerinjan, 1964).

The capillary depression of oil can be calculated by using Blaisdell's (1940) numerical equations provided that density, surface tension, and contact angle are known. Because of the capillary forces, the oil meniscus is concave. This constitutes a drawback when measuring the height of the liquid column by noncontacting techniques. The effect of capillarity can be reduced by using large-bore tubes, very clean conditions for both the glass vessel accomodating the manometer fluid and the fluid itself, and careful manipulation techniques for determinations.

Usually, U-tube manometers have glass tubes of about 38-mm bore. Cleaning procedures for glass (Espe, 1966, vol. II, p. 270), mercury (Espe 1966, vol. I, p. 602), and oil (Poulter and Nash, 1979) are available. Tilford (1973) covered the glass walls of a U-tube manometer with a thin, optically transparent layer (chromium or nickel), which permitted reproducible contact angles for mercury.

The distillation of the manometer fluid into the arms of the instrument (Thomas *et al.*, 1963) assures the cleanness of the fluid, but the procedure may change its density when the fluid is oil.

(iii) *The vapor pressure*. The relatively high pressure (p_v) of mercury vapor (about 0.13 Pa or 1×10^{-3} Torr at room temperature) may be significant compared to the pressure to be measured, especially in absolute measurements where $P_l = 0$. Special techniques to maintain the mercury vapor pressure at reduced values so that the last term of Eq. (3.8) should be negligible were used by Ruthberg (1975, p. 233).

The vapor pressure of mercury can be calculated according to Espe (1966, vol. I, p. 612) as

$$\log p_v = 10.377 - 0.8254 \log T - 3285/T \qquad (3.12)$$

where T is the thermodynamic temperature.

Accurate data for mercury vapor pressure within a large range of temperatures are available (Honig and Kramer, 1969).

Mercury vapor is usually prevented from entering the pumping system by interposing a cold trap between the U-tube manometer and the system. Owing to the drag of gas molecules by the mercury vapor from the manometer to the cold trap, a pressure gradient and diffusion current build up. Unless readings are not corrected for this effect, errors as high as 100% might be obtained in pressure measurement (see Section 3.3.5).

Since mercury is poisonous, the health hazard associated with its high vapor pressure constitutes a serious drawback to its use as manometer fluid. Smoking should be prohibited where mercury manometer are used since mercury might come in contact with the cigarette, evaporate, and be inhaled.

The vapor pressure of oils used for manometric work is extremely low (about 1×10^{-4} Pa or 1×10^{-6} Torr at 40°C).

(b) *Temperature*. The disturbance introduced by variations from the average temperature (T_m) in the measurement of pressure with U-tube manometers is rather critical, producing uncertainties proportional to the reading [first term of Eq. (3.8)]. Temperature differences between the fluid columns in the two arms of the manometer cause errors due to changes both in the volume expansion coefficient β [second term of Eq. (3.8)] and in the refractive index, where oil is the manometer fluid.

The effect of temperature variations on the micrometers used for the determination of the height of the manometer column is discussed in the following paragraph.

(c) *Measurement of the fluid-column heights*. This measurement is accomplished by determining the surface of the menisci. Two techniques are available to date, for this purpose: (i) *contacting techniques* in which a mechanical contact is established at the surface of the menisci and the height is measured between this contact and a fiducial line or point, and (ii) *non-contacting techniques* in which light or ultrasound reflected from the menisci provide the required information.

(i) *Contacting techniques.* A micrometer-driven index point is used to detect the surface of the meniscus by touching it from below or from above according to whether the manometer fluid wets or does not wet the glass tube. Figure 3.3 schematically shows the approach used for measuring the height of the mercury columns in a mercury U-tube manometer. Figures 3.4 and 3.5 illustrate a mercury and an oil micrometer-driven point for measuring the height of the liquid columns. The actual position of the surface is indicated by a dimple at the contact point between the micrometer spindle and the meniscus. The dimple can be easily detected when mercury is the manometer fluid due to the optical reflectance of the contact surface. For both mercury and oil U-tube manometers, a "zebra" pattern consisting of sharply defined black parallel bands drawn on a white card placed behind the tubes greatly enhances observations. For accurate location of the contact point, the pattern is displace relative to the menisci until the dimple is viewed at the edge of a black band. Using optical magnification a skilled operator can set the thimble of the micrometer to about 0.9 μm.

The length of the liquid columns can be deduced from the average value of the zero level measurement. This technique reduces the measurement uncertainty. Inaccuracies in the measurement of the liquid column height arise when the manometer is leveled and from temperature effects on micrometers.

The manometer leveling consists in setting at vertical position the geometrical axes of the micrometers. Cosine errors occur when these axes

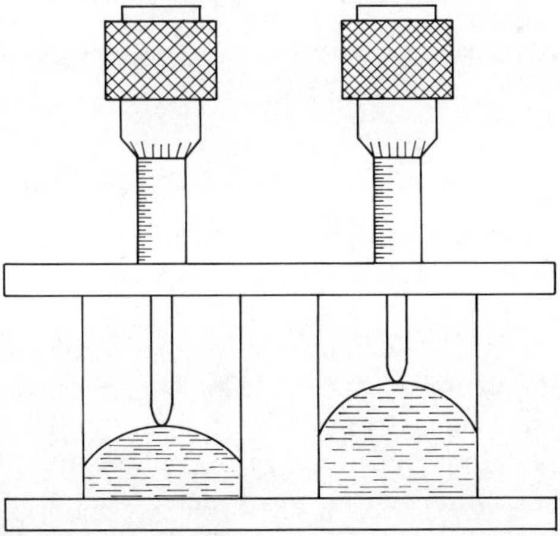

Fig. 3.3 Arrangement for measuring mercury column heights with micrometers.

Fig. 3.4 Photograph of a U-tube mercury point-contact manometer, with valves, cold traps, and optical background (after Ruthberg, 1975).

deviate from the vertical. Sine errors result from changes in the instrument level in the plane of the micrometer axes during the measurement of the fluid surfaces.

The cosine error (Fig. 3.6) can be kept negligible by strictly maintaining the verticality of the micrometer axes. The sine error can be reduced to a minimum by setting the instrument on a solid mount equipped with a sensitive level. Optical magnification of the bubble and its maintenance at a fiducial line are essential for the manometer operation. In addition the vacuum system must be damped for vibrations and mechanically isolated (bellows) from the manometer.

Fig. 3.5 Photograph of an oil U-tube micrometer point contact, with valves and optical background (after Ruthberg, 1975).

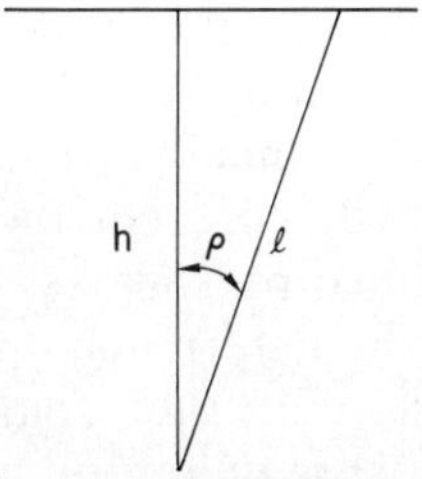

Fig. 3.6 Illustration of the cosine error. h and l are height and length of the manometric column.

Ruthberg (1969) used the reversal technique to reduce leveling errors. The technique consists of reestablishing an identical test pressure in the arm where the reference pressure was previously. The technique is, however, time-consuming since it involves manipulation of valves and pump-down procedures.

Temperature effects on the micrometers arise as result of temperature differences in the micrometer parts (e.g., between the spindle in contact with the manometer fluid and the barrel) and drifts from the average temperature of the thimbles due to handling during determinations. The magnitude of the temperature effect can be calculated according to Ruthberg (1969). The drifts in the temperature of the thimble can be measured by tiny thermocouples in good thermal contact with the thimble and reduced using adequate insulation (Ruthberg, 1975, p. 238). Usually, thermometers calibrated to 0.1°C, fitted in the bottom of the base plate of the manometer are used to monitor the temperature of the instrument. Thermostating the manometer is also a common practice in manometric work.

Table 3.2 lists the errors encountered with mercury and oil U-tube manometers.

The accuracy and performance of contacting techniques can be evaluated by considering that the sensitivity ($\Delta h/\Delta P$) of oil manometers is more than one order of magnitude higher than that of mercury manometers due to the difference in the densities of the two manometer fluids. However, the use of oil presents some drawbacks, such as loss (or gain) of fluid on the walls of the instrument arm during subsequent determinations, thus increasing the time needed for determination; the formation of gas bubbles or loss of oil as a result of loose gaskets; or changes in oil density due to absorbed gas.

The measurement resolution of the liquid-column heights limits the high pressure range of precision liquid-column manometers.

Manual operations make it difficult to maintain the mechanical and thermal stability of the instrument.

The reading accuracy of the liquid-column height varies from operator to operator. The reported accuracy obtained by different operators in taking repeated readings with manually driven micrometers is controversial, and the figures advanced represent the best expectation for ideal operating conditions and skilled persons.

A change in pressure during determinations requires resetting the micrometers, thus involving shifts in their temperature from the ambient, hence, loss of accuracy. Pressure therefore, cannot be continuously measured.

As a result of the high dynamic viscosity of mercury, ripples caused by

Table 3.2

SUMMARY OF ANALYSES OF ERRORS FOR U-TUBE MICROMETER POINT-CONTACT MANOMETERS WITH 38-MM DIAMETER AND 40-MM LENGTH COLUMNS AT $(P_1 - P_2) = 0$

	Error E	
	Ruthberg (1969) Oil (DC 704)	Thomas and Cross (1967) Mercury
Errors not proportional to reading[a]		
Level change	0.03	0.74
Capillarity	0.01	0.3
Setting and reading	0.10	2.0
Density	0.06	1.5
Micrometer temperature	0.02	2.6
Total theoretical $(\Sigma E^2)^{1/2}$	0.12	3.7
Observed 3σ	0.07–0.13	—
Errors proportional to reading[b]		
Density	0.4	0.4
Micrometer temperature	0.0	0.2
Manometer temperature	0.2	—
	0.6	0.6
Total (not proportional and proportional) for 1 Torr[a]	1.8×10^{-4}	6.7

[a] Error in millitorr.
[b] Error in parts per 10,000 of the reading.

the slightest vibration make awkward the assessment of the meniscus. Hence, determinations become time-consuming.

(ii) *Noncontacting techniques.* The displacement relative to a fiducial level of either the meniscus or a float resting upon the meniscus can be measured by using various optical, ultrasonic, or capacitance techniques. Optical and ultrasonic fringe-counting interferometry are extensively used in vacuum metrology owing to their high resolution and accuracy. Instruments utilizing these techniques are usually referred to as *micromanometers.*

Optical fringe-counting interferometry based on Fizeau (Thomas *et al.*, 1963), Fabry–Perot (Aubry and Delbart, 1965) or Michelson (Stevenson and McFadden, 1965) interferometric methods has been utilized for producing interference fringes for the measurement of changes in the height of the manometer columns.

Thomas *et al.* (1963) measured changes of the meniscus surfaces produced by pressure changes in a micromanometer having a W configuration (Fig. 3.7). This design permits minimizing leveling errors. The same fluid, di-octyl sebacate, was used in the manometer and in the diffusion pump which evacuated the instrument in order to reduce the vapor species in the measurement setup.

Collimated monochromatic light (mercury green line $\lambda_0 = 0.5461\ \mu m$) obtained after filtering the radiations emitted by a mercury lamp was directed by adequate optics on the menisci. Changes in the columns heights resulting from pressure changes were measured from the interference fringes between the light reflected from the upper surface of the oil and the lower surface of a quartz optical flat (F in Fig. 3.7). The fringe pattern obtained while the instrument was at zero pressure differential was watched and compared with the relative fringe shift resulting when the fluid was pressurized. A sensitivity of 2.6×10^{-4} Pa (2×10^{-6} Torr) was reported as having been attained by detecting a shift of $\frac{1}{10}$ of a fringe.

Aubry and Delbart (1965) designed a micromanometer of the fixed-cistern type (i.e., an instrument in which the height of *only one* liquid column is measured) in which both mercury and oil (type 704) have been utilized as manometer fluid (Fig. 3.8). When the manometer fluid was mercury, a thin layer of oil, floating on the surface of the meniscus, was used to damp the ripples generated by vibrations transmitted to the instrument. The authors claim that the thickness of the oil layer is critical: less than 0.3 mm yield a big damping of the ripples, thus enhancing the time response of the measuring system; more than 0.03 mm results in disturbances due to oil capillarity.

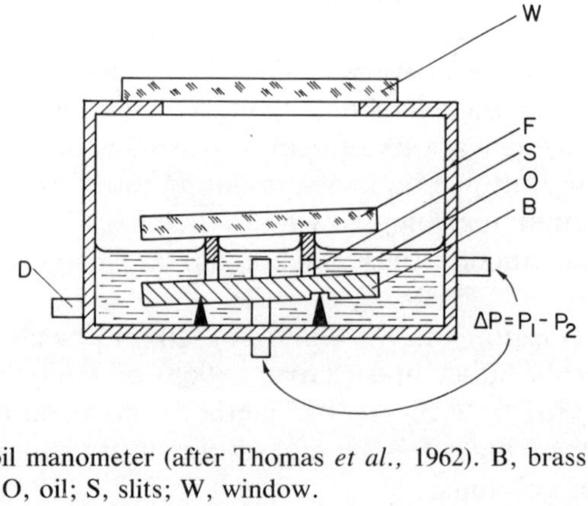

Fig. 3.7 W-type oil manometer (after Thomas *et al.*, 1962). B, brass block; D, drain; F, quartz optical flat; O, oil; S, slits; W, window.

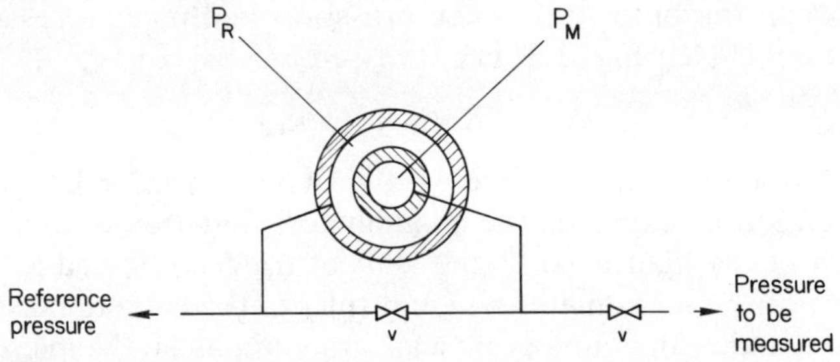

Fig. 3.8 Diagram of the micromanometer used by Aubry and Delbart (1965). P_M, measuring pool; P_R, reference pool; V, valve.

In this technique monochromatic light (cadmium line $\lambda_0 = 0.585\ \mu m$) emitted from a source (S in Fig. 3.9) is directed at the surface of the meniscus and reflected either by the mercury or by a semireflecting mirror (M) consolidated inside the pool, when oil is the manometer fluid. The optical path difference for fringe formation is yielded in the oil wedge between the meniscus and the mirror. The differential pressure ΔP mea-

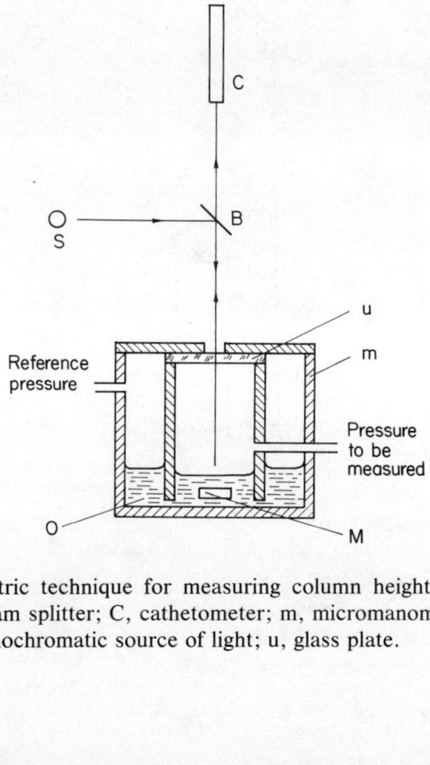

Fig. 3.9 Interferometric technique for measuring column heights, used by Aubry and Delbart (1965). B, beam splitter; C, cathetometer; m, micromanometer; M, semireflecting mirror; O, oil; S, monochromatic source of light; u, glass plate.

sured between the outer (reference pressure) and inner (pressure to be measured) pools (see Fig. 3.8) is expressed as

$$\Delta P = K\rho(\lambda_0/n)(\psi/\psi_o) \tag{3.13}$$

Here K is the ratio of the surfaces of the two pools, ρ and n the density and the refraction index of the manometer fluid, respectively, λ_0 the wavelength of the light used, ψ the shift of fringes referred to the scale graduations of the micrometer which displaces the cathetomer for fringe observation, and ψ_0 the number of scale graduations on the micrometer of the cathetometer which separate two adjacent fringes.

The reported accuracy of this technique is 3.23×10^{-3} Pa (2.43×10^{-5} Torr) and 2.62×10^{-4} Pa (1.96×10^{-6} Torr) for mercury and oil, respectively. The range of pressure of Aubry and Delbart's micromanometer is 10^{-3}–30 Pa (10^{-5}–3×10^{-1} Torr).

Stevenson and McFadden (1965) utilized coherent light emitted from a laser (L in Fig. 3.10) ($\lambda_0 = 0.6328\ \mu$m) and two fully reflecting mirrors (M) located horizontally below the menisci of the U-tube manometer to produce interference fringes. A change in the pressure difference between

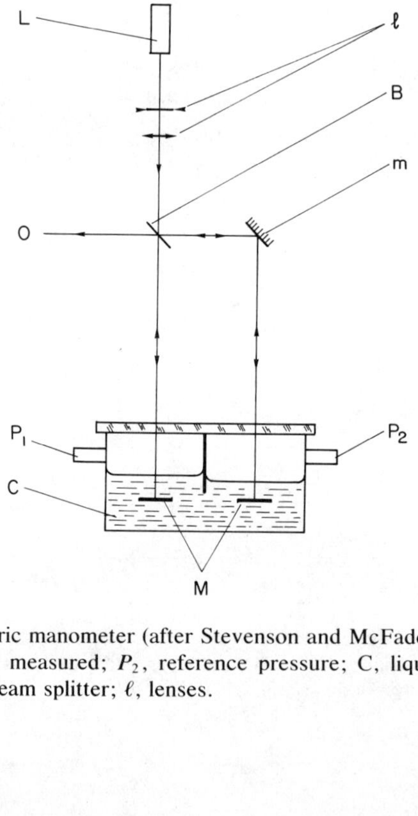

Fig. 3.10 Interferometric manometer (after Stevenson and McFadden, 1965). L, laser; O, observer; P_1, pressure measured; P_2, reference pressure; C, liquid column; M and m, reflecting mirrors; B, beam splitter; ℓ, lenses.

the two pools creates a difference in the optical path between the two incident light beams as a result of the change produced in the level of the columns. Fringes thus produced are controlled by adjusting the beam splitter (B) and the front mirror surface m.

The shift S of a fringe resulting from the change in the level h of the manometer columns is expressed as

$$S = (2h/\lambda_0)(n_o - n_g) \tag{3.14}$$

Here λ_0 is the wavelength (in vacuum) of the light used and n the refractive indices of the oil (subscript o) and the gas (subscript g) in the manometer. The manometer fluid was galvanometer damping oil (specific gravity 0.907). The probable error in determining h is about $+0.1\ \mu$m.

Tilford (1973) designed a fringe-counting laser interferometer manometer permitting continuous tracking and measurement of the column heights during pressure changes. The radiation of a CO_2 laser (L) ($\lambda_0 = 10.6\ \mu$m) (Fig. 3.11) is passed through a beam expander (E) and a beam splitter (B) to the mercury menisci. The resulting reflected beams are recombined at the beam splitter and focused through a lens (ℓ) on a pyroelectric detector (D). Upon modulating the optical path of the beams at a certain frequency it is possible to obtain signals out of phase by 90°.

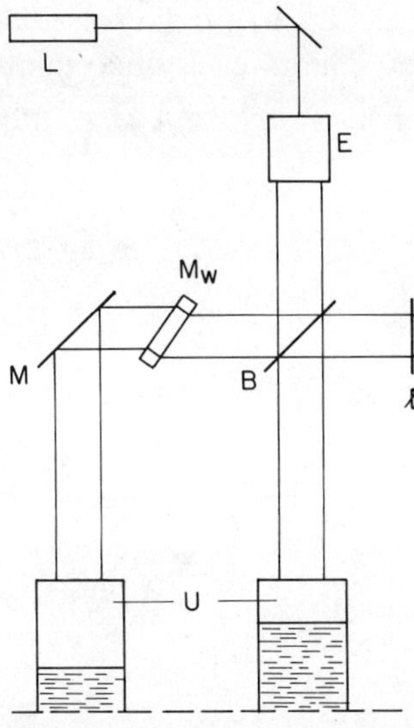

Fig. 3.11 Diagram of a fringe-counting laser interferometer manometer (after Tilford, 1973). B, beam splitter; D, pyroelectric detector; E, beam expander; ℓ, focusing optic; L, laser; M, folding mirror; M_w, modulating mirror; U, manometer.

These signals, containing the quadrature information necessary to discriminate between increasing or decreasing heights of the manometer columns, are fed to an up–down counter, thus making possible automatic up–down counting of the fringes.

Owing to the long wavelength of the emitted radiation from the laser and to the optically transparent metallic layer covering the inner side of the cisterns, ripples produced at the menisci surfaces do not disturb the measurement. However, to avoid vibrations, the U-tube manometer assembly is attached to a 800-kg mass contained in a special construction. Measurements with this manometer where temperature is not controlled have a resolution better than 0.1 μm.

Laser interferometry applied to mercury surfaces was also used by Bennet *et al.* (1975) to measure the height of a manometer column. An annular support (F) (Fig. 3.12) supporting a lens (ℓ) is floated on the meniscus and the radiation from a laser focused on the mercury surface. The incident and reflected radiations are then combined for interferometric analysis. The reflected beam is affected by both rocking and rotation of the lens support.

Truffier and Choumoff (1974) used a photomultiplier and an electronic system permitting the recording of the fringe motion versus time. With their method oil-column heights as small as those corresponding to $\frac{1}{100}$ of a fringe (i.e., 1.72×10^{-7} cm) can be accurately measured. Mechanical and thermal stability are achieved by surrounding the micromanometer by a sand mass of 1.5 m^3 volume. The overall uncertainty in the differential pressure measurement if $(\Delta P/P) < 10^{-7}$.

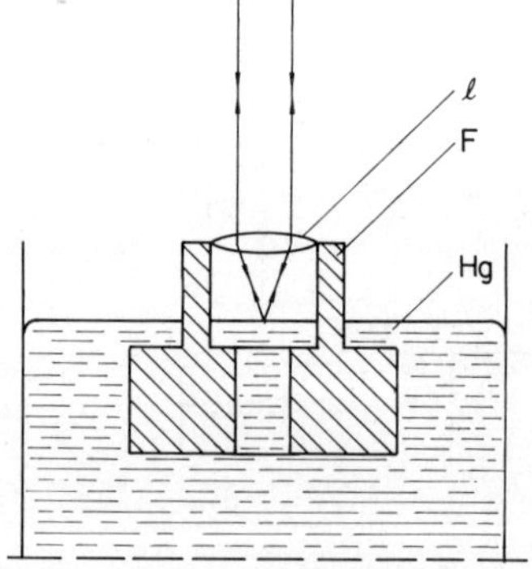

Fig. 3.12 Catseye float. F annular support, *l* lens. Reprinted with permission from *J. Phys. E., Sci. Instrum.*, **8,** 5. Bennet *et al.* Copyright 1975 The Institute of Physics.

Poulter and Nash (1979) designed a polarizing interferometer and fringe-counting electronics. The light source used was a neon–helium laser. The range of their oil micromanometer is 0.3–6 Pa (3×10^{-3}–4.5×10^{-2} Torr), and the total uncertainty of the measurement is $\pm$ (1.4×10^{-3} Pa + 0.5% of reading).

Pulsed ultrasonic fringe-counting interferometry consists of sending through the liquid column an ultrasonic signal whose wavelength and frequency are known and registering the resulting echo. The signal traverses the length of the liquid column several times, undergoing multiple reflections at the meniscus. The *n*th echoes are amplified and demodulated in phase by sensitive detectors with a 90° phase shift between their reference signals. The amplitude of the phase-sensitive signals is proportional to the amplitude of the received signal and to the phase shift between signal and reference. The phase of received echoes varies with changes in the column length. A fringe-counting pulsed interferometer extracts the transit time changes between the primary and reflected signals, thus enabling the measurement of the column length. For length measurement with the ultrasonic interferometer to be feasible, pulses have to be short enough so that the returning echoes do not overlap.

The principle of the pulsed ultrasonic fringe-counting interferometer for the automatic measurement of transit times (Heydemann, 1971) was used in the construction of two ultrasonic manometers for low and medium vacua (Heydemann *et al.*, 1977). The U-type manometers could be used with mercury or DEHS oil (see Table 3.1). Electrical pulses with fast rise and fall times are injected into a transducer located at the bottom of the manometer columns. The electrical pulses are transformed into mechanical waves which travel up and down the column until all the ultrasonic energy of the signal is lost by absorption and diffraction. An automatic fringe-counting pulsed interferometer interconnected with a microcomputer for data collection makes possible very accurate measurements for both absolute and differential pressures to 10^4 Pa (10^2 Torr).

Table 3.3 summarizes the uncertainties in pressure measurement with an interferometric oil micromanometer (Poulter and Nash, 1979) and an ultrasonic interferometer manometer (Heydemann *et al.*, 1977).

The main advantages and drawbacks of noncontacting techniques are remote and automatic determinations permitting a high degree of temperature stability, continuous tracking and measurement of the column heights, potentially high resolution and accuracy owing to the traceability of the sources emitting radiations (light, ultrasound) to wavelength standards. The greatest disadvantage of the technique is the mechanical instability during determinations.

Table 3.3

SUMMARY OF UNCERTAINTIES AFFECTING PRESSURE MEASUREMENT WITH THE INTERFEROMETRIC OIL MANOMETER AND THE ULTRASONIC INTERFEROMETER MANOMETER

	Interferometric oil micromanometer[a] Oil (MS 704)		Ultrasonic fringe-counting interferometer manometer[b]			
			Oil (DEHS)		Mercury	
Source of uncertainty	Random (Pa)	Systematic (%)	Random	Systematic (ppm)	Random	Systematic (ppm)
Uncertainties proportional to pressure						
Density		±0.01		100		2
Cosine		±0.0001		<1		<1
Gravity		±0.0002	Estimated to be less than 1 ppm	1	Estimated to be less than 1 ppm	1
Refractive index		±0.03				
Laser wavelength		±0.0003				
Laser wavelength		±0.0003				
Total	±0.04					
Temperature				10		5
Speed of sound				No estimate available		70
Carrier frequency				1		1
Flexure of mounting plate				<1		<1
Total				—		81

	Random (Pa)	Systematic (%)	Random	Systematic (mPa)	Random (1σ)	Systematic (mPa)
Uncertainties independent of pressure						
Capillarity	$\pm 5.2 \times 10^{-4}$		—	<0.01		0.1
Temperature gradients	$\pm 2.0 \times 10^{-4}$		0.01		3	
Residual pressure	$\pm 1.0 \times 10^{-3}$					
Resolution of counters	$\pm 4.5 \times 10^{-4}$					
Refractive index	$\pm 0.01\%$					
Tilt	$\pm 1.0 \times 10^{-4}$		0.01		0.1	
Total	$\pm 1.4 \times 10^{-3}$ Pa $+0.01\%$ of reading					
Mechanical and electrical noise			0.1		1	
Demodulation				0.7		10
3 times the square root of the sum of the squares random			0.10		9.6	
Sum of systematic uncertainty				0.7		10.1
Total			0.80		19.7	

[a] After Poulter and Nash (1979).
[b] After Heydemann *et al.* (1977).

3.3 The Compression Manometer—McLeod Gauge

3.3.1 Principle

Pressure measurement with the McLeod gauge, is based on Boyle's law,

$$PV = \text{const} \tag{3.15}$$

which expresses the relationship between the pressure P and volume V of an ideal gas at constant temperature.

In principle, a McLeod gauge (Fig. 3.13) consists of a glass bulb (1), also called a compression bulb, to which a closed-end capillary (2), referred to as compression capillary, is attached. The compression bulb is connected to a mercury reservoir (not shown in Fig. 3.13) and to a side arm (4) with a sealed bypath capillary (3), referred to as a comparison capillary. Capillaries (2) and (3) run parallel and have identical bores. The side arm (4) connects the instrument to the system containing the gas of unknown pressure.

To operate the gauge the glassware shown in Fig. 3.13 and the reservoir are first thoroughly evacuated. The mercury is held at its highest level in the reservoir, and gas from the system is permitted to enter into the side arm (4). Then the mercury from the reservoir is forced to enter the side arm (4). When mercury reaches the level X–X′, the communication between the side arm and the compression bulb is cut off, thus trapping the amount of gas to be measured in the volume of the compression bulb and compression capillary. By further forcing the rise of mercury in the manometer, gas is compressed in the closed-end capillary, until a certain height h of the gas column is reached. The pressure of the compressed gas is then

$$PV = (P + \Delta h\rho g)(Ah) \tag{3.16}$$

or

$$P = (\rho g)(Ah\,\Delta h/V - Ah) \tag{3.17}$$

Here P is the unknown pressure of gas in the system, V the volume into which the gas is compressed after the mercury has passed the cutoff level, Δh the manometric height of the compressed column of gas measured as the difference between the mercury menisci in the compression and comparison capillaries, ρ the density of mercury, A the average cross section of the compression capillary, and h the height of the compressed gas measured between the mercury meniscus in the compression capillary and the top end of this capillary.

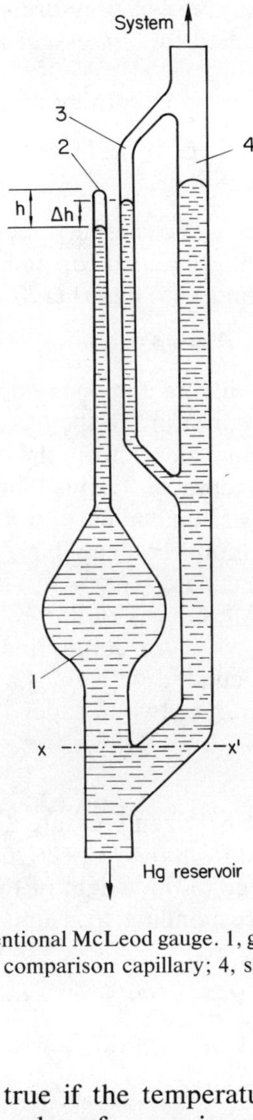

Fig. 3.13 Upper part of a conventional McLeod gauge. 1, glass compression bulb; 2, closed-end (compression) capillary; 3, comparison capillary; 4, side arm; X–X′, cutoff level.

Equation (3.17) holds true if the temperature is constant during the operation, and hence the value of ρ remains constant.

Since the value of Ah is small compared to V, Eq. (3.17) can be rewritten to a close approximation as

$$P = (A/V)(h\,\Delta h\rho g). \tag{3.18}$$

Provided that ρ and g are accurately determined, the calibration of the manometer can be calculated from the geometrical dimensions of the instrument. Practically, the range of pressure which can be measured is limited by the ratio A/V, called the *gauge constant.*

3.3.2 Scales Used for Pressure Determination

Pressure determination is made by compressing gas according to either *quadratic* or *linear scale* techniques.

(a) *The quadratic scale.* The mercury meniscus in the comparison capillary is brought level with the inner top end of the compression capillary (i.e., $\Delta h = h$). Equation (3.18) can therefore be written

$$P = (A/V)(h^2 \rho g) \tag{3.19}$$

Significant inaccuracies will affect pressure determination due to differences between the radii or curvature of the inner top end of the compression capillary and the mercury meniscus in the comparison capillary. The practical way to reduce inaccuracies of this kind is to allow the meniscus of the mercury in the comparison capillary to rise less than 0.5 cm below the inner top end of the compression capillary. This scale covers a pressure range of about four decades.

(b) *The linear scale.* Mercury is raised in the compression capillary up to an arbitrary level, *always the same,* and heights h and Δh determined. Pressure is then calculated with Eq. (3.18). The linear scale permits pressure measurement over two decades.

3.3.3 Sensitivity

The qualitative difference between pressure determination according to these scales can be obtained from the respective sensitivities. If we define sensitivity as the unit change in the height of the mercury column in the compression capillary corresponding to a unit change of pressure, the sensitivities of these scales are

$$dh/dP = (V/A)/(1/2h\rho g) \qquad \text{(quadratic)} \tag{3.20}$$

$$d(\Delta h) = (V/A)/(1/\rho g) \qquad \text{(linear)} \tag{3.21}$$

If v is the final volume of the compressed gas in the compression capillary, V/v expresses the *compression ratio* which shows the degree of pressure amplification achieved in the closed-end capillary.

Equations (3.20) and (3.21) can be rewritten as functions of the compression ratio as

$$dh/dP = (V/v)(1/2\rho g) \qquad \text{(quadratic)} \tag{3.22}$$

$$d(\Delta h)/dP = (V/v)(h/\rho g) \qquad \text{(linear)} \tag{3.23}$$

An analysis of Eqs. (3.20–3.23) shows that sensitivity

(a) increases when using the quadratic scale (Eq. 3.20) as h decreases (low pressure),
(b) is independent of the value of h when using the linear scale (Eq. 3.21), thus remaining constant at all points in the range, and
(c) is related to the compression ratio (V/v) of the gauge [Eqs. (3.22) and (3.23)] no matter which scale is used.

Sensitivity [Eqs. (3.22) and (3.23)] can be increased by either of the procedures shown below or by a combination of them by

(a) increasing the volume V, or
(b) reducing the volume v by using small bore capillaries.

Increased values for V involve special constructions necessary to sustain the increased weight of the glass bulb [(1) in Fig. 3.13)] filled with mercury as well as long times for pressure determination.

The drawback of reducing the capillary bores (0.075 cm) is that the mercury column tends to "stick" to the walls or to "break," leaving a hanging thread when the level of liquid is raised and subsequently lowered. In addition, capillary depression of mercury gets larger with decreasing capillary bore.

3.3.4 Systematic Errors Arising from the Instrument

The following assumptions have tacitly been admitted in the deduction of Eq. (3.16):

(a) The gas whose pressure is measured obeys Boyle's law.
(b) The temperatures of the gas compressed in the volume V and of the gas in the system are the same.
(c) The pressure of gas in the whole measuring assembly (gauge and system) is negligibly small compared with that of the gas compressed in the compression capillary.
(d) The volume in which the gas is compressed in the compression capillary is that of a right circular cylinder.
(e) The height of the mercury in the gauge is controlled *only* by the hydrostatic forces that balance the pressure exerted by the gas.
(f) The quantity of gas trapped in the compression volume V is not altered by subtraction or addition of gas molecules.

In practice, these assumptions are never met. The following analysis is devoted to the major dissimilarities between actual and ideal measurement conditions and therefore will provide correction factors for Eq. (3.18).

(a) *Invalidation of Boyle's law.* Boyle's law, which expresses the relationship between the pressure P, volume V, and temperature T of an ideal gas, cannot be used for a *real gas* (a gas whose molecules or atoms are of finite size and interact with each other). For such a gas, various theoretical or empirical equations (Van der Waals, virial coefficient, Keyes, etc.) relate the physical quantities P, V, and T more accurately than does Boyle's law.

Jansen and Venema (1959) used the *Van der Waals* equation

$$(P + a/V^2)(V - b) = R_0 T \tag{3.24}$$

(where a and b are constants and R_0 is the universal gas constant) for pressure determination with a McLeod gauge of special construction. They demonstrated, however, that Boyle's law gives sufficiently accurate results for

(i) all gases where

$$T_c < T_m \tag{3.25}$$

the subscripts c and m referring to the critical and measurement temperatures, respectively;

(ii) other gases where $T_c > T_m$ but where the saturated pressure in the compression capillary at temperature T_m is higher than the sum of pressures of gases in the system and in the gauge column (Δh in Fig. 3.13). After checking 10 gases fulfilling condition (i) and 9 gases for condition (ii), Jansen and Venema concluded that the error in pressure determination by using Boyle's law instead of Van der Waals' equation is less than 0.5%.

(b) *Temperature differences between the gas in the gauge and in the system.* The average error produced as a result of the difference in the temperatures of the compressed gas in the volume V and in the system is less than 0.15% with the normal use of the instrument (Jansen and Venema, 1959).

The temperature of the gas in the gauge is maintained equal to that in the system by means of adequate thermostating and by raising the mercury in such a way as to avoid heating the compressed sample. Thus, before the mercury column in the side arm reaches the cutoff level X–X′ (Fig. 3.13) the mercury rise is stopped for a while (about 5 min) to allow the gas in the instrument to reach temperature equilibrium with the system. Then the mercury is *slowly* raised to the point of pressure determination.

The reduced rate of rise discussed has the following advantages:

(i) It permits easier stopping of the mercury inflow, such that differences in level of less than 0.1 cm can be obtained between the menisci in the compression and comparison capillaries.

(ii) It prevents the erratic behavior of the contact angle θ owing to the lack of electric charges on the glass walls (Brombacher *et al.*, 1960). Electric charges build up on the walls when mercury is rapidly raised or lowered. Thus, Podgurski and Davis (1960) noticed that a mercury thermometer suspended near the gauge is attracted to the instrument when the mercury level is rapidly shifted up and down. At certain pressures, a luminous electrical discharge can even be noticed. It is advised *always* to take readings corresponding to the *rise* of the mercury column since its downward movement in the capillaries yields erratic behavior of the contact angle θ (Löcherbach and Schitko, 1962; Porter, 1933).

For large temperature differences, temperature corrections of Eq. (3.7) are necessary.

(c) *The gas pressure in the system is small compared with that in the gauge.* Practically, the dimensions of the gauge are chosen such that the pressure of gas in the system is negligible compared with that in the compression volume.

(d) *The gas is compressed in a volume different from that of a right circular cylinder.* This occurs because

(i) the cross section of the capillary bore is not uniform along the length considered for the calculation of the compressed gas

(ii) the top surface of the mercury column assumes the shape of a convex meniscus, and

(iii) the top end of the compression capillary is not flat.

An analysis of these causes show the following:

(i) The accuracy of glass capillary tubing manufactured with modern techniques is about 4 parts in 10^4 in 1 cm for diameters ranging between 0.1 and 0.05 cm. With such capillaries the error introduced by the variation of the diameter of the bore is negligible.

(ii) The mercury column in the tubings of the manometer terminates with a convex-shaped meniscus owing to the action of both gravity and surface tension (Fig. 3.14). The height h_v of the meniscus, measured as the distance from its top to the plane of the contact ring between mercury and the glass tubing, changes as a function of the contact angle θ. The value of θ is determined by the equilibrium of the energies of the surfaces in contact, i.e., liquid, solid, and gas (Kistemaker, 1945; Whitlaw-Gray and Teich, 1948). Fluctuations of the contact angle θ, even under the same

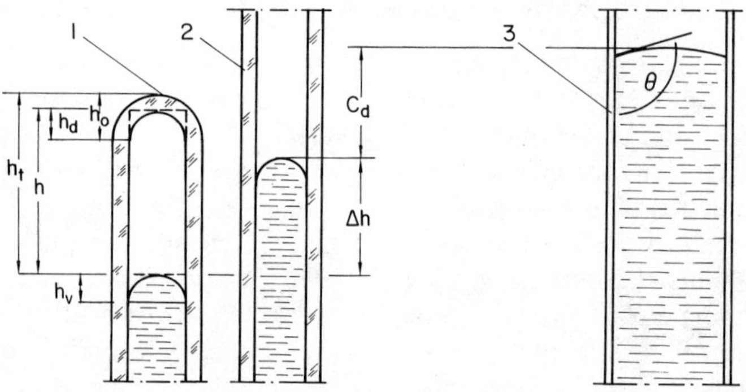

Fig. 3.14 The mercury level in the capillaries and the side arm of a McLeod gauge. 1, compression capillary; 2, comparison capillary; 3, side arm; h_v, height of the mercury miniscus; h and h_t, heights of the compressed gas measured with reference to the inner and outer effective end of the compression capillary, respectively; h_o and h_d, heights of the outer and inner effective end of the compression capillary, respectively; Δh, height of the mercury in the comparison capillary measured with reference to the mercury meniscus in the compression capillary; C_d, height of the mercury in the side arm measured with reference to the mercury meniscus in the comparison capillary; θ, contact angle between mercury and glass.

experimental conditions, have been attributed to various causes such as the cleanliness of the mercury or the glass tubing, irregularities in the glass surface (Rosenberg, 1939), and aging (Keevil *et al.*, 1941). However, a definite relationship between these factors has not yet been proved (Nakayama *et al.*, 1968).

Thorough cleaning of the mercury (Davidson and Rinker, 1974) and glass (Podgurski and Davis, 1960; Berman, 1973) combined with thorough outgassing of the gauge help to diminish the instability of θ (see Section 3.2.2).

Contamination-free conditions for mercury are obtained if the reservoir of the gauge is kept under clean, dry nitrogen or inert gas. The mercury column is raised by exerting pressure in the reservoir with one of these gases at atmospheric pressure. Although several techniques are available for raising the mercury from the reservoir into the glass work (Roth, 1982, p. 293) the procedure based on the contraction or expansion of the gas above the mercury pool has the advantage of avoiding contamination of the manometer fluid (Fletcher and Butcher, 1969). The gas volume in the reservoir is changed by allowing the gas to reach either the temperature of liquid nitrogen or that of the room. The drawback of this technique is that it is time consuming.

Chemical treatment of the inner surface of the capillaries results in an increase in the contact angle θ, the capillary depression becoming relatively less sensitive to a change of θ. Rosenberg (1938) reported that grinding the inner surface of the capillary to such a degree as to leave enough transparency for the observation of the mercury column reduces irreproducibility of the mercury meniscus height by about 5%. This technique, however, increases the capillary depression (Klemperer, 1944).

Despite all the procedures discussed for improving the cleanliness of both the mercury and the glass, the spread in the values of the contact angle is considerable even for surfaces having identical microscopic properties. The claim that capillary depression is eliminated by using the same piece of capillary tubing (identical bore) for the compression and comparison capillaries has also been demonstrated to be untrue (Jansen and Venema, 1959; Podgurski and Davis, 1960; Löcherbach and Schitko, 1962; Nakayama *et al.*, 1968).

The error produced by the instability of the contact angle increases as the diameter of the capillary bore is reduced and the height of the compressed gas in the compression capillary is decreased (Nakayama *et al.*, 1968). Figure 3.15 illustrates the error caused by the variation of the meniscus volume as a function of the length l of the compressed gas.

The contact angle and consequently the capillary depression may be stabilized by tapping the capillary. It is assumed that the mechanical energy thus transmitted to the liquid helps in creating more stable equilibrium conditions, but no quantitative explanation is yet available. Jansen and Venema (1959) experimentally demonstrated that the average error in observing mercury menisci is much smaller with tapping against the capillary wall than with no tapping. However, Kreisman (1961) states that hand tapping gives rise to poor reproducibility of mercury menisci. He advises raising the mercury in the capillaries of the manometer under conditions of minimum vibration.

To correct the volume of the compressed gas for the error introduced by the curvature of the mercury meniscus, the supplementary volume corresponding to the height h_v must be added to the effective height h of the compressed gas (Fig. 3.14).

(iii) Several constructive solutions have been designed which permit the compression of gas in the compression capillary in a volume approximating as much as possible that of a right circular cylinder. Barr and Anhorn (1946) welded a plug at the top end of the compression capillary in order to obtain a "flat" surface. The drawbacks to this solution are twofold: the rough surface of glass obtained in a previous operation with a chemical treatment is removed in the portion where the plug is welded, thus enhancing the sticking of the mercury to glass, and the refractive index of the glass capillary varies in the region of the seal due to nonuniformities of the wall produced at the plug weld.

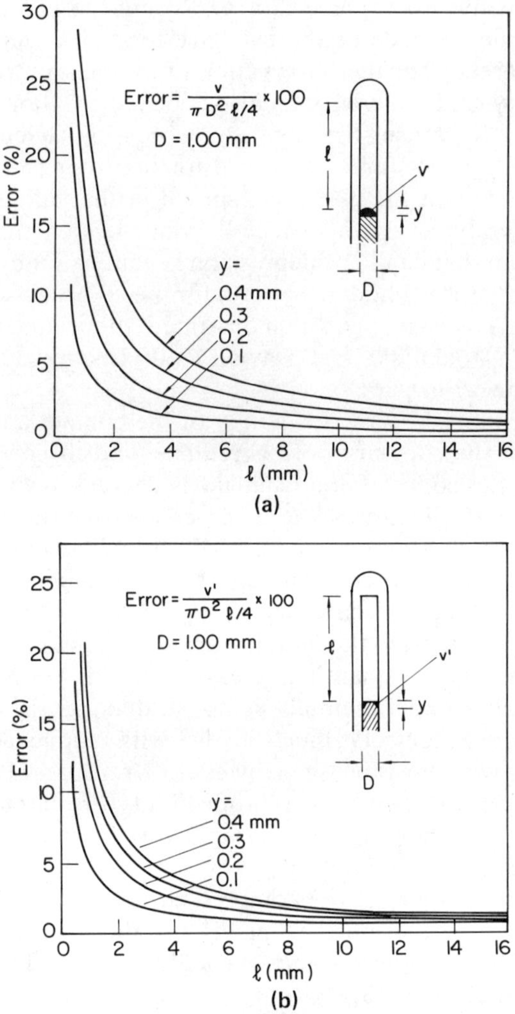

Fig. 3.15 Error caused by the variation of the meniscus volume. Reprinted with permission from *Vacuum* **18,** 71. Nakayama *et al.* Copyright 1968 Pergamon Press.

To avoid these drawbacks, Reid (1968) cemented a flat glass disk at the top end of the compression capillary polished flat to within one light fringe (Fig. 3.16a). The diameter of the glass disk was larger by 3 mm than the outer diameter of the capillary tubing.

Podgurski and Davis (1960) sealed a tapered plug at the top end of the compression capillary (Fig. 3.26b) of their gauge which had a special

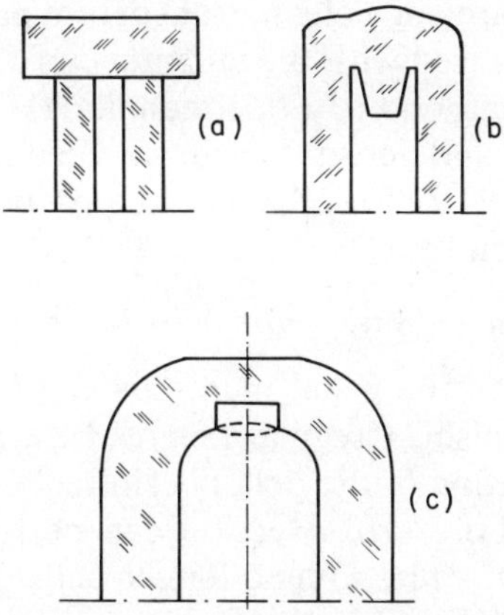

Fig. 3.16 Designs for flattening the end of the compression capillary, (a) Reid (1968), (b) Podgurski and Davis (1960), (c) Moser and Poltz (1958).

construction. The conical shape of the plug permitted conserving the rough finish of the inner wall of the capillary, but did not fulfill the requirement of terminating "flat" the top end of the tubing due to the annular clearance left around the plug.

Moser and Poltz (1957) designed a rectangular-shaped compression chamber at the end of the compression capillary (Fig. 3.16c). The lower side of this chamber forms a sharp edge with the glass walls, and mercury sticks to that edge. It is possible to adjust the position of the mercury level plane at this edge and detect the mercury surface by special methods of illumination (see Section 3.3.6).

The accuracy determination of the heights of the inner and outer ends of the compression capillary is used to correct the error arising in pressure measurement due to the problems discussed above. Two approaches are available according to whether determinations are made during the manufacture of the manometer, when the capillary can be handled as a separate part, or after the manometer is assembled. In the first case the capillary is sealed, and a fiducial line tangent to the inner top end of the capillary bore is marked on glass. A small amount of mercury is then introduced in the bore, and the mercury pellet which is formed is moved to fill in the closed end. After measuring the length of the pellet the mercury is removed and weighed. The volume corresponding to the pellet can thus be calculated. If the bore cross section is accurately known, the volume corresponding to a right circular cylinder with height h_d is then easily computed (Dunoyer, 1926).

The cross-sectional area of a clean, transparent piece of capillary tubing can be accurately measured with the technique of Ruthberg (1975, p. 250) before sealing the tubing bore. In this technique a small amount of mercury of known mass m and density ρ is introduced into the capillary bore. The cross-sectional area a of the bore in the portion where the mercury pellet is located is given by

$$a = (3m - \pi\rho h_v^3)/3\rho(l - h_v) \tag{3.26}$$

where h_v is the height of the pellet meniscus and l the pellet length measured between the menisci. Irregularities of the cross-sectional area can be revealed as the position of the pellet is shifted along the capillary. The average value of the bore cross section can be found by using a large amount of mercury, thus providing a longer pellet.

In the second approach, gas at low pressure is admitted into the assembled manometer and compressed in the closed-end capillary. The height of the compressed gas column is then measured from the top of the mercury meniscus to a fiducial line tangent to the outer top end of the capillary (h_t in Fig. 3.14). Pressure P is calculated as

$$P = (A/V)(h_t - h_o)(\Delta h - \Delta h_c)\rho g \tag{3.27}$$

Here, in addition to the symbols used for writing Eq. (3.18), h_o is the height of the effective outer top end of the compression capillary and Δh_c the difference in the capillary depressions between the compression and the comparison capillaries.

The value of h_t calculated from Eq. (3.27) is

$$h_t = C/\Delta h_r + h_o \tag{3.28}$$

where

$$C = P(V/A)(1/\rho g) = \text{const} \tag{3.29}$$

and

$$\Delta h_r = \Delta h - \Delta h_c \tag{3.30}$$

Equation (3.28) represents a straight line in a plot of h_t versus $1/\Delta h_r$, with a h_o intercept at the h_t axis, (Fig. 3.17). Usually, a set of readings of h_t and Δh at constant pressure yield the average value of h_o, within the limits of the experimental error. Systematic differences in the value of h_o obtained indicate a dirty or a nonuniform capillary bore, or both. The accuracy of the interpolation of h_o can be checked by plotting the results obtained for several pressures (P_1, P_2, P_3). This technique has been used by Gotto (1960), Nottingham and Tornery (1961), and Eschbach (1962).

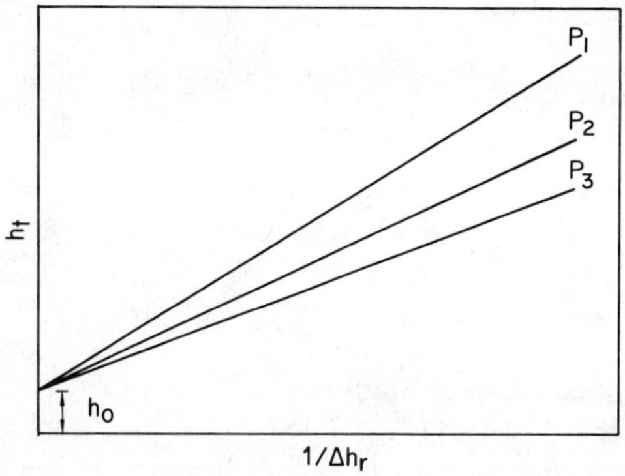

Fig. 3.17 Graphical determination of the effective top end of the compression capillary.

(e) *Capillary forces.* The height reached by the mercury columns within the compression and comparison capillaries is actually lower than that which would correspond to the hydrostatic equilibrium of these columns. This effect is produced as a result of the capillary depression which varies with temperature, local variations of the boundary angle between mercury and glass, etc. (for the quantitative aspect of capillarity, see Section 3.2.2).

Capillary depression is quantitatively determined by the following techniques.

For instruments in which the comparison and compression capillaries have the same bore, gas is trapped above the cutoff level (X–X′ in Fig. 3.13) at the instant when the pressure in the system is about 10^{-1} Pa (10^{-3} Torr) or less and then compressed. The difference between the tops of the mercury menisci in the compression capillary (1) and the side arm (2) (Fig. 3.14) gives the value of the capillary depression.

For instruments in which the compression capillary is sealed by a tapered or cylindrical glass plug and the comparison capillary is replaced by a wide-bore tube (Podgurski and Davis, 1960; Eliott *et al.*, 1967), the technique is different. Assuming that gas is compressed within a right circular cylinder, Eq. (3.18) can be written

$$P = (A/V)(h + h_v - h_d)(\Delta h - \delta h_c)(\rho g) \tag{3.31}$$

or

$$P = (A/V)(h_t + h_v - h_o)(\Delta h - \Delta h_c)(\rho g) \tag{3.32}$$

Here h_v is the height of the mercury meniscus and h_d the virtual height of the compression capillary if it were flat (see Fig. 3.14). The height h_i of the right circular cylinder is

$$h_i = (h + h_v - h_d) = (h_t + h_v - h_o) \tag{3.33}$$

Substituting Eq. (3.33) in Eqs. (3.31) and (3.32) gives

$$P = (A/V)h_i(\Delta h - \Delta h_c)(\rho g) \tag{3.34}$$

For P and T constant, ρ does not vary and Eq. (3.34) becomes

$$h_i(\Delta h - \Delta h_c) = \text{const} \tag{3.35}$$

This equation represents a straight line in a plot of Δh versus $1/h_i$, with a Δh_c intercept at the Δh axis (Fig. 3.18).

Plotting Δh_c as a function of different heights of the mercury column permits both rapid correction of the capillary depression at any point on the capillaries and checking of the manometer stability during determinations. If the values of Δh are corrected for the errors introduced by Δh_c, pressure determinations are affected only by reading errors, random in nature. Podgurski and Davis (1960) recommend a *variable compression technique* for the assessment of Δh_c. This technique consists in *taking several readings at different compression ratios for the same pressure.* The uncertainty arising from this technique is due to the difference in the temperatures of the manometer during the determination of Δh_c and during the determination of the actual pressure.

(f) *Alteration of the gas amount in the gauge.* Assuming a leak-free connection between the manometer and the system, the following sinks and sources may alter the amount and chemical composition of the gas being measured.

(i) *Sorption at the glass walls of the measuring system.* Gas can be sorbed at the glass walls of the gauge or at the cold trap connected between the instrument and the system when the coolant level is raised in

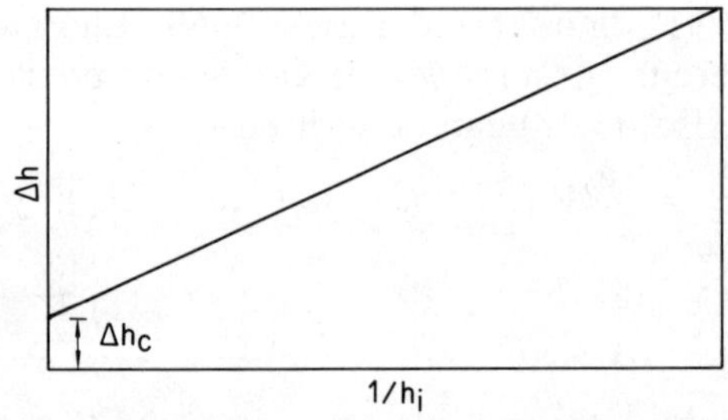

Fig. 3.18 Graphical determination of the difference Δh_c in capillary depression.

the trap. Carr (1964) discussed the influence of gas sorption on the measurement and calculated the sorption constant of the gas.

(ii) *Gas released from the cold trap.* When the coolant level of the refrigerant in the cold trap decreases as a result of evaporation, gas formerly adsorbed at the walls of the trap is released. This leads to pressure differences in the tubing connecting the instrument with the system. The value of the pressure difference depends on the conductance of the tubing. Gas release from the cold trap can be reduced by utilizing a coolant with gaseous cesium can reduce-permeation to negligible rates (Bryant and ide wetted with alcohol) instead of one with low evaporation temperature (liquid nitrogen).

(iii) *Gas permeation.* The permeation of gas (usually He through fused silica) is proportional to pressure and amenable to calculation (Roth, 1980, p. 145). Adequate treatment of the inner surface of the glass with gaseous cesium can reduce permeation to negligible rates (Bryant and Gosselin, 1966).

(iv) *Spontaneous gas release from the glass walls of the gauge.* The rate of gas release can be reduced to negligible values by outgassing the instrument and the tubing connecting it to the system. This is usually achieved by either torching or bake-out procedures. Podgurski and Davis (1960) recommend flushing the gauge with O_2 at about 660 Pa (about 5 Torr) after bake-out in order to burn out nonvolatile carbonaceous products.

3.3.5 Systematic Errors Arising from the Measurement Process

Pressure readings with the McLeod manometer are accurate for gases obeying Boyle's law within the limits discussed in Section 3.3.4. A mixture of gas and vapor which does not saturate during compression in the manometer will approximate this law as well. The manometer will indicate in this case the total pressure due to both gas and vapor at the measurement temperature. If condensation takes place during the measurement, the instrument will read the pressure due to the gas plus the saturation vapor pressure due to the condensed vapor.

Condensation can be detected by using the methods of Pirani and Yarwood (1961) or Wear (1968). According to the first method the height Δh of the mercury in the comparison capillary is varied, and the corresponding changes of the mercury heights h in the compression capillary are noted (see Fig. 3.14). If no condensable vapor is present or if the compression is insufficient to produce condensation, the plot of $h \, \Delta h$ versus h will be a straight line (Fig. 3.19a). In the case of a gas or a gas–vapor mixture,

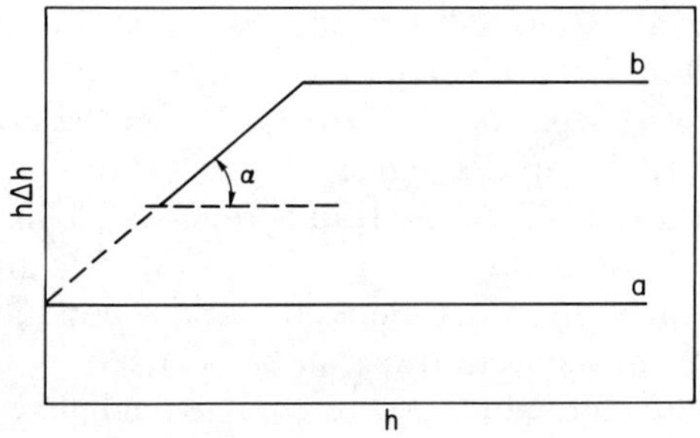

Fig. 3.19 Graphical determination of condensable vapor. a, Noncondensable vapor; b, condensable vapor.

as soon as condensation begins, the plot of the same variables is a straight line with slope α to the h axis (Fig. 3.19b). The slope α gives the magnitude of the pressure produced by the saturated vapor.

The entry of condensable vapor into the manometer is prevented by connecting the instrument with the measuring system via a cold trap. The presence of the cold trap introduces, however, supplementary errors to pressure measurement due to (a) thermal transpiration and (b) the mercury pumping effect.

(a) *Thermal transpiration.* A gas flow builds up in the tubing which connects the manometer with the cold trap because of the temperature difference between these parts. This gas flow generates a pressure gradient. The relationship between the pressures P and temperatures T at the location of the manometer (1) and cold trap (2) is

$$P_1/P_2 = (T_1/T_2)^{1/2} \tag{3.36}$$

and holds true until and equilibrium state is established.

Deviations from this relationship are introduced by the geometry and the reflectivity of the tubing walls. Hobson (1970, 1973) proved that Eq. (3.36) is valid for tubings with a rough surface. In order to use this equation for tubings with a vitreous (smooth) surface, the term containing the square root ratio of temperatures must be multiplied by a term $a \cong 1.1$–1.3. Siu (1973) has shown that Eq. (3.36) holds true for short tubes only for specular reflection and for long tubes only for diffuse reflection of molecules.

The pressure gradient which is a sensitive function of the dimensions of the cold trap can be reduced to negligible values by using a molecularly rough (Ruthberg, 1975, p. 254) tubing symmetrically bent in a U shape (Rush and Bunge, 1932).

(b) *Mercury pumping effect.* Because of mercury vapor flow to the cold trap, molecules inside the compression volume of the gauge are dragged away. Therefore the gas pressure in the McLeod manometer is lower than that in the system. This effect, first described by Gaede (1915) and later by others, is well known, and corrections can be made in the McLeod readings.

Several proposed methods to diminish or to avoid the mercury vapor pumping effect either introduced a length of capillary tubing in the line connecting the gauge to the cold trap (Tunniclife and Rees, 1967; Sharma *et al.,* 1980) or cooled the cutoff section (Ishii and Nakayama, 1962). Also, special McLeod constructions were suggested (Meineke and Reich, 1963b; Colgate and Genre, 1968; Reichman and Wharton, 1970), but none were reported in routine practice.

The magnitude of mercury vapor pumping can be computed. The considerations for computation may be classified according to one of two theoretical hypotheses, either the molecular transport theory of diluted gases or the kinetic theory of gases in the region of molecular flow.

Gaede (1915), taking into account the molecular transport theory of diluted gases, presented a mathematical model expressing the ratio between the pressures of gas and mercury vapor in the system. In accordance with the theory, an increasing pressure gradient of the gas in the system builds up along the tubing connecting the McLeod manometer to the cold trap because the mercury vapor flow. Owing to its high density at the cold trap, the gas diffuses back through the mercury vapor towards the manometer. Expressing the two phenomena in the stationary state, Gaede reached a mathematical expression which he integrated over the distance traveled by the gases and mercury vapor between the manometer and the cold trap. The velocity of mercury vapor was considered constant and independent of the pressure of the gas in the system.

Using Gaedes equation, many workers (Ishii and Nakayama, 1962; Meineke and Reich, 1963a,b; de Vries and Rol, 1965; Elliott *et al.,* 1967) determined the errors arising in McLeod manometer readings, but their results are slightly inconsistent. An analysis of their reports indicates that there are three reasons for the discrepancies:

(i) errors in calculating the numerical coefficient in Gaede's equation;
(ii) calculation of the diffusion coefficient by unlike methods;
(iii) different values chosen for the mercury pressure at the same temperatures.

It is worth mentioning that the experimental work reported was carried out within pressure ranges below 10^{-5} Pa (10^{-3} Torr).

In criticizing Gaede's equation as not representing the physical meaning of the phenomena, Takaishi (1965) studied the mercury pumping effect in the light of the kinetic theory of gases in the region of molecular flow. Owing to the difficulties in solving nonlinear differential equations for density gradients, he made use of two approximations, one underestimating and the other overestimating the effect of mercury drag. In a more recent work Takaishi and Sensuy (1970) undertook an experimental study to verify the results obtained by Takaishi to explain the mercury vapor pumping effect. They concluded that the experimental results are well described by his theory but could not eliminate the uncertainty in determining the value of one parameter in the Takaishi equation. Neither was any progress made in Sensui's (1970) analysis regarding the accurate determination of this parameter.

Rambeau (1969) also made use of the kinetic theory of gases to study the errors associated with the mercury vapor pumping. He deduced three expressions, each giving the magnitude of the error for a definite range of pressures.

Colgate and Genre (1968) tried to eliminate the error due to the mercury pumping effect by means of a device added to the McLeod manometer. They also obtained a theoretical solution of the problem by integrating Gaede's transport equation for mercury vapor velocity considering the pressure of the gas in the system to be variable. The experimental results they obtained are different, however, from those found by Takaishi in the range 10^{-2}–6.66 Pa (10^{-4}–5×10^{-2} Torr).

The effect of mercury vapor pumping on the McLeod readings is still controversial. Each mathematical model proposed gives, for the same range of pressures and temperatures, results depending upon the value chosen for the mercury vapor pressure.

The error arising from the mercury vapor pumping depends on the absolute temperature (which affects both mercury vapor pressure and the diffusion coefficient of the measured gas in mercury vapor) and on the inner radius of the tubing connecting the manometer to the cold trap. If the McLeod manometer is not thermostated, error calculations for each pressure reading become a laborious task, especially since for each pressure several readings must be taken at different compression ratios. Even for routine calibration work, where the connecting tube radius and the calibration gas remain the same for all measurements, error calculation is still complicated.

With a view to providing the possibility of a fast and accurate evalution of the errors associated with mercury vapor pumping for routine work, calculations were made (Berman, 1974) for a large range of temperatures and radii of connecting tubing. These calculations hold true in the range 10^{-5} Pa (10^{-3} Torr) down to the pressures where capillary depression errors limit the accuracy of the manometer readings. Since in gauge calibration work pure dry nitrogen is used, calculations were made for this gas. For convenience, the calculated results were plotted.

For the range of pressures below 10^{-5} Pa (10^{-3} Torr), all the mathematical models agree, so Gaede's equation was chosen for computations. Also it was assumed that no thermal transpiration effect exists, due to temperature nonuniformity between the manometer and the remainder of the vacuum equipment.

In the stationary state, the rates at which the gas in the system and mercury vapor flow along the tubing connecting the McLeod manometer to the cold trap are expressed by

$$dP/P = -(p_v u/D_{12})\,dx \tag{3.37}$$

where P is the pressure of the gas, p_v the pressure of mercury vapor, u the velocity of mercury vapor, and D_{12} the diffusion coefficient of the gas against the mercury vapor at 133.332 Pa (1 Torr).

The velocity of mercury vapor is assumed to be constant (i.e., not modified by the gas diffusing in the opposite direction). Equation (3.37) can be integrated over the length of the connecting tubing, from $x = 0$ and $P = P_1$ (pressure at the cold trap) to $x = l$ and $P = P_2$ (pressure at the McLeod gauge).

$$\ln(P_1/P_2) = (p_v u/D_{12})\,l \tag{3.38}$$

The velocity of the mercury vapor can be written

$$u = Q/(\pi r^2 p_{v2}) = C(p_{v2} - p_{v1})/(\pi r^2 p_{v2}) \tag{3.39}$$

where l and r are the length and inner radius, respectively, of the tubing connecting the gauge to the cold trap, Q is the throughput of mercury vapor along the tube, C the conductance of the connecting tube for mercury vapor, T the thermodynamic temperature, and M the molecular weight. The mercury vapor pressure at the manometer and at the trap are p_{v2} and p_{v1}, respectively. As $p_{v1} \ll p_{v2}$

$$u = C/(\pi r^2) \tag{3.40}$$

For the molecular range, where $l/r \geq 100$ (Dushman, 1962, p. 88),

$$C = 30480(r^3/l)(T/M)^{1/2} \tag{3.41}$$

Substituting into Eq. (3.38) the values of u and C obtained from expressions (3.40) and (3.41), and taking the diffusion coefficient at atmospheric pressure,

$$\begin{aligned} \ln P_1/P_2 &= [30480/(3.14 \times 14.18 \times 760)](p_v r T^{1/2}/D_{12}) \\ &= 0.901(p_v r T^{1/2}/D_{12}) \end{aligned} \tag{3.42}$$

Solving Eq. (3.42) gives

$$P_1/P_2 = \exp(0.901 p_v r T^{1/2}/D_{12}) \tag{3.43}$$

or approximately

$$(P_1 - P_2)/P_2 = P/P_2 = 0.901(p_v r T^{1/2})/D_{12} \tag{3.44}$$

The error introduced in expression (3.44) due to disregarding the second term in the e^x series is 2×10^{-2} and 5×10^{-4} at $T = 303$ K and $T = 278$ K, respectively.

The mercury vapor pressure p_v and the diffusion constant D_{12} were calculated for each K in the range 278–303 K. The mercury vapor pressure was calculated using Kelley's expression (Dushman, 1962, p. 736), and the value of the diffusion constant was calculated using the Hirshfelder formula and tables 1-A and 1-M of Hirshfelder *et al.* (1964, pp. 539, 1111–1112, 1126). Inner radii of the connecting tubing ranging from 0.2 to 1 cm were considered. The results are plotted in Figs. 3.20 and 3.21.

For the experimental work, a standard calibration system [tentative standard AVS 6.4 (1969)] (Fig. 3.22) was used. The stainless steel dome (D) is pumped down by a mercury diffusion pump. The mercury vapors

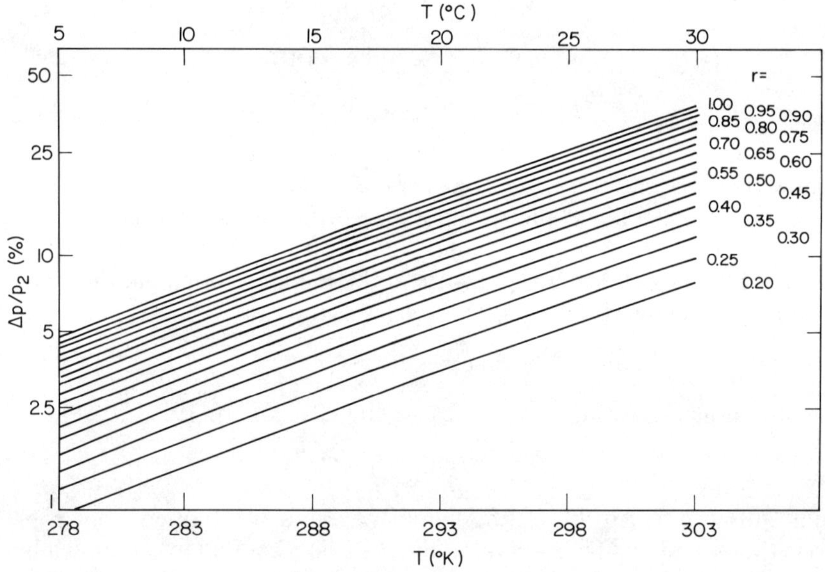

Fig. 3.20 $\Delta P/P_2$ as a function of temperature for connective tubing with inner radii ranging from 0.2 to 1.0 cm, in 0.05 cm intervals.

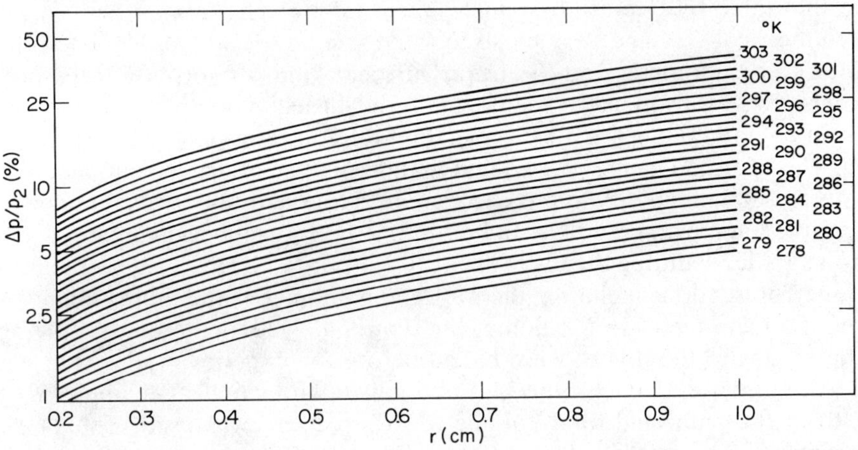

Fig. 3.21 $\Delta P/P_2$ as a function of the tubing radius for temperatures ranging from 278 to 303 K, in 1 K intervals.

are trapped in a liquid nitrogen trap (LN). A butterfly valve (not shown in Fig. 3.22) and a needle valve V_1 offer the possibility of generating fixed points of pressure by balancing the gas flow in and out of the test dome (see Section 5.5.2). An ionization manometer of the Bayard–Alpert (BA) type and a capacitance manometer (CM) are used to monitor and read pressure, respectively, inside the dome. The McLeod gauge is connected

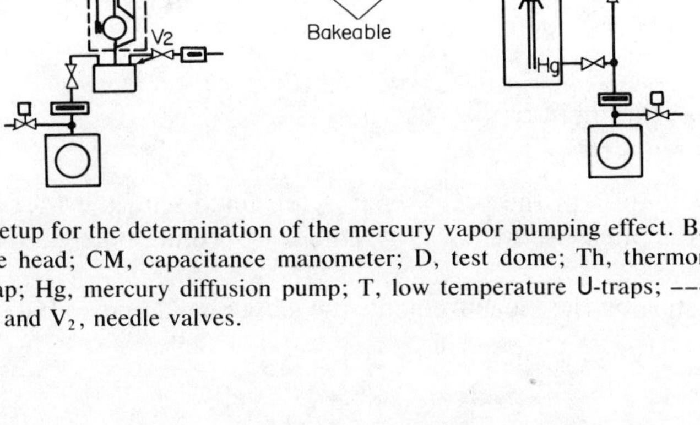

Fig. 3.22 Setup for the determination of the mercury vapor pumping effect. BA, Bayard–Alpert gauge head; CM, capacitance manometer; D, test dome; Th, thermometer; LN, liquid N_2 trap; Hg, mercury diffusion pump; T, low temperature U-traps; ---, bakeable sections; V_1 and V_2, needle valves.

to the dome by 6-mm i.d. stainless steel tubing. The tubing, bent into a symmetrical U shape very close to the dome, is 63-cm long between the McLeod manometer and the trap. Different kinds of sorption traps prevent oil vapor from backstreaming into the diffusion pump or water vapor from contaminating the mercury in the McLeod reservoir. The trap backing the McLeod gauge was cooled with a mixture of dry ice and alcohol. Any possible humidity in the nitrogen used for calibration was retained in a trap cooled with liquid N_2. In both traps the coolant was maintained at a constant level during the experiment. High-purity N_2 was used for calibration. The sections including the McLeod manometer, the tubing connecting the manometer to the dome, the dome, as well as the line supplying the N_2 gas to the dome, were baked before the experiments.

A commercial bench-type McLeod gauge with capillary diameter $d = 1.01 \pm 0.01$ mm and total volume of 412 ± 2 cm^3 was used. The gauge constant was therefore $K = 1.94 \pm 0.03 \times 10^{-6}$ mm^{-1}.

The estimated uncertainty of the McLeod gauge measurements was found to range between ±2.7% at 0.042 Pa (3.15×10^{-4} Torr) and ±5.8% at 4.2×10^{-3} Pa (3.15×10^{-5} Torr) (Simons, 1963; Carr, 1964). The standard deviation of the pressure readings was about 2.6%.

The capacitance manometer was calibrated in the system shown in Fig. 3.22 according to the method of Utterback and Griffith (1966). The accuracy of the capacitance manometer was ±0.25%.

Before connecting the McLeod gauge into the system, the glass was specially treated (Berman, 1973). In order to avoid mercury contamination resulting from glassblowing in joining the manometer to the system, the connection was made by means of a metallic bellows and a demountable union screw.

Pressure readings were taken with the aid of a cathetometer according to the linear scale. After determining the location of the top of the sealed capillary, 20 measurements were taken for each pressure. During these measurements, the temperatures indicated by the thermometers located at the bulb of the gauge and at the test dome were the same. The results are plotted in Fig. 3.23. Good agreement between calculated and experimental results is evident.

3.3.6 Measurement of the Mercury Column Heights

The calculation of the pressure P determined with the McLeod gauge involves the measurement of the mercury column heights in the instrument.

Before starting the measurement, the gauge has to be stabilized against

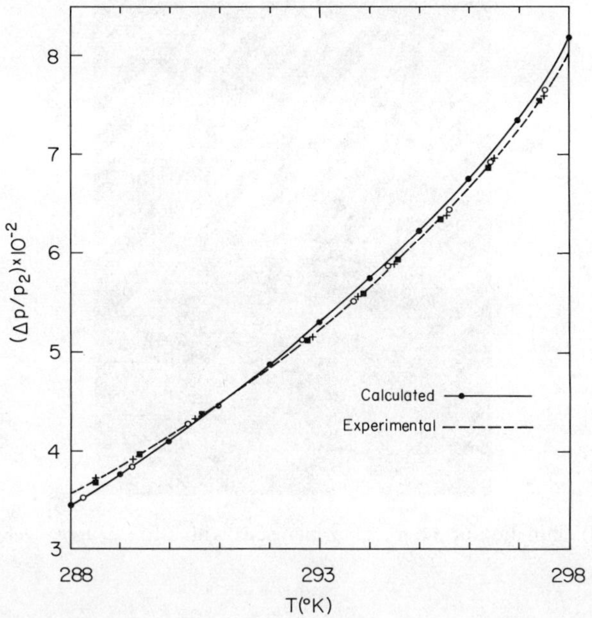

Fig. 3.23 Calculated and experimental curves of $\Delta P/P_2$ as a function of temperature for $r =$ 0.30 cm and a measuring pressure of pure N_2.

vibrations and leveled. The cosine error (see Section 3.2.2) can be reduced to a negligible value much more easily than in the case of U-tube manometers. The sine error resulting from the misalignment of either the scale used for observation or the cathetometer column from the vertical is, however, more critical than in the case of the U-tube manometer.

The height of the mercury columns in the manometer glassware is measured on a ruled scale placed close to the compression capillary. For accurate readings, a telescope (equipped with a cross-hair eye-piece), sliding by means of a vernier along a graduated column (cathetometer), is used. The precision of measurement depends on whether a ruled scale or a cathetometer column is used to take readings (Ruthberg, 1975, p. 249). In the latter case the telescope must be located at such a distance that the presence of the operator does not affect the manometer temperature.

Sighting of the menisci tops needs adequate illumination. For this purpose, optical devices which collimate light by means of a slit and lens system are used (Collins and Blaisdell, 1936). The drawback of the method is that for each meniscus shift the optical device must be readjusted. Good illumination (Zeleny, 1936) is attained by locating behind the

Fig. 3.24 Illumination of the mercury meniscus with a translucent "zebra" pattern.

meniscus a translucent screen on which a "zebra" pattern is drawn (Fig. 3.24). The screen is illuminated from behind, thus transmitting diffuse light. The setting is made on the sharp junction between the mercury meniscus top and the reflected pattern on the mercury surface. Regardless of the system of illumination, care must be taken that the temperature of the gas in the compression volume not be changed as a result of thermal radiation emitted from the illumination source (Miller, 1972; Cespiro, 1973).

3.4 Pressure Generators

3.4.1 Range of Pressure Generators

Pressure measurement in the pressure range where direct comparison with a primary standard is inaccurate is accomplished by resorting to the indirect method of measuring pressure (see Section 2.4.2). Basically, a mass of gas, the pressure of which can be accurately measured, is used as a means of generating a rarefied environment in an evacuated volume to which, usually, a secondary standard (see Section 3.1) is attached. The readings of the instrument are then related to the gas pressure which has generated the environment. During measurements, the volume accommodating the rarefied environment (referred to as calibration volume in the following discussion) is isolated from the pumping system (hence, the name *static method*). The pressure range of this method is $\sim 10^3$ Pa (10 Torr) $> P >$ 10^{-6} Pa (10^{-8} Torr).

Although density essentially is the physical quantity which accurately characterizes a gas at equilibrium below a certain degree of rarefaction, for convenience, pressure will be used instead throughout this section.

3.4.2 Principle

Gas is admitted into a previously evacuated volume V_1 in such an amount as to generate a pressure P_1 large enough to be accurately measured with a primary or secondary standard (Fig. 3.25). After measuring P_1, the gas is expanded in an evacuated volume V_2, notably large compared with V_1. For a perfect gas, expanded under isothermal conditions such that the quantity of gas remains constant during the process, Boyle's law applies and

$$P_1 V_1 = P_2 V_2 \tag{3.45}$$

From Eq. (3.45) P_2 can be determined if the values of P_1, V_1, and V_2 are known. The readings of the secondary standard M_2 are then compared with the values of P_2 calculated according to Eq. (3.45), and the required pressure point is inferred after suitable corrections. Reasonable pressure points on the pressure scale can be obtained by selecting appropriate values for P_1, V_1, and V_2.

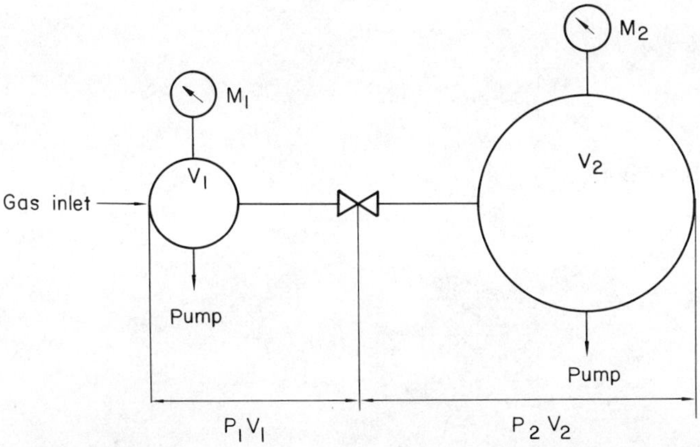

Fig. 3.25 Principle of gas expansion. V_1, V_2, enclosures with volume V_1 and V_2, respectively; M_1, primary or secondary pressure standard; M_2, secondary pressure standard.

The method of expanding a known mass of gas to generate low pressures, which suggests the reverse operation of the McLeod gauge, was first used by Knudsen (1910) to calibrate his radiometer gauge.

3.4.3 Basic Operation Procedures

For generating the rarefied environment in volume V_2, two approaches have evolved: (a) the single- and (b) the multiple-stage expansions.

(a) *Single-stage expansion.* A typical schematic diagram of a pressure generator system utilizing this technique is shown in Fig. 3.26. The system is evacuated and degassed such that the base pressure is much lower than that necessary for the pressure point requested, and the rate of rise of the residual gas is low. The system is then isolated from the pump, and gas is admitted in the gas container G and the transfer sampler Tr. The pressure of the gas is measured with the primary or secondary gauge M_1 and then G is isolated from Tr. Finally, the gas contained in the transfer sampler is allowed to expand in the calibration chamber C. In admitting or expanding gas, a certain time must elapse before the pressure is measured to permit the gas mass to reach temperature equilibrium with the container. [Equilibrium conditions prevail when the readings of each instrument involved in pressure measurement remain constant. Thus, Elliott and Clapham (1978) allow 1 min elapsed time between successive

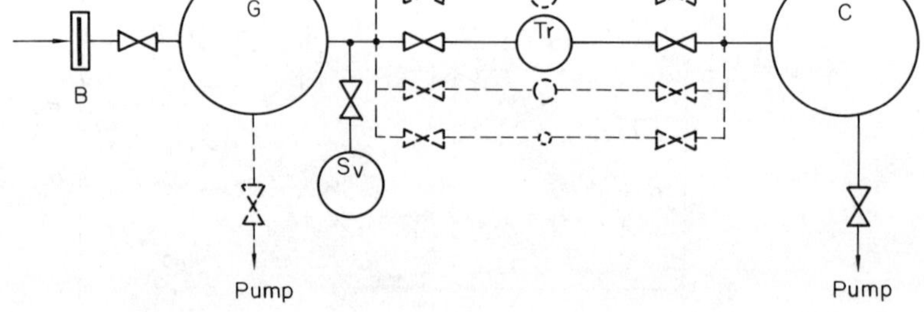

Fig. 3.26 Diagram of a system used in the single-stage expansion technique. G, gas container; Tr, transfer sampler; C, calibration chamber; M_1, primary or secondary pressure standard; M_2, secondary pressure standard; B, cold trap; S_v, standard volume.

steps, although according to their experience equilibrium is reached in less than 10 s]. If the sequence of operations leading to a pressure point is conventionally called a step, a series of calibration points can be generated by completing a number of steps:

(i) with various inlet pressures,
(ii) with transfer samplers of different sizes (broken lines in Fig. 3.25), or
(iii) by cascading, i.e., by pumping the calibration chamber after the first step with the adjacent transfer sampler isolated and then once again expanding the gas from the transfer sampler to the calibration chamber.

The required pressure point is obtained by comparing the readings of the gauge M_2 for a certain pressure, with the calculated value of that pressure. The value of the first pressure point P_{1c} generated in the calibration chamber after completing a step is expressed by

$$P_{1c} = (P_{1c} V_{Tr})/(V_{Tr} + V_c) \tag{3.46}$$

or by

$$P_{1c} = (P_{1G} V_{Tr})/(V_{Tr} + V_c)^2 \tag{3.47}$$

after a cascade expansion, where P_{1G} is the initial pressure of the gas in the container G, before completing the first step and V_{Tr}, V_c are the volumes of the transfer sampler and calibration chamber, respectively. In writing Eqs. (3.46) and (3.47) the volumes of the connecting tubing (including valves), as well as the volumes of the cold traps connected to the system, were disregarded.

After the first step, the pressure in the transfer sampler and the calibration chamber will differ from the initial value and Eqs. (3.46) and (3.47) will consequently have to be corrected. Smetana and Carley (1966), who treated the gas transfer in a single-stage expansion system, more rigorously express the pressure generated in the calibration chamber after the first step as

$$\begin{aligned} P_{1c} = {} & (P_{1G} V_G V_{Tr})/|(V_G + V_{Tr})(V_{Tr} + V_c)| \\ & + P_{bc}|1 - (V_G V_{Tr})/[(V_G + V_{Tr})(V_{Tr} + V_c)]| \end{aligned} \tag{3.48}$$

and after the nth step as

$$\begin{aligned} P_{nc} = P_{bc} + {} & \left| \frac{(P_{1G} - P_{bc})}{(1 + V_{Tr}/V_G + V_c/V_G)} \right| \\ & \times \left[1 - \left| \frac{1}{(1 + V_{Tr}/V_c)(1 + V_{Tr}/V_G)} \right|^n \right] \end{aligned} \tag{3.49}$$

Here P_{bc} is the pressure in the calibration chamber due to the background of residual gas as measured by the gauge M_2 after separating the chamber from the pump. Using Eq. (3.49) and assuming that the volumetric ratios V_{Tr}/V_G and V_{Tr}/V_c as well as the ratio P_{bc}/P_{1G} are much less than unity, Holanda (1969) expressed the pressure generated in the calibration chamber after the nth step by

$$P_{nc} = P_{n-1,c} + (V_{Tr}/V_c)P_{nG} \tag{3.50}$$

where P_{nG} is the pressure in the container G after the nth step. Equation (3.50) is a close approximation to Eq. (3.49) and is more convenient to use when a large number of data points (usually more than 100) are necessary for determinations.

(b) *Multiple-stage expansion.* This technique permits generating a larger range of pressure points than that obtained with the single-stage expansion technique by utilizing a relatively smaller number of steps. Basically the gas is transferred to the calibration chamber by expanding it through transfer samplers of different sizes, alternated such that a large-volume sampler is always connected to a small-volume sampler (Fig. 3.27). Each large transfer sampler as well as the calibration chamber is pumped separately.

After processing the system so as to insure both a background pressure at least two decades lower than the smallest calibration pressure and a reasonable rate of rise of the residual gas, the system is isolated from the pumps. Two operation modes are then available.

(i) *Sequential expansion mode.* Gas is expanded in the same manner as in the single-stage expansion. If the circuit drawn in solid lines in

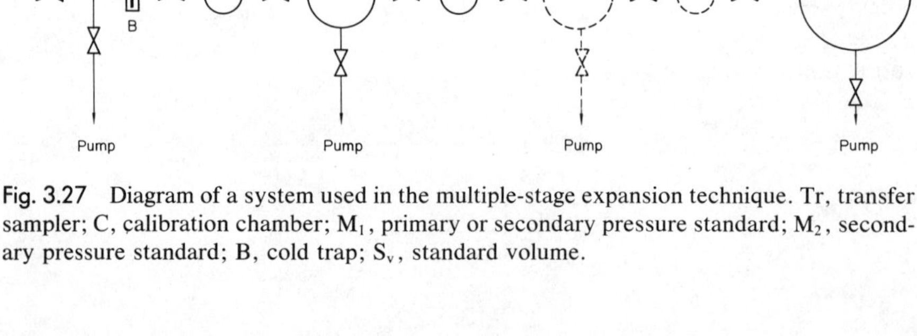

Fig. 3.27 Diagram of a system used in the multiple-stage expansion technique. Tr, transfer sampler; C, calibration chamber; M_1, primary or secondary pressure standard; M_2, secondary pressure standard; B, cold trap; S_v, standard volume.

Fig. 3.27 is considered, the pressure generated in the calibration chamber after the first step will be

$$P_{1c} = P_{M1}[V_{Tr_1}/(V_{Tr_1} + V_{Tr_2})][V_{Tr_2}/(V_{Tr_2} + V_{Tr_3})][V_{Tr_3}/(V_{Tr_3} + V_c)] \tag{3.51}$$

(ii) *Cascaded expansion mode.* Gas is admitted into the transfer sampler Tr_1, which is isolated from the gas supply after measuring the gas pressure. The gas is then expanded from Tr_1 to Tr_2 and allowed to reach temperature equilibrium with the sampler. Subsequently, Tr_2 is isolated from Tr_1 and pumped down by the pump directly connected to it. After reaching a convenient base pressure in Tr_2, the pump is isolated from the transfer sampler, and the gas from Tr_1 is again expanded in the volume Tr_2. The remaining expansion is then performed in the usual manner previously described. For this situation, Eq. (3.51) becomes

$$P_{1c} = P_{M1}|V_{Tr_1}/(V_{Tr_1} + V_{Tr_2})|^2 \times |V_{Tr_2}/(V_{Tr_2} + V_{Tr_3})||V_{Tr_3}/(V_{Tr_3} + V_c)| \tag{3.52}$$

A three-fold expansion from V_{Tr_1} to V_{Tr_2} will lead to

$$P_{1c} = P_{M1}|V_{Tr_1}/(V_{Tr_1} + V_{Tr_2})|^3 \times |V_{Tr_2}/(V_{Tr_2} + V_{Tr_3})||V_{Tr_3}/(V_{Tr_3} + V_c)| \tag{3.53}$$

Equations (3.51–3.53) must be corrected to account for the fact that after the first expansion, the pressure changes in both V_{Tr} and V_c.

Meineke and Reich (1967) used a calibration system built according to the typical schematic diagram shown in solid lines in Fig. 3.26. The expansion ratios of their system were

$$\begin{aligned} V_{Tr_1}/(V_{Tr_1} + V_{Tr_2}) &= 2.224 \times 10^{-2} \\ V_{Tr_2}/(V_{Tr_2} + V_{Tr_3}) &= 9.774 \times 10^{-1} \\ V_{Tr_3}/(V_{Tr_3} + V_c) &= 2.091 \times 10^{-3} \end{aligned}$$

which provide the following values for the sequential or the cascaded modes:

$$P_{1c} = (4.545 \times 10^{-5} \pm 0.47\%)P_{M1} \quad \text{calculated according to Eq. (3.51)} \tag{3.54}$$

$$P_{1c} = (1.011 \times 10^{-6} \pm 0.52\%)P_{M1} \quad \text{calculated according to Eq. (3.52)} \tag{3.55}$$

$$P_{1c} = (2.248 \times 10^{-8} \pm 0.57\%)P_{M1} \quad \text{calculated according to Eq. (3.53)} \tag{3.56}$$

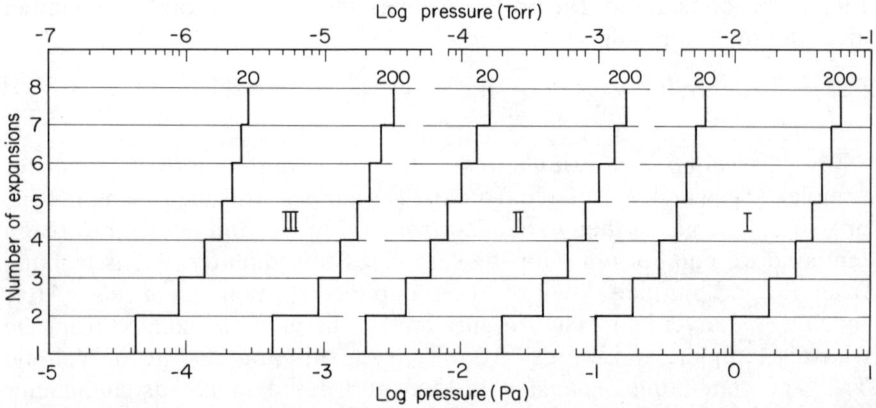

Fig. 3.28 The pressure range of the static expansion method (after Meineke and Reich, 1967).

The range of pressures they could generate in the calibration chamber of their system is shown in Fig. 3.28, where sections I–III refer to the procedures expressed by Eqs. (3.51–3.53), respectively. With a starting pressure of about 2.7 kPa (20 Torr) and a two-fold expansion of the gas from V_{Tr_1} to V_{Tr_2}, Eq. (3.55) gives $P_{1c} \simeq 2.7$ mPa (2.02×10^{-5} Torr) after the first step and $P_{1c} \simeq 21$ mPa (1.6×10^{-4} Torr) after the eighth step (II in Fig. 3.28).

Equations (3.46–3.53) were obtained assuming that the generation of pressure points develops under isothermal conditions. Since in practice the temperature of different parts of the calibration setup rise above ambient, a correction factor to account for temperature variations becomes necessary. Thus, the value of P_{1c} (Eq. 3.46) must be reduced by a factor $(T_n - T_1)/2T_1$, where subscripts to T refer to the temperature of the expansion vessel after the first and the nth step, respectively (Elliott *et al.*, 1967). This correction must be calculated for each expansion vessel in the setup.

Theoretically, in the static method of gas expansion the value of the required pressure point is determined only from the inlet pressure of the gas and geometric considerations, without the necessity of utilizing other assumptions or measurements. Practically, however, this is not the case, because factors such as adsorption and desorption phenomena in the system during a drug or a sequence of steps, replacement phenomena which change the nature of the gas used, etc., cannot accurately be taken into account.

3.4.4 Accuracy and Traceability of Pressure Points Determination

For accurate determinations, the pressure P_G and the volume ratio V_{Tr}/V_c or $V_{Tr,n}/(V_{Tr,n} + V_{Tr,n+1})$ must be known precisely.

In addition, a correct evaluation of uncertainties in the experimental techniques is necessary.

(a) *Measurement of the high pressure P_G.* This measurement is performed with a primary or secondary standard which meets the requirements with regard to pressure range and accuracy. To illustrate how such an instrument is selected, the following example is given.

A calibration system of the single-stage expansion type is required to generate pressures P_c ranging from about 1 Pa (10^{-2} Torr) down to about 7×10^{-5} Pa (5×10^{-7} Torr) in the calibration chamber. The volumes selected for the three vessels V_G, V_{Tr}, and V_c are 10, 10^{-3}, and 10 L, respectively. These values represent the usual dimensions met in practical systems (Schuman, 1962; Arney and Henderson, 1967). The pressures on the high-pressure side of the system calculated by Eq. (3.46) are ~10 kPa (100 Torr) to ~0.7 Pa (5×10^{-3} Torr). To span this range two standard instruments are chosen, one to measure from ~13 kPa (100 Torr) to ~13 Pa (10^{-1} Torr) and the other from ~13 Pa (10^{-1} Torr) to ~0.7 Pa (5×10^{-3} Torr). If the precision of the first instrument is 0.01% full scale at the lower limit of the range, the precision at the high limit of the range will be 1%. The same reasoning applies with regard to the accuracy of the instrument which measures the lower range. Regardless of the type of standard instrument, it is necessary to make sure that the range of pressures it spans lies within its optimum limits of accuracy.

(b) *Determination of the volume ratio.* This measurement refers to the volumes involved in gas expansion as well as those parts which contribute additional volumes, such as connecting tubing, valves, cold traps, and measuring instruments. Two techniques are available: gravimetric expansion and gas expansion, frequently used in combination with each other.

(i) *Gravimetric technique.* Small and moderate sized volumes, such as those corresponding to the transfer samplers, are measured before assembling the system by weighing them empty, filling them with water or mercury (Schuman, 1962), and reweighing. When determinations are made with water, measurement errors are possible owing to trapped air bubbles which are difficult to remove. Mercury circumvents this drawback, but special care is required in its handling to avoid spillage, which would constitute hazardous contamination. However, use of mercury is compulsory in determining minute volumes, especially in cavities shaped in bulky parts, where water fails to provide accurate results.

(ii) *Gas expansion techniques.* The gravimetric method is impractical for the measurement of large or very large volumes or for the determination of the volume ratio of assembled parts as they actually exist in a calibration system. For such measurements, the gas expansion technique is used as previously described with the volume ratio as the unknown variable. The gas container G, or a special standard volume S_v, (Figs. 3.25 and 3.25) supplies the gas which is then expanded into different sections, the volume ratios of which are to be determined (Bobenrieth, 1946; Vermandé, 1952; Elliott *et al.*, 1967; Holanda, 1969). It is recommended that the pressure of the gas after expansion be high enough both to be accurately measured with a standard instrument and also to minimize the gas adsorption in the system.

When the expansion technique is utilized for the determination of the calibration chamber volume, allowance must be made for the volumes added to the chamber by the measuring instruments (such as RGA), the cold traps, and the valve to the transfer sampler. These volumes must be estimated at their operating temperature during measurement. Thus, the volume V_g of a gauge head to be calibrated measured at room temperature T_L must be increased by the ratio $(T_w - T_L)/T_w$, where T_w is the temperature of the gauge during operation. The volume of the valve which connects the calibration chamber with the transfer sampler is to be determined at room temperature (if hand operated), and the volume of the cold trap is to be determined at the temperature of the coolant, with the trap filled to operating level.

Elliott and Clapham (1978) measured volume ratios on the order of 100 of an apparatus for vacuum gauge calibration by the static method. They utilized the successive expansion technique and determined pressure with instruments which do not require the presence of a cold trap. The uncertainty affecting their measurements, after taking into account all the factors discussed above, was about $\pm 0.1\%$.

3.4.5 Uncertainties of the Method

Uncertainties in the use of pressure generators result from (a) the invalidation of Boyle's law and (b) systematic errors in the measurement of the different quantities involved in the evaluation of the pressure generated in the calibration chamber.

(a) *Invalidation of the Boyle's law.* Boyle's law remains valid when (i) the molar quantity of gas transferred through the system is maintained constant, (ii) the process of pressure generation develops under isothermal conditions, and (iii) the gas used for pressure generation obeys the ideal gas laws.

(i) *Alteration of the molar quantity of the calibration gas.* The molar quantity of the gas used for pressure points generation may be altered as a result of *gas pumping* by measuring instruments. Ionization gauges and other measuring instruments, which generate ions from the gas they sense, pump that gas by several mechanisms. The rate of change of the pressure in the calibration chamber P_{EC} due to the electronic pumping of the measuring instrument depends on the pressure P_c in the calibration chamber and is an exponential function of the time at which the instrument was switched on (see Section 7.5). The value of the ratio P_{EC}/P_c is controlled to a large extent by the volume of the calibration chamber. Thus, for instance, if the measuring instrument is of the Bayard–Alpert (BAG) type ($S_E \simeq 0.1$ L/s for N_2) and the time interval required to generate a pressure increment is 20 s, the ratio P_{EC}/P_c will assume the values 0.2, 2, and 20% for values of V_c equal to 1000, 100, and 10 L, respectively. As a result, S_E becomes a significant factor since it alters the pressure generated in calibration chambers of small volume. Nevertheless, the problem has been satisfactorily solved by either correcting for, or minimizing, the pumping effect of the measuring instrument (see Section 7.5.6). Thus, Barton and Chubb (1965) noticed strong pumping effects produced by the calibrated gauge (BAG type) when hydrogen was used as the gas for generating the pressure points in their relatively small volume (10 L) calibration chamber. They recorded an oscillogram of the response of the gauge to the injection of hydrogen and corrected for gauge pumping by intersecting the tangent to the curve of the ion current as a function of time with the axis of the ion current (Fig. 3.29). The same technique of correcting for gauge pumping for argon and nitrogen was used by Elliott *et al.* (1967) in their system where a calibration chamber of about 6 L was used. Schuman (1962) reported having minimized the influence of a calibrated BAG pumping by intermittently energizing the grid of the gauge and by maintaining the emission current at a low level.

The molar quantity of the gas may also be altered as a result of *gas desorption due to temperature*. After the ultimate pressure (also called *blank-off*) is reached in all the enclosures and the system has been separated from the pumps, the partial pressure of noncondensable gases bound in the system starts to rise. In a carefully designed system, this occurs as a result of the spontaneous gas desorption from the walls of the calibration chamber, transfer sampler, tubing, etc. The largest amount of desorbed gas is generated in the calibration chamber, which usually has a

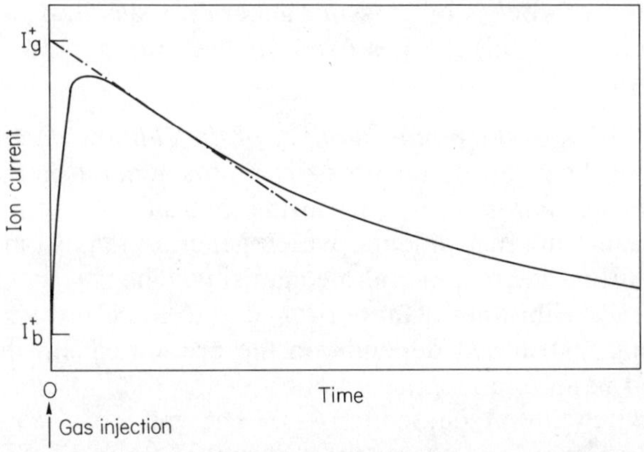

Fig. 3.29 Correction of the ion current variation for gauge pumping during hydrogen injection. I_b^+, current due to the background gas; I_g^+, current corrected for gauge pumping. Reprinted with permission from *Vacuum* **15,** 307. Barton and Chubb. Copyright 1965 Pergamon Press.

large internal wall area. In order to evaluate the contribution of the outgassing from the calibration chamber to the ultimate pressure, it is necessary to consider the ratio of the surface area of the walls A_c to the volume V_c, the time necessary to complete a step, and the outgassing rate of the chamber material. Thus, for a stainless steel calibration chamber with $A_c/V_c = 10\ m^{-1}$, the pressure rise due to outgassing during a step is calculated as follows: The outgassing rate of the material after treatment and bakeout procedures (Power and Robson, 1962) is $Q \cong 10^{-9}$ Pa m³/s/m² (10^{-12} torr L/s/cm²); the pressure rise due to outgassing is $Q(A_c/V_c) \cong 1.3 \times 10^{-8}$ Pa/s (1×10^{-10} Torr/s); the pressure rise during a step ($t_s = 20$ s) is $Q(A_c/V_c)t_s \cong 3 \times 10^{-7}$ Pa (2×10^{-9} Torr). If the smallest pressure point generated by a step is 1.3×10^{-5} Pa (10^{-7} Torr), the pressure rise contributed by outgassing of the walls is 2%.

The effect produced by outgassing on the pressure measured after the gas injection can be assessed graphically. For this purpose, the curve of the variation of the ion current due to background gas, as measured with a residual gas analyzer (RGA), is subtracted from that of the variation of the total current, as measured with the instrument connected to the calibration chamber where both are plotted as a function of time. Figure 3.30 shows how this correction is made when the expanded gas is hydrogen, which is assumed not to interact with the outgassing yield (Barton and Chubb, 1965).

The contribution of the transfer samplers due to outgassing can be measured in a dummy gas-sample injection (Barton and Chubb, 1965). In

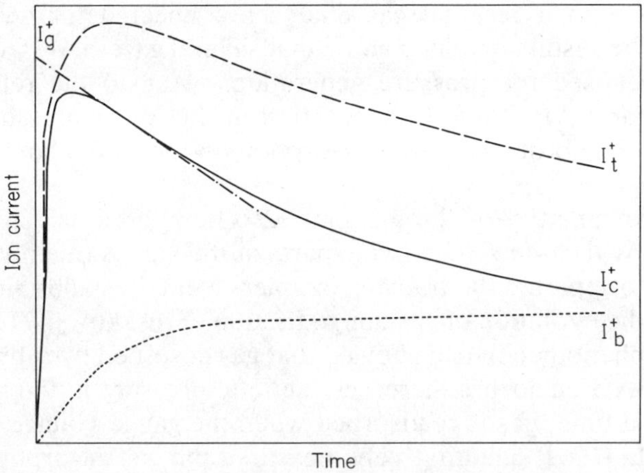

Fig. 3.30 Calibration technique for hydrogen. I_t^+, variation with time of the total current; I_c^+, variation with time of the current following hydrogen injection; I_b^+, variation with time of the current due to background gas; I_g^+, current corrected for gauge pumping. Reprinted with permission from *Vacuum* **15,** 307. Barton and Chubb. Copyright 1965 Pergamon Press.

such a test, the step is performed in the same sequence as in actual calibration, but without transferring any gas.

The molar quantity of the gas may also be altered as a result of *gas adsorption.* Gas is adsorbed on the walls of the pressure generator system owing to physi- and chemisorption phenomena, which in turn depend on several factors, such as the chemical composition of the gas, the wall condition, and the range of pressures at which determinations are made.

Adsorption used to be considered one of the main factors contributing to the limitation of the method (Roehring and Simons, 1962). However, more recent theoretical and experimental research has supplied quantitative evidence that adsorption contributes errors of less than 1 part in 10^4 (Poulter, 1977). Poulter (1977) calculated that the uncertainty due to physical adsorption of inert gases in a system with a volume-to-surface-area ratio of 200 : 1 does not exceed 1 part in 4×10^4 for argon or 1 part in 5×10^5 for nitrogen. Experimental data obtained by Holanda (1965, 1969) and Bottorf and Chuan (1966) show that the gas used for pressure points generation is adsorbed (chemisorbed) on the walls of the calibration chamber in nondetectable amounts. Messer (1977) also presented experimental evidence showing the influence of adsorption phenomena on the accuracy of the static method of generating pressure points. He considerably increased the surface-to-volume ratio of the calibration chamber of his system by connecting it to an additional volume filled with hollow tubes. The experiment was performed with several gases and the readings were taken with a residual gas analyzer connected to the calibration chamber. The results obtained show that several gases (Ar, CH_4, He, N_2, Xe) may be used for pressure generation owing to the relatively low pressure drop ($-\Delta P/P$ due to adsorption in the lower pressure range of calibration. The problem of gas adsorption on the walls, however, is still controversial.

The molar quantity of the gas may also be altered as a result of *gas desorption by friction*. The moving parts of the valves (mainly of the gate type) used to separate the transfer samplers from the calibration chamber may cause the evolution of gas due to friction. Groszkowsky (1962) investigated the phenomenon and showed that gas desorbed by rubbing moving parts in a BAG enclosure increases the total pressure in the gauge envelope. After a time, gas is readsorbed when the gauge voltage is switched off. Holanda (1969) quantitatively measured the gas desorption resulting from repeated operation of the gate valves of his calibration system in a dummy gas-sample injection. He found a random uncertainty of $\pm 1\%$ at the lower pressure limit of the system (1.3×10^{-6} Pa, 10^{-8} Torr). At higher pressures the uncertainty diminishes to a value which may be disregarded.

Finally, the molar quantity of the gas may also be altered as a reuslt of *replacement of the gas for pressure points generation by other gas species*. The chemical composition of the gas admitted into the system may be changed as a result of the release and replacement of previously adsorbed gas species. Replacement phenomena (Robins, 1962; Brown and Leck, 1955) occur with gases physically or chemically adsorbed as well as with gas ionically trapped (Carmichael and Trendelenburg, 1958; Burt *et al.*, 1961; James *et al.*, 1964; Kornelsen and Sinha, 1966; Erents and Carter, 1967; Grant and Carter, 1967) (see Section 7.5.1).

(ii) *Pressure points generation under nonisothermal conditions*. Gas expansion under isothermal conditions may be disturbed by the combined effect produced by warming of the pressure generator setup during determinations and by temperature gradients in different parts of the system.

Drifts in the temperature of some part of the system during determinations are due to variations in room temperature or to variations in the temperature of that part during a sequence of steps. Thus, a change in room temperature of only 1°C introduces an error of 1 part in 300 on the absolute temperature scale. Elliott and Clapham (1978) measured temperature variations of 0.60 and 0.40°C in two adjacent transfer volumes during the 80 min necessary for determinations. Upon correcting the ratio R of these volumes by a factor $(T_n - T_1)/2T_1$, they found the corresponding uncertainty of the measurement to be inconsequential.

Differences in the temperature of dissimilar parts of the system may occur, even in a laboratory where temperature is closely controlled, as a result of thermal energy transfer to adjacent parts; Holanda (1969) mentioned local heating effects due to heat transfer from the electrical coils activating the solenoid valves of the system. He estimated the error due to both types of temperature effects to be $\pm 0.9\%$.

The effect of cooling generated by adiabatic expansion of gas is negligible, even for expansions performed at large pressure differences.

(iii) *Failure to obey the ideal gas law.* The ideal gas law $PV = RT$ (for 1 mole) relates the measurable quantities P, V, and T of a perfect gas at low pressures. For pressures approaching the high range at which gas is admitted into the system and for real gases such as argon, hydrogen, and nitrogen, other relations more accurately approximate the behavior of the gas. Among these, the equation of state of Kamerling Ones and Holborn, expressing the product PV as a function of a power series in $1/V$ or in P with virial coefficients of state, offers the most accurate approach (Himmelblau, 1967, p. 142). Poulter (1977) expressed the relation which predicts the pressure P_{1c} in the calibration chamber after the first step by utilizing both Holborn's equation of state and the ideal gas law, for $P =$ 100 kPa (760 Torr) and $T = 273$ K. He found that the values of P_{1c} calculated using the ideal gas law have to be multiplied by a correction factor in order to get those resulting from Holborn's equation. The value of the correction factor increases with the molecular mass of the gas used for the generation of the pressure points (e.g., 1.0004 for N_2 and 1.0066 for Xe).

(b) *Systematic errors in the measurement.* The calculation of the pressure P_c generated in the calibration chamber requires the accurate determination of the high pressure P_G of the gas. The systematic uncertainty in the determination of P_c, inherent in the standard measurement instrument, is frequently due to mercury vapor flow to the cold trap, capillary depression (mercury manometers), and thermal transpiration (manometers which have a large transducer surface area exposed to the gas). The uncertainty due to these sources, which constitutes the major part of the error in the measurement, can be minimized by the appropriate experimental procedures discussed above.

The total error resulting in the generation of pressure points by static methods can be calculated as the root mean square of the limits of the uncertainties discussed. Plots showing various analyses of uncertainties, determined by the root-mean-square summation method, for the different techniques of the static method are given in Fig. 3.31.

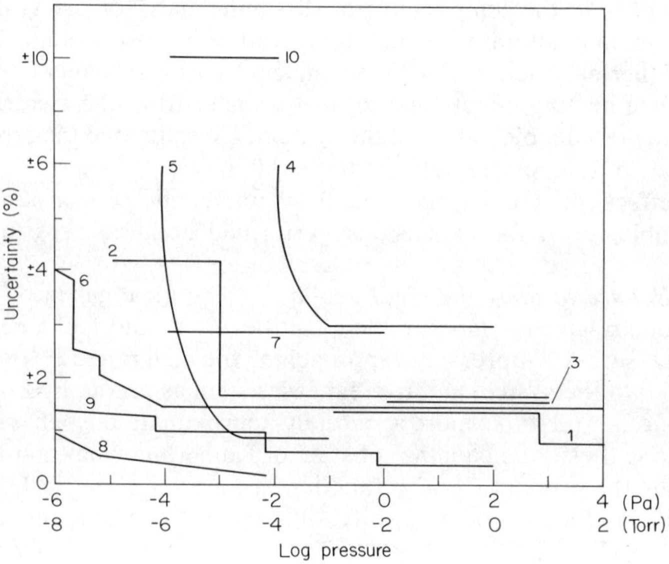

Fig. 3.31 Estimated uncertainty (percentage) in pressure reading as a function of pressure (pascals) for different techniques of the static method. The letter in parenthesis following the reference denotes the technique of the gas expansion (single, i; multiple, ii). 1, Arnery and Henderson (1967) (i); 2, Bottorf and Chuan (1966) (ii); 3, Flanick and Ainsworth (1961) (i); 4, Holanda (1963) (i); 5, Holanda (1965) (i); 6, Holanda (1969) (i); 7, Meineke and Reich (1967) (ii); 8, Messer (1977) (i); 9, NPL (ii) (see Poulter, 1977); 10, Schuman (1962) (i).

3.4.6 Upper and Lower Limits of the Generated Pressure Range

Theoretically, the upper limit of the pressure generated in using the static methods discussed is not restricted by physical bounds. Practically, this limit depends on the pressure at which the gas used for pressure points generation is admitted in the first transfer sampler, the volume ratios involved in the performance of a step, and the number of steps necessary to achieve the required pressure in the calibration chamber.

The lower limit of the pressure range is set by the cumulative effect of the uncertainties discussed. Practically, the method fails if the pressure point to be generated is not at least one order of magnitude higher than that of the system background and if the pumping speed is such that the time constant (see Section 2.3.6) is less than 10^3 s (Redhead *et al.*, 1968, p. 256). The time constant in this case can be expressed as

$$t \gg V_c/C_T \tag{3.57}$$

Table 3.4

THE MAIN CHARACTERISTICS OF SOME PRESSURE GENERATORS BASED ON STATIC METHODS[a]

Reference	Pressure range [Pa (Torr)]		Volume ratio V_{Tr}/V_c		Calibration chamber volume	Gas expansion technique		Accuracy
	Calibration chamber	Calibration gas	Minimum	Maximum	L	Single	Multiple	(Fig. 3.31)
Holanda (1969)	10^{-6}–10^{3} (10^{-8}–10)	9–10^{5} (7×10^{-2}–8×10^{2})	1.44×10^{-7}	1.36×10^{-4}	1444	X		Curve 6[b]
Schumann (1962)	26×10^{-5}–1.3×10^{-2} (2×10^{-7}–1×10^{-4})	$\sim 6.6 \times 10^{3}$ ~ 50	1×10^{-4}	1×10^{-2}	10	X		Curve 10[b]
Arney and Henderson (1967)	1.3×10^{-1}–4×10^{3} (1×10^{-3}–3×10)		18×10^{-5}	1.8×10^{-3}	30.85	X		Curve 1
Meinke and Reich (1967)	6.5×10^{-5}–9.6 (4.9×10^{7}–7.2×10^{-2})	2.6×10^{3}–2.6×10^{4} (20–200)	2.06×10^{-3}	9×10^{-2}	12.974		X	Curve 7

[a] After Berman (1979).

[b] The volume of the transfer samplers was measured after assembling the system.

where C_T is the conductance of the valve, located between the last transfer sampler and the calibration chamber, when it is open.

The main characteristics of pressure generators reported in the literature are given in Table 3.4.

3.5 Accuracy of Primary Standards

All known factors which might contribute uncertainties to pressure measurement with the primary standards selected for discussion in this section have been examined in Sections 3.2–3.4. The plot of the estimated uncertainty versus pressure of some usual primary standards (Fig. 3.32) shows that

(a) liquid column manometers (curves 1 and 2) provide accuracies of better than 1% in the pressure range of about 1 to 10^2 Pa (10^{-2} to 1 Torr);

(b) interferometric micromanometers using noncontacting techniques for the determination of the manometric column height are characterized by accuracies better than 1% up to pressures of about 10^{-2} Pa (10^{-4} Torr) (curve 3);

(c) under favorable conditions and with expert handling, the accuracy of the conventional McLeod gauge (curve 4) is better than 1% at about 1 Pa (10^{-2} Torr); it may be possible to improve the accuracy (<1% in the

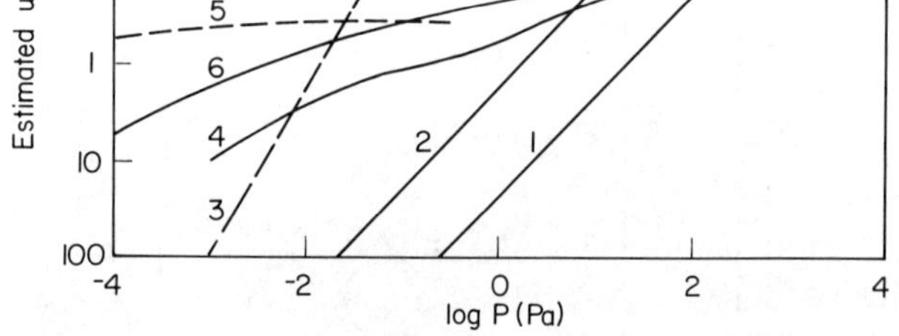

Fig. 3.32 Accuracy of primary standards versus pressure. 1, mercury U-tube micrometer point-contact manometer; 2, oil (DC 704) U-tube point-contact micromanometer; 3, oil interferometric micromanometer; 4, conventional McLeod gauge; 5, modified Moser–Poltz McLeod gauge; 6, pressure generators (static). Dashed curves denote moderately optimistic hopes for techniques under development. Curves 1–4 and 6, Ruthberg (1975); Curve 5, Lloyd (1967).

range 10^{-4} to ~1 Pa, 10^{-6} to ~10^{-2} Torr) of this gauge (curve 5) by utilizing the construction proposed by Moser and Poltz (1957) (see Section 3.3.4); and

(d) static-type pressure generators are accurate to better than 1% in the pressure range 10^{-2} Pa (10^{-4} Torr) (curve 6).

The accuracy required for pressure measurements in vacuum metrology usually ranges between 1 and 10%, and only a few special cases require accuracies as low as 0.1%. Achievement of measurements with critical accuracies at this lower limit is possible in a few national laboratories which maintain reference standards corresponding to the international specifications. Some of these laboratories in countries of the European community as well as Japan, U.S.A., and U.S.S.R. are Laboratoire de Metrologie, Leybold-Heraeus-Sogev, France; Physikalisch-Technische Bundesanstalt, West Germany; Instituto di Metrologia "G. Colonetti," Italy; National Physical Laboratory, United Kingdom; National Research Laboratory of Metrology, Japan; National Bureau of Standards, U.S.A.; All Union Research Institut of Metrology (Mendeleev), U.S.S.R.

Reliability in pressure standards maintained by each of these laboratories as well as in the methods used for operating them is established by international comparison. Thus, the intercomparison of the equipment maintained at the national laboratories in France, West Germany, Italy, and United Kingdom has proved that the pressure standard of each laboratory does not differ from the weighted mean of all the laboratories by more than 2% in the range 8×10^{-5}–8×10^{-2} Pa (6×10^{-7}–6×10^{-4} Torr) (Poulter *et al.*, 1980). The intercomparison of pressure standards and calibration methods is made feasible by the use of suitable secondary (transfer) standards.

3.6 The Conventional Triode

Hot-cathode ionization gauges, specially designed and built, were utilized as secondary pressure standards by Meineke and Reich (1967), Choumoff and Iapteff (1977), and Gentsch and Messer (1980). These gauges required nonstandard control units to supply the electron emission and the electrode potentials and were utilized only for industrial calibration work. The performance of these gauges was better than that of Bayard–Alpert gauges.

Sutton and Poulter (1982) designed a triode gauge head similar to a conventional triode (see Section 4.4.7) whose sensitivity changed by less than 0.1% per 100 h and could therefore be used as a secondary pressure standard. Studies at NBS, however, demonstrate that *unknown system-*

atic effects may influence the performance of such gauges used as transfer standards (Tilford, 1983).

3.7 The Capacitance Manometer

An elastic metallic diaphragm (D in Fig. 3.33) separates two cavity volumes subjected to a known reference pressure P_r and the unknown pressure P_x to be measured. The diaphragm deflects due to the pressure differential, thus modifying the capacitance of an electrical circuit where the diaphragm (grounded to the box B) constitutes one electrode, and two metallic insulated plates or rings constitute the other electrode. The deflection of the diaphragm, hence the change in the capacitance of the electrical circuit, is a function of the force exerted by the gas per unit area of the diaphragm.

The capacitance of a system of two parallel electrodes is proportional, in turn, to the dielectric constant of gas separating the electrodes and to a geometric factor expressed as the ratio of the electrode area to the distance separating the electrodes.

The capacitance manometer consists of a gauge head containing the capacitance transducer and an electronic unit to both energize and display the transducer's output. The transducer may be of the double-sided or single-sided type according to the relative position of the elastic (diaphragm) and fixed electrodes. The double-sided type, an early type of instrument, contains two fixed electrodes (E) located on each side of the diaphragm (Fig. 3.33a). For absolute pressure determinations, the reference side is continuously evacuated. Except for gases at high pressure or those with a dielectric constant significantly different from air (e.g., conductive, radioactive gases), pressure measured is independent of the gas

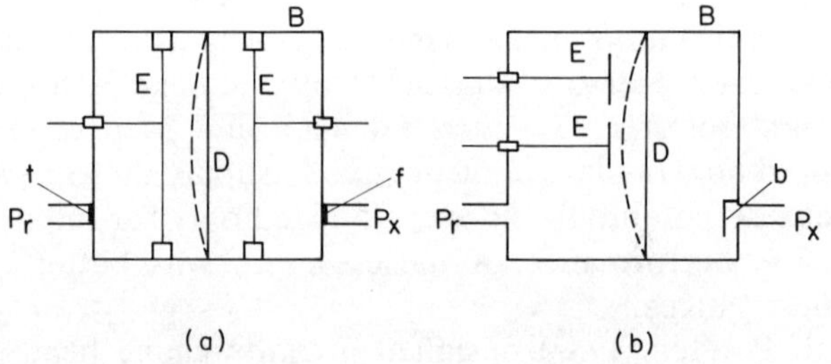

Fig. 3.33 Diagram of a capacitance diaphragm manometer. (a) Double-side construction, (b) single-sided construction. E, electrode; D, elastic diaphragm; B, leak-tight box; P_x, pressure to be measured; P_r, reference pressure; b, baffle; f, filter.

composition. The double-sided location of electrodes provides for a good zero stability but has the drawback that the gas whose pressure is being measured may be contaminated by the presence of fixed electrodes. Filters (f) at the entrance to the cavity volumes keep particulate matter from entering the sensor assembly but considerably lengthen the response time of the gauge head. The decontamination of the gauge head is difficult because of the ceramic insulators which are the substrate for metallic electrodes.

In the single-sided construction (Fig. 3.33b) there is no electrode in the cavity volume containing the gas at the unknown pressure. Two electrodes, one at the center of the diaphragm for sensing large displacements of this diaphragm and the other (ring shaped) around the first electrode for sensing small displacements, are located in the cavity volume sealed at the reference pressure P_0. A baffle (b) keeps gas from entering the cavity volume containing gas at unknown pressure and from directly impinging on the diaphragm.

Gauge heads are available with the reference side P_0 open (for differential determinations) or permanently sealed under vacuum (10^{-7} Pa, 10^{-5} Torr). In the latter case a getter pellet, fired during the capacitance transducer manufacture, permits the maintenance of the sealed vacuum through the rated life time of the transducer under proper operating conditions.

Electronic input circuitry applies an ac signal to the electrodes at some frequency (typically 10 kHz), referred to as the carrier frequency (Sullivan, 1979). The deflection of the diaphragm modulates an ac input drive. A synchronized phase-sensitive detector circuit generates positive and negative signals according to the direction of the diaphragm deflection, hence according to the sign of the applied pressure. These signals are then demodulated, amplified in a very stable output, and displayed in analog or digital form.

Gauge heads of the capacitance type span pressures in the range from 10^{-2} Pa (10^{-4} Torr) to above atmospheric pressure.

The main systematic errors which affect pressure determination with capacitance-type transducers are produced by variations of the ambient temperature, nonlinearity, and hysteresis. Other errors are due to the gas composition, the adsorption, and ionization during the transducers operation. These errors are usually very small and will not be discussed further.

Temperature effects produce changes in the capacitance of the electrodes, the diaphragm stiffness, and the dielectric constant of the gas (Brombacher, 1970, p. 31). Among these, the change in the capacitance of the electrodes accounts for the largest source of error. It can be reduced both by a proper choice of materials (inconel) for the construction of the diaphragm and the box and by special techniques to compensate for or to control the variations in ambient temperature (Sullivan, 1967, 1979).

In the temperature compensation technique a sensor, in good thermal contact with the box accomodating the electrode system, continuously tracks the temperature of the box. The signals from the sensor are fed to electronic circuitry which compensates for the changes in the zero value.

The temperature control technique consists in maintaining the transducer at constant temperature (usually 45°C) within very close limits of variation (±0.1°C). The control ratio, i.e., the ratio of the transducer temperature to the change in the ambient temperature, is usually 1/10. Proper temperature regulation of the transducer with either technique can improve the zero changes with ambient temperature (zero stability) and the span variations by about one order of magnitude.

According to Sullivan (1977), nonlinearity of readings accounts for the largest error apart from temperature effects, which account for about 80% of the total error. Errors due to nonlinearity can be reduced by choosing an adequate transducer design (Brombacher, 1970, p. 24) and suitable electronic components of the control unit.

Hysteresis is significant for sensors where the diaphragm is simply stretched and the assembly of the cavity volumes accommodating gas at P_x and P_r is clamped together with metal clips under large forces. This construction was used from 1961 to 1966 and was then abandoned. In modern constructions the error due to hysteresis is negligible.

3.8 The Spinning-Rotor Viscosity Gauge

The spinning-rotor viscosity gauge is based on the drag due to the momentum transfer effect of a gas at reduced pressure (see Section 4.3.4) on a spinning bearing ball suspended by levitation in a magnetic field. A steel ball at rest (1 in Fig. 3.34) in a tubular extension (2) of a flange (7) connected to the volume containing gas at unknown pressure is suspended vertically by two permanent magnets (3) and maintained in the center of the magnetic field by the field produced by a pair of coaxially arranged control coils (4). The ball is laterally stabilized by a field produced by four coils (not shown in Fig. 3.34). Driving coils (5) produce a magnetic rotating field which spins the ball to a set frequency of 400 Hz. After attaining the required frequency, the ball is allowed to coast, and the decay of speed with time due to the gas friction is measured.

The pressure can be calculated from the drag resulting from the momentum transfer between the gas molecules and the ball, the moment of

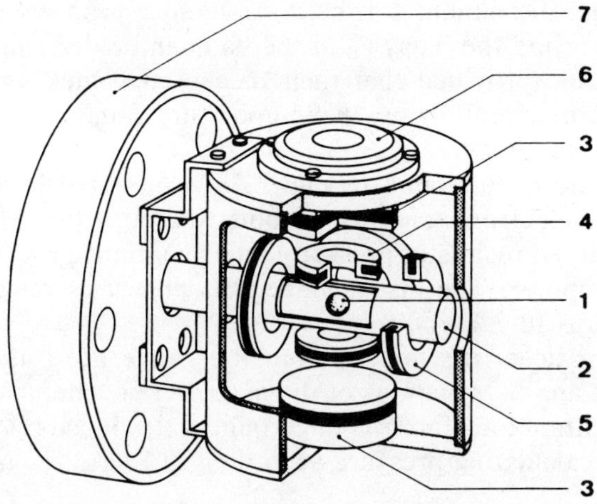

Fig. 3.34 Cross section of the spinning-rotor viscosity gauge head (industrial reproduction of the original KFA-Jülich design). 1, ball; 2, evacuated tubular extension of 7; 3, permanent magnets; 4, coils for vertical stabilization; 5, driving coils; 6, water level; 7, connecting flange. (Not shown are the four coils for lateral stabilization and the two pickup coils.) Connection flange and tubular extension are bakeable up to 400°C (after Reich, 1982).

inertia of the ball of radius r, and the times t_n and t_{n-1} required for completing a given number of revolutions of the ball as

$$P = \frac{(\pi \bar{c} r \rho_B / 10\sigma)(t_n - t_{n-1})}{(t_n t_{n-1})} \tag{3.58}$$

where $\bar{c}$ is the average speed of the gas molecules, ρ_B the density of the ball, and σ the macroscopic coefficient of tangential momentum transfer.

Equation (3.58) holds true provided that

(a) the mean free path of gas particles is large compared with the radius of the ball,
(b) thermal equilibrium conditions prevail during measurements, and
(c) the residual drag (due to eddy currents) associated with the magnetic suspension is accounted for.

One of the most important parameters is the surface condition of the ball, which affects the tangential momentum transfer (Comsa *et al.*, 1980).

The sensitivity s of the gauge

$$s = (t_n - t_{n-1})(P t_n t_{n-1}) = (40\sigma r^2)/(3\bar{c} g_B) \tag{3.59}$$

shows that the instrument can be considered a primary standard since both the radius and the mass g_B of the ball can be determined by fundamental methods, provided that their measured values remain constant during long term operation and the temperature T and the molecular mass M of the gas are known.

The gauge head illustrated in Fig. 3.34 is operated by a control unit containing the electronic circuity for both operating the sensor (ball) and displaying printed results of pressure determinations processed by a programmed computer. The instrument spans pressures ranging from 1 to 10^{-5} Pa (10^{-2} to 10^{-7} Torr).

The systematic errors arising from the gauge head are due to time measurement and to variations of the macroscopic coefficient of the tangential momentum transfer. The uncertainty introduced by time measurement (Δt) in calculating pressure according to Eq. (3.58) is

$$\Delta P = (\pi \bar{c} r / 10\sigma)(2\Delta t / t_N^2) \tag{3.60}$$

where t_N is the time taken for completing N revolutions.

The macroscopic coefficient of tangential momentum transfer σ between the gas molecules and the tangentially moving surface of the rotor (ball) as measured by Comsa *et al.* (1980) for a single 4.5-mm diameter ball and several gases is close to unity in the pressure range 10^{-7}–10^{-4} Pa (10^{-9}–10^{-6} Torr). Determinations at PTB (Physikalisch-Technische Bundesanstalt, W. Germany) made by Messer (Table 3.5) (private communication) show that the maximum variation of σ_{eff} (4.5 mm of stainless steel) during the years 1977–1984 is about 3%. Different values ($\sigma_{eff} \simeq 1.17$)

Table 3.5

VALUES OF σ_{eff} MEASURED WITH A SRG ROTOR[a]

Date of recalibration	σ_{eff} for Ar	Date of recalibration	σ_{eff} for Ar
25-07-77	1.00_3	04-02-81	1.00_2
26-07-77	1.00_4	02-03-81	1.00_1
17-07-78	1.00_6	03-03-81	1.00_3
18-07-78	1.00_5	19-08-81	1.01_0
19-07-78	1.00_1	26-04-82	1.00_4
12-11-79	1.00_6	25-03-83	0.99_4
06-12-79	1.00_5	29-03-83	0.98_4
07-03-80	1.00_6	30-03-83	0.98_5
		03-04-84	0.98_1

[a] Bearing ball made of normal steel, 4.5 mm ϕ.

have been obtained for balls made of bearing ball steel with rough surface. The long-term behavior of such balls requires determination over a period of a few years.

References

Adam, H. (1971). "Some unsolved problems in modern vacuum technology." Lecture given at the *2nd Israel Vacuum Congress* (unpublished).

AVS 6.4 (1969). *Procedure for Calibrating Hot Filament Ionization Gauges Against a Reference Manometer in the Range 10^{-2}–10^{-5} Torr.*

Arney, G. D., and Henderson, W. F. (1967). *AEDC-Tr-67-83.*

Aubry, B., and Delbart, R. (1965). *Vide* **20,** 194.

Barr, W. E., and Anhorn, V. J. (1946). *Instruments* **19,** 666.

Barton, R. S., and Chubb, J. N. (1965). *Vacuum* **15,** 113.

Bennet, S. J., Clapham, P. B., Daborn, J. E., and Simpson, D. I. (1975). *J. Phys. E.,* **8,** 5.

Berman, A. (1973). *Vacuum* **23,** 61.

Berman, A. (1974). *Vacuum* **24,** 241.

Blaisdell, B. E. (1940). *J. Math. Phys.* **19,** 186.

Bobenrieth, A. (1946). *Vide* **2,** 62.

Bottorf, M. R., and Chuan, R. I. (1966). *Res./Dev. Aug.,* 60.

Boyer, D. W., and Pankhurst, R. C. (1971). "Pressure probe methods for determining wind speed and flow direction HMSO." London.

Bradshaw, P. (1965). *J. Sci. Instrum.* **42,** 677.

Brombacher, W. G. (1967). *NBS Technical Note 298.*

Brombacher, W. G. (1970). *NBS Monograph 114.*

Brombacher, W. G., Johnson, D. P., and Cross, J. L. (1960). *NBS Monograph 8.*

Brown, E., and Leck, J. H. (1955). *Brit. J. Appl. Phys.* **6,** 161.

Bryant, P. J., and Gosselin, Ch, M. (1966). *NASA Brief 66-10372.*

Burt, R. B., Colligon, J. S., and Leck, J. H. (1961). *Brit. J. Appl. Phys.* **12,** 396.

Calcatelli, A., Ferrero, C., and Rumiano, G. (1975). *Vuoto* **8,** 136.

Carmichael, J. H., and Trendelenburg, E. A. (1958). *J. Appl. Phys.* **29,** 1570.

Carr, P. (1964). *Vacuum* **14,** 37.

Cespiro, Z. (1973). *Vacuum* **23,** 277.

Choumoff, P., and Iapteff, B. (1974). *Proc. 6th Int. Vac. Congr. Kyoto 1974,* p. 143. *Jpn. J. Appl. Phys. Suppl. 2* (Part 1).

Colgate, S. O., and Genre, P. A. (1968). *Vacuum* **18,** 553.

Collins, S. C., and Blaisdell, B. E. (1936). *Rev. Sci. Instrum.* **7,** 213.

Comsa, G., Fremerey, J. K., Lindenau, B., Messer, G., and Rohl, P. (1980). *J. Vac. Sci. Technol.* **17,** 642.

Cook, A. H. (1961). *Philos. Trans. R. Soc.* **254,** A 1038, 125.

Cook, A. H., and Stone, W. B. (1957). *Philos. Trans. R. Soc.* **250,** A 978, 279.

Davidson, H. L., and Rinker, R. (1974). *Rev. Sci. Instr.* **45,** 134.

Dunoyer, L. (1926). "Vacuum Practice." D. Van Nostrand, New York.

Dushman, S. (1962). "Scientific Foundations of Vacuum Technique." J. Wiley, London.

Elliott, K. W. T., and Clapham, P. B. (1978). *NPL Rep. MOM 28.*

Elliott, K. W., Woodman, D. M., and Dadson, R. S. (1967). *Vacuum* **17,** 439.

Erents, K., and Carter, G. (1967). *Vacuum* **17,** 97.

Eschbach, H. L. (1962). "Praktikum der Hochvakuumtechnik." Akademische Verlagsgeselshaft, Leipzig.

Eschbach, H. L. (1976). *CBNM, EUR 4797 d.*
Espe, W. (1966). "Materials of High Vacuum Technology," Vols. Pergamon Press, New York.
Flanick, A. P., and Ainsworth, J. E. (1961). *Rev. Sci. Instrum.* **32,** 408.
Fletcher, B., and Butcher, G. R. (1969). *J. Sci. Instrum. Series* **2,** 2, 740.
Fryburg, G. C., and Simons, J. H. (1949). *Rev. Sci. Instrum.* **20,** 541.
Gaede, W. (1915). *Ann. Phys.* **46,** 354.
Gentsch, H., and Messer, G. (1980). *Proc. 8th. Int. Vac. Congr. Cannes 1980,* p. 203. *Suppl. Le Vide, les Couches Minces No. 201.*
Gotto, M. (1960). *Proc. 1st Int. Congr. Vac. Technique Belgium 1958,* p. 266. Pergamon Press, New York.
Gould, F. A., and Vickers, T. (1952). *J. Sci. Instrum.* **29,** 85.
Grant, W. A., and Carter, G. (1967). *Brit. J. Appl. Phys.* **18,** 527.
Groszkowsky, J. (1962). *Trans. 8th Natl. Vac. Symp. USA 1961,* p. 186. Pergamon Press, New York.
Hales, J. L. (1970). *J. Phys. E.* **3,** 855.
Heydemann, P. L. M. (1971). *R. Sci. Instrum.* **42,** 983.
Heydemann, P. L. M., Tilford, C. R., and Hyland, R. W. (1977). *J. Vac. Sci. Technol.* **14,** 597.
Himmelblau, D. M. (1967). "Basic Principles and Calculations in Chemical Engineering." Prentice-Hall, Englewood Cliffs, New Jersey.
Hirschfelder, J. O., Curtiss Ch., F., and Bird, R. B. (1964). "Molecular Theory of Gases and Liquids." Wiley, New York.
Hobson, J. P. (1970). *J. Vac. Sci. Technol.* **7,** 351.
Hobson, J. P. (1973). *J. Vac. Sci. Technol.* **10,** 73.
Holanda, R. (1963). *NASA Technical Note NASA TN D-1729.*
Holanda, R. (1965). *NASA Technical Note TN D-3100.*
Holanda, R. (1969). *NASA Technical Note TN D-5406.*
Honig, R. E., and Kramer, D. A. (1969). *RCA Rev.* **30,** 285.
Huang, R. J., Demirel, T., and McGee, T. D. (1972). *J. Phys. E.* **5,** 707.
Ishii, H., and Nakayama, K. (1962). *Trans. 8th Natl. Vac. Symp. USA 1961,* p. 519. Pergamon Press, New York.
Jansen, C. G., and Venema, A. (1959). *Vacuum* **9,** 219.
James, L. H., Leck, J. H., and Carter, G. (1964). *Brit. J. Appl. Phys.* **15,** 681.
Kaye, G. W. C., and Laby, T. H. (1973). "Table of Physical and Chemical Constants," 14th ed. Longman, London.
Keevil, N. B., Errington, R. F., and Neuman, L. T. (1941). *Rev. Sci. Instrum.* **12,** 609.
Kistemaker, J. (1945). *Physica* **11,** 270, and 277.
Klemeperer, O. (1944). *J. Sci. Instrum.* **21,** 88.
Knudsen, M. (1910). *Ann. Phys.* **31,** 633.
Kornelsen, E. V., and Sinha, M. K. (1966). *Appl. Phys. Lett.* **9,** 112.
Kreisman, W. (1961). *Trans. 7th Natl. Symp. Vac. Technol. USA 1960,* p. 75. Pergamon Press, New York.
Laplace, P. S. (1816). *Phyl. Mag.* **47,** 102.
Leck, J. H. (1964). "Pressure Measurement in Vacuum Systems." Chapman and Hall Ltd., London.
Löcherbach, E. O., and Schitko, F. J. (1962). *Vak. Tech.* **11,** 51.
Lloyd, E. C. (1967). *Instrum. Control Syst.* **40,** 105.
Marton, L. L. (1974). *Acad. R. Belg. Cl. Sci. Mem. June 8,* 676.
Meinecke, Ch., and Reich, G. (1962). *Vak. Tech.* **11,** 86.

Meineke, Ch., and Reich, G. (1963a). *Vacuum* **13,** 579.
Meineke, Ch., and Reich, G. (1963b). *Vak. Tech.* **12,** 70.
Meineke, Ch., and Reich, G. (1967). *J. Vac. Sci. Technol.* **4,** 356.
Messer, G. (1977). *Phys. Bl.* **33,** 343.
Miller, III, J. R. (1972). *J. Vac. Sci. Technol.* **9,** 201.
Miller, III, J. R., and Jerinjan, J. E. (1964). *J. Vac. Sci. Technol.* **1,** 81.
Moser, H., and Poltz, H. (1957). *Z. Instrkd.* **65,** 43.
Nakayama, K., Akiyama, Y., and Hashimoto, H. (1968). *Vacuum* **18,** 65.
Nottingham, W. B., and Torney, Jr., F. L. (1961). *Trans. 7th Natl. Symp. Vac. Technol. USA 1960,* p. 117. Pergamon Press, New York.
Orcutt, R. H. (1973). *J. Vac. Sci. Technol.* **10,** 506.
Peggs, G. N. (1980). *J. Phys. E.* **13,** 1254.
Pirani, M., and Yarwood, J. (1961). "Principles of Vacuum Engineering." Chapman and Hall, London.
Podgurski, H. H., and Davis, F. N. (1960). *Vacuum* **10,** 377.
Porter, A. W. (1933). *Trans. Faraday Soc.* **29,** 702.
Poulter, K. F. (1977). *J. Phys. E.* **10,** 112.
Poulter, K. F., and Nash, P. J. (1979). *J. Phys. E.* **12,** 931.
Poulter, K. F., Calcatelli, A., Choumoff, P. S. Iapteff, B., Messer, G., and Grosse, G. (1980). *J. Vac. Sci. Technol.* **17,** 679.
Power, B. C. (1966). "High vacuum pumping equipment," p. 66. Chapman and Hall, London.
Power, B. D., and Robson, F. C. (1962). *Trans. 8th Int. Vac. Symp. USA 1961,* p. 1175. Pergamon Press, New York.
Rambeau, G. (1969). *Vide* **19,** 219.
Redhead, P. A., Hobson, J. P., and Kornelsen, E. V. (1962). *In* "Advances in Electronics and Electron Physics" (L. Marton, ed), vol. 17. Academic Press, New York.
Reich, G. (1982). *J. Vac. Sci. Technol.* **20,** 1148.
Reich, G. (1982). Gasreibungs-Vakuummeter VISCOVAC VM 210 Leybold-Heraeus GMBH.
Reich, G. (1984). Gasrei bungsrakummanometer VISCOVAC VM 211 Leybold-Heraeus, 3 veränderte Auflage.
Reichman, E., and Wharton, L. (1970). *R. Sci. Instrum.* **41,** 402.
Reid, G. R. (1968). *J. Sci. Instrum. Series 2* **1,** 60.
Robins, J. L. (1962). *Trans. 9th Natl. Vac. Symp. USA 1962,* p. 510. Macmillan, New York.
Roehring, J. R., and Simons, Jr., J. (1962). *Trans. 8th Natl. Vac. Symp. USA 1961,* p. 511. Pergamon Press, New York.
Rosenberg, P. (1938). *Rev. Sci. Instrum.* **9,** 258.
Rosenberg, P. (1939). *Rev. Sci. Instrum.* **10,** 131.
Roth, A. (1982). "Vacuum Technology." North-Holland, Amsterdam.
Rush, M., and Bunge, O. (1932). *Z. Tech. Physik.* **13,** 77.
Ruthberg, S. (1969). *J. Vac. Sci. Technol.* **6,** 401.
Ruthberg, S. (1975). *Pressure Measurements for the Range 1 kPa to 100 Pa. In* "Experimental Thermodynamics," vol. II. (B. Le Neindre and B. Vodar, eds.). Butterworth, London.
Schuman, S. (1962). *Trans. 9th Natl. Vac. Symp. USA 1962,* p. 463. Macmillan, New York.
Sellenger, F. R. (1968). Vacuum **18,** 646.
Sensui, Y. (1970). *Vacuum* **20,** 539.
Sharma, J. K. N., Dwivedi, H. K., and Sharma, D. R. (1980). *J. Vac. Sci. Technol.* **17,** 820.

Simons, Jr, J. C. (1963). *Trans. 10th Natl. Vac. Symp. USA 1963,* p. 246. Macmillan, New York.
Siu, M. C. I. (1973). *J. Vac. Sci. Technol.* **10,** 368.
Smetana, F. O., and Carley, Jr., C. T. (1966). *J. Vac. Sci. Technol.* **3,** 49.
Smith, A. M. O., and Murphy, J. S. (1955). *Rev. Sci. Instrum.* **26,** 775.
Steckelmacher, W. (1969). *Vuoto* **2,** 189.
Stevenson, W. H., and McFadden, P. W. (1965). *Rev. Sci. Instrum.* **36,** 1272.
Sullivan, J. J. (1967). *Res./Dev.* **27,** 41.
Sullivan, J. J. (1979). *Transducer Technol.* **8,** 22.
Sutton, C. M., and Poulter, K. F. (1982). *Vacuum* **32,** 247.
Takaishi, T. (1965). *Trans. Faraday Soc.* **61,** 840.
Takaishi, T., and Sensui, Y. (1970). *Vacuum* **20,** 495.
Thomas, A. M., and Cross, J. L. (1967). *J. Vac. Sci. Technol.* **4,** 1.
Thomas, A. M., Johnson, D. P., and Little, J. W. (1962). *Trans. 9th Vac. Symp. USA 1962,* p. 468. Macmillan, New York.
Thomas, L. B., Harris, R. E., and Kreuger, C. L. (1974). *Proc. 6th Int. Vac. Congr. Kyoto 1974,* p. 151. *Jpn. J. Appl. Phys. Suppl. 2* (Part 1).
Thomas, L. B., Kreuger, C. L., and Harris, R. E. (1977). *Proc. 7th Int. Vac. Congr. Vienna 1977.* Berger and Sohne Horn, Austria.
Tilford, Ch. R. (1973). *Rev. Sci. Instrum.* **44,** 180.
Tilford, Ch. R. (1983). *J. Vac. Sci. Technol. A* **1**(2), 152.
Truffier, J. L., and Choumoff, P. S. (1974). *Proc. 6th Int. Vac. Congr. Kyoto 1974,* p. 139. *Jpn. J. Appl. Phys. Suppl. 2* (Part 1).
Tunniclife, R. J., and Rees, A. J. (1967). *Vacuum* **17,** 457.
Utterback, N. G., and Griffith, Jr, T. (1966). *Rev. Sci. Instrum.* **37,** 866.
Vermandé, M. J. (1962). *Vide* **7,** 1145.
de Vries, A. E., and Rol, P. K. (1965). *Vacuum* **15,** 135.
Warshawsky, I. (1972). *J. Vac. Sci. Technol.* **9,** 1.
Wear, K. B. (1968). *Rev. Sci. Instrum.* **39,** 245.
Whitlaw-Gray, R., and Teich, N. (1948). *Trans. Faraday Soc.* **44,** 774.
Zeleny, J. (1936). *Rev. Sci. Instrum.* **7,** 289.

CHAPTER 4

Gauges for Low-Pressure Measurement

4.1 Classes of Instruments

Modern vacuum technology techniques permit the measurement of total pressures down to 10^{-12} Pa (10^{-14} Torr). Still lower pressures than this can be measured with ionization gauges, provided that special means of decreasing spurious currents and enhancing conductivity are utilized. Thus, Blechschmidt (1974) describes an ionization gauge which can measure pressures estimated to be below 10^{-13} Pa (10^{-15} Torr). This gauge, which has a sensitivity independent of pressure of some 10^{-7}/Pa (10^{-9}/Torr) and an x-ray limit below 10^{-14} Pa (10^{-16} Torr), uses a channel electron multiplier to collect electrons. However, it is useful to consider whether the technical or financial effort required to utilize such a gauge is worthwhile or whether the use of a partial pressure analyzer would not be a better approach.

To cope with the entire spectrum of pressures mentioned, several types of instruments and measurement methods, overlapping in range, are used. These instruments are classified according to their operating principle in four fundamental classes (see Fig. 4.1).

Instruments in the first class directly measure pressure as force per unit area. Pressure measurement in the low (10^5–10^2 Pa, 760–1 Torr) and medium (10^2–10^{-1} Pa, 1–10^{-3} Torr) vacuum ranges is not particularly difficult since it involves the determination of relatively large mechanical forces. Usually, no energy input to the instrument is necessary. Below 10^{-1} to about 10^{-3} Pa (10^{-3} to about 10^{-5} Torr) the instrument requires the supply of energy to convert, amplify, and display the input signal.

The second class consists of instruments which determine pressure indirectly by measuring a pressure-dependent property of gas. They exploit gas kinetic effects such as transport of momentum or of energy through the rarefied gas. The pressure range they span extends from 10^5 to about 10^{-5} Pa (760 to about 10^{-7} Torr). Instruments based on the transport of momentum were utilized in earlier investigations concerning the measurement of low gas pressures. Their delicate suspension, long time response, noncontinuous indication of pressure, and other drawbacks have impeded their development in forms suitable for industrial applications. It is only several years ago that advanced electronics have made possible the industrial construction of vacuum gauges utilizing magnetic suspension of a rotating body whose deceleration permits the measurement of gas pressure in terms of the force produced by molecular drag. Instruments using the transport of energy, known as thermal conductivity gauges, are in turn developed on a large industrial scale.

In the third class are instruments which relate pressure to phenomena produced by charged particles which cross the gas, such as fluorescence, energy loss, scattering, and ionization. The pressure range they span extends from several hundred down to about 10^{-13} Pa (1 to about 10^{-15} Torr) for instruments of the ionization type.

Finally, in the fourth class is the Brownian-motion gauge which has not yet been developed as a practical instrument but which has possibilities in the future if it is combined with the latest advances in electronics.

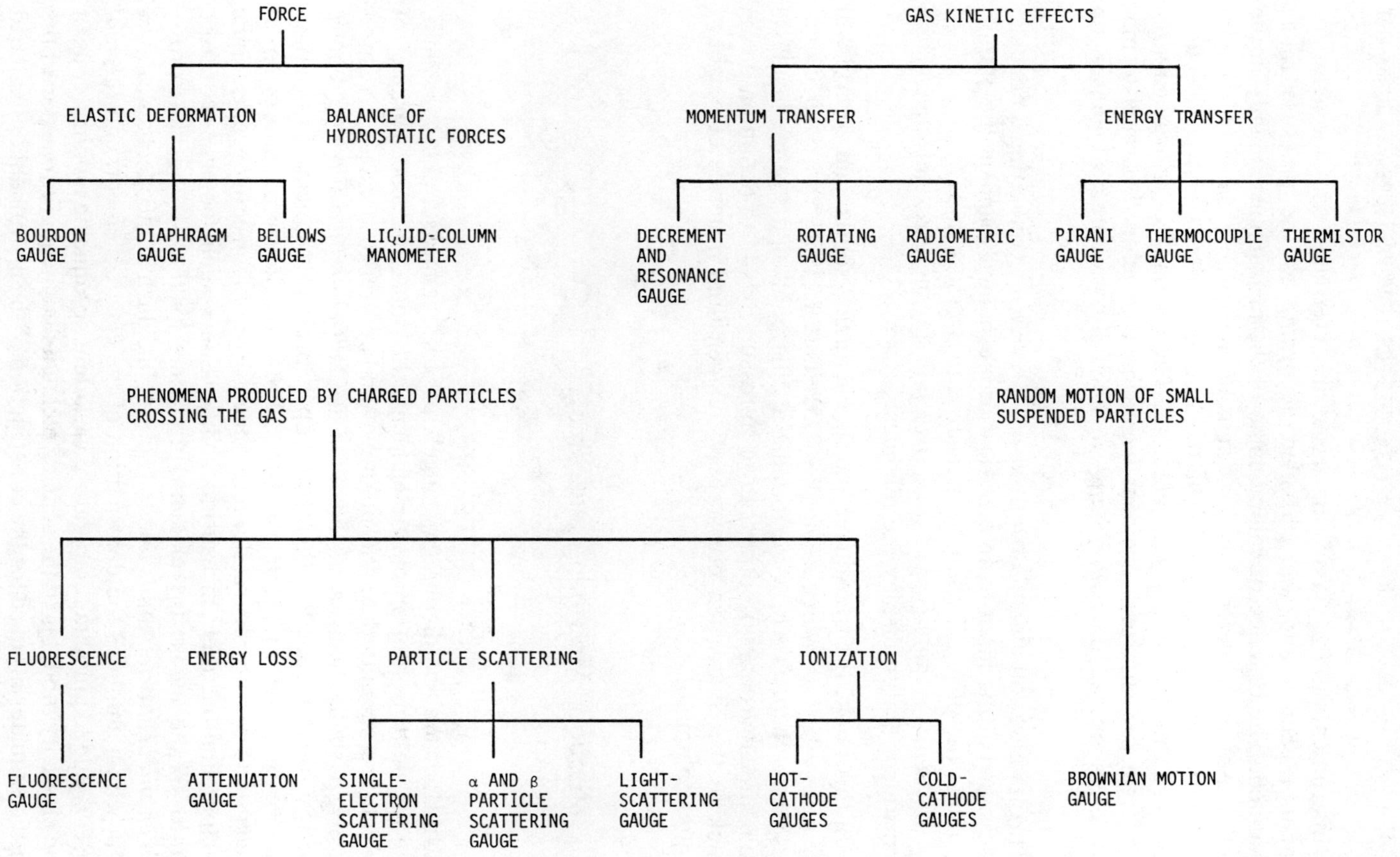

Fig. 4.1 Physical quantities measured in gas for low total pressure determination.

Instruments in the classes discussed are referred to as manometers or vacuum gauges and are classified according to the pressure range spanned. The practice of calling some of the instruments (mainly in the first class) manometers has its origin in the earliest times of vacuum application when instruments of the liquid-column type were used to measure the degree of gas rarefaction. The diversity of instruments for low total pressure measurement which have developed parallel to manometers are called *vacuum gauges*. Generally, a vacuum gauge is made up of

(a) a part containing a *transducer* (a device for converting a nonelectric signal at the input to an electrical or mechanical signal at the output) to relate the physical quantity measured to pressure and
(b) a *power supply and a control unit* to energize and control the transducer and to display the readings.

In this book, the transducer and its energizing source and control will be referred to as *gauge head* (BS 2951: Part 1: 1958) and *control unit,* respectively. It is worth mentioning at this point that "gauge head" is sometimes wrongly used instead of "pressure head" (a definition dating back to the time when pressures were measured in terms of height of liquid columns).

4.2 Gauges Measuring the Force Effect of Gas

4.2.1 Bourdon and Diaphragm Gauges

In Bourdon and diaphragm gauges, pressure is measured by determining the deflection of an elastic element, the outer and inner surfaces of which are submitted to the unknown and reference pressures, respectively. Usually, gauges of this type are *self-contained* (do not require energy input to operate them) and span pressures ranging from atmosphere to about 133.32 Pa (1 Torr). Pressures lower by one order of magnitude can be accurately determined, by compensating against some factors limiting their measuring ability. Some such factors are *elastic faults* of the measuring element producing hysteresis and zero drifts, *mechanical shortcomings* due to friction, backlash of the mechanical system which amplifies the deformation of the elastic element, and *thermal effects* due to ambient temperature variations, producing modifications of the elasticity modulus of the sensing element and inequal expansion of the different materials used for the construction of the instrument (unmatched temperature coefficients).

The *Bourdon gauge* consists of a hollow metallic tubing with a flat or elliptical cross section bent in various shapes (Fig. 4.2a). In the case of a coil-shaped tubing, one of its ends, consolidated to a box, can be connected to the vacuum system. The other end is sealed and attached mechanically to a pointer moving over a calibrated scale (Fig. 4.2b). If the pressure inside the hollow tube varies with respect to the atmospheric pressure outside (reference pressure), the coil deflects. The curvature of the coiled tubing tends to become larger or smaller according to whether pressure in the system is higher or smaller than the reference pressure. Variations in atmospheric pressure yield important errors mainly at the end of the low pressure scale.

Ordinarily, the hollow tubing is hardened to avoid errors due to elastic defects. Bourdon tubes of quartz (fused silica) are used instead of metallic tubes in applications where hysteresis of the sensing element has to be held at lower values. The response to vibrations of quartz tubing is more pronounced compared with that of metallic tubings. Dressler (1965) set the theoretical basis of Bourdon tubes.

The *diaphragm gauge* uses as a pressure-responsive element, a thin, circular, (usually) metallic plate, either flat or corrugated to avoid buckling (Fig. 4.3). For the same diameter and thickness, flat plates are more sensitive than corrugated ones (Brombacher, 1970, p. 18).

With *aneroid* instruments two diaphragms, leaktight and joined around their periphery, are placed in what is called a *capsule*. The capsule, evacuated to a low pressure (reference), is sealed and placed in a box where it is subjected to the pressure which is being measured. The deflection of the capsule is mechanically amplified. Increased deflections can be obtained by stacking several capsules.

The pressure range of diaphragm gauges can be substantially lowered by using electrical transducers of the capacitance or inductance type.

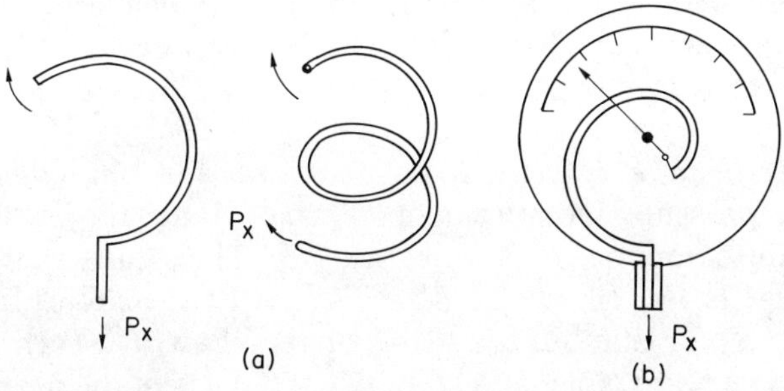

Fig. 4.2 (a) Usual shapes of bending Bourdon tubes; (b) diagram of a Bourdon manometer. P_x is unknown pressure.

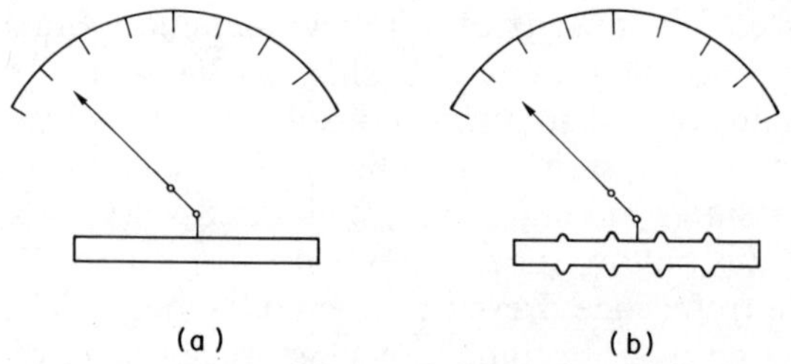

Fig. 4.3 Capsule diaphragm. (a) Flat; (b) corrugated.

4.2.2 Liquid-Column Manometers

Absolute or differential pressures can be measured in terms of the height of a liquid column.

The *open-end* manometer (Fig. 4.4a) is a glass tube, usually U-shaped, filled with the manometer fluid (mercury or oil) (see Section 3.2). One end of the tube is open to atmosphere (reference pressure P_r), the other end is connected to a system containing the gas whose pressure P_x must be determined. The height h, measured between the menisci of the manometer fluid in the two limbs of the manometer, is related to pressure by Eq. (1.2). The drawback of the open-end manometer is that pressure determination depends, under constant temperature conditions, upon the knowledge of atmospheric pressure. This drawback is circumvented by closing one limb under low pressure (reference pressure) (Fig. 4.4b).

The pressure range of the U-shaped liquid manometer is from 1×10^4 Pa (100 Torr) to about 10 Pa (10^{-1} Torr), the lower limit being set by the lack of resolution in distinguishing with the naked eye between the levels of the liquid columns in the two limbs (less than 0.1 mm). Lower pressures can be measured either by using optical magnification or by tilting the closed-end limb to a certain angle, thus increasing the number of divisions on the manometer scale (Fig. 4.4c). The last approach is feasible if the ratio of the cross sections of the closed and open limbs is larger than 1/200.

The *McLeod gauge,* referred to as a *compression manometer,* permits the accurate pressure measurement of gases and vapor which do not saturate during compression (see Section 3.3). The measuring range of the McLeod gauge is 10^2–10^{-3} Pa (1–10^{-5} Torr) with accuracies from 10% at 10^{-1} Pa (10^{-3} Torr) to more than 50% at 10^{-3} Pa (10^{-5} Torr). Trends in modern vacuum technology aim at avoiding the use of the McLeod gauge for the reasons discussed in Section 3.3.

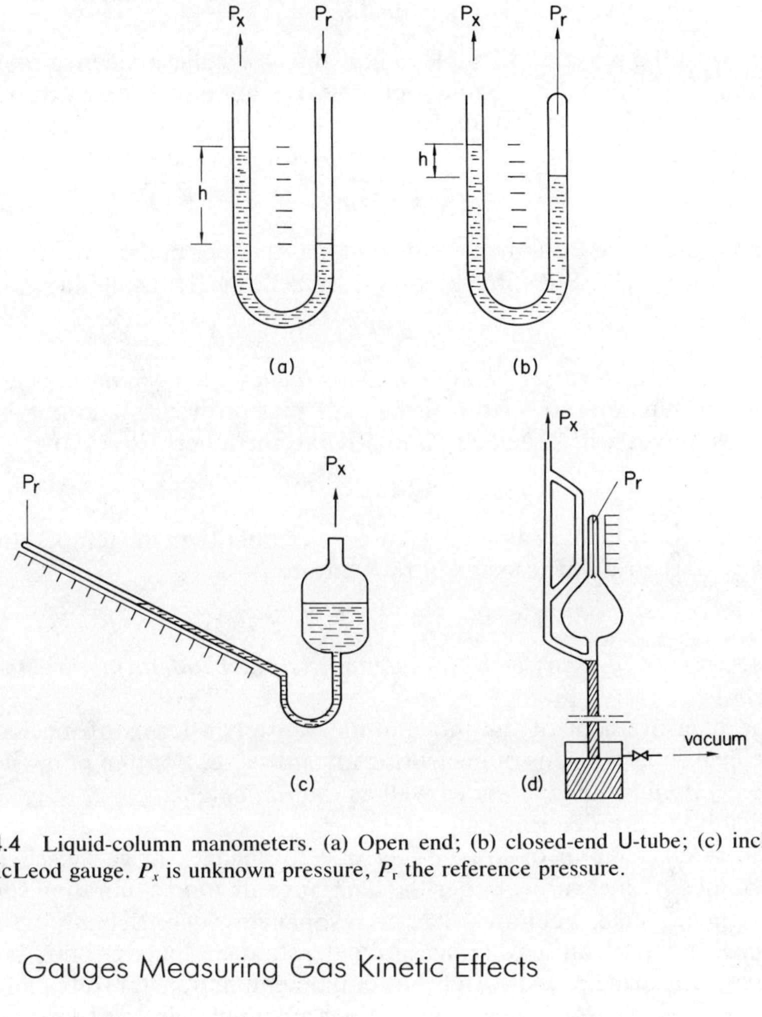

Fig. 4.4 Liquid-column manometers. (a) Open end; (b) closed-end U-tube; (c) inclined; (d) McLeod gauge. P_x is unknown pressure, P_r the reference pressure.

4.3 Gauges Measuring Gas Kinetic Effects

4.3.1 Principle

Among the transport phenomena in gas at low pressure, only those involving (a) *momentum transfer* and (b) *energy transfer* are used for indirect pressure determination. The main theoretical aspects of transport phenomena are briefly discussed.

(a) *Momentum transfer.* According to the kinetic theory of rarefied gases, the rate of transfer of momentum of mass motion (q_m) by molecules between two surfaces 1 and 2 separated by a distance much less than the molecular mean free path is

$$q_m = m\phi(v_1 - v_2) \tag{4.1}$$

where m is the mass of the molecule, ϕ the molecular incidence rate (see Section 2.3), and v the rebound velocity. Equation (4.1) holds true if the surfaces 1 and 2 are at rest and at the temperature of the surrounding gas.

The molecular incidence rate ϕ is related to pressure P as

$$\phi = P/(2\pi mkT)^{1/2} \tag{4.2}$$

where k and T are Boltzman's constant and the thermodynamic temperature, respectively. Substituting Eq. (4.2) in Eq. (4.1) results in

$$q_m = P(m/2\pi kT)^{1/2}(v_1 - v_2) \tag{4.3}$$

Thus *the rate of transfer of momentum is linearly dependent on pressure*. When the temperature of the surfaces is that of the gas and one of the surfaces moves with a velocity u relative to the other, Eq. (4.3) becomes

$$q_m = P/(m2\pi kT)^{1/2}u \tag{4.4}$$

If one of the surfaces is at temperature T_1, greater than the temperature of the gas T, the rate of transfer of momentum is

$$q = P(3/2\pi)^{1/2}\,[(T_1/T)^{1/2} - 1] \tag{4.5}$$

Thus *the rate of transfer of momentum is independent of the mass of molecules*.

The measurement of the momentum transfer in terms of mechanical force is used as a means of measuring low pressure. For that purpose, the *viscous damp* or *drag effect* as well as the *radiometric effect* of gas on a moving body are determined.

The *viscous damp* or *drag effect* due to changes in gas viscosity is determined by measuring either the *damping* effect on a vibrating surface (decrement gauge, Section 4.3.2, or resonance gauge, Section 4.3.3) or the *drag* exerted on a rotating surface (rotating gauge, Section 4.3.4). Gauges measuring pressure in terms of momentum transfer (referred to as molecular or friction gauges) are rather unpopular in modern vacuum technology. Their drawbacks are the hours required for reading at the lower limit of pressure, disturbance of determinations by vibrations and extraneous magnetic fields, difficulties in maintaining constant the temperatures of both gas and vibrating surface, noncontinuous readings, and unsuitability for operation by unskilled workers. Pressure measurement is dependent on gas species and becomes easier with heavier gases (Eq. 4.3) (Anderson, 1958; Austin, 1969; Austin and Leck, 1972; Carter, 1972; Dimeff *et al.*, 1962; Hobson, 1970; Jones and Roberts, 1978; Neher, 1962; Novak, 1972; Robinson, 1962; Schwarz, 1960, 1962).

The *radiometric effect* arises as follows: Under certain conditions, a surface free to move in a rarefied environment (temperature T_1 of gas and surface are equal) is repelled with a force F_r by another stationary surface (at temperature $T_2 > T_1$) a small distance apart. The force F_r, referred to as radiometric force, proportional to the pressure P of the gas is

$$F_r = (P/2)[(T_1/T_2)^{1/2} - 1] \tag{4.6}$$

In writing Eq. (4.6), the thermal accommodation coefficient of the surfaces is assumed to be unity. The *thermal accommodation coefficient* α_T stands for the transfer of energy to a surface immersed in gas by impinging molecules of gas. It is defined as the ratio of the energy actually transferred between impinging molecules and the surface to the energy which would be transferred if the molecules attained thermal equilibrium with the surface. That is

$$\alpha_T = (T_r - T_i)/(T_s - T_i) \tag{4.7}$$

Here the subscripts to temperature T refer to the temperatures of reflected (re-emitted) molecules (r), incident molecules (i), and the surface (s). The values of the thermal accommodation coefficient α_T for different gases are listed in Table 4.1.

Vacuum gauges using the radiometric effect for measuring pressure can be used as primary standards (see Section 3.1) provided that the molecu-

Table 4.1

VALUES OF THE THERMAL ACCOMMODATION COEFFICIENT FOR SEVERAL GASES

	Gas									
Surface	Air	Ar	CO_2	H_2	He	Kr	N_2	Ne	Hg	O_2
Fe	—	—	—	—	—	—	—	0.1[a]	—	—
Ni	—	0.93[b]	—	0.29[b]	—	—	0.82[b]	0.82[b]	—	0.86[b]
Pt (ordinary)	0.90[e]	0.89[c]	—	0.36	—	0.69[e]	0.89[c]	—	—	—
Pt polished	—	—	0.868[c]	0.358[c]	—	—	—	—	—	0.835[c]
Pt slightly coated with black	—	—	0.945[c]	0.556[c]	—	—	—	—	—	0.927[b]
Pt heavily coated with black	—	—	0.975[c]	0.712[c]	—	—	—	—	—	0.956[c]
W	—	0.85[c]	—	0.20[c]	0.016[d]	—	0.57[c]	—	0.95[c]	0.9[b]

[a] Eggelton and Tompkins (1952).
[b] Amdur and Guildner (1957).
[c] Dushman (1949, p. 55).
[d] Thomas and Shofield (1955).
[e] Thomas and Brown (1950).

lar weight of gas M is known and the thermal accommodation coefficient α_T is unity.

(b) *Energy transfer.* Heat is transferred in gas by molecular collisions from hotter to colder regions. This process, called *conduction,* is independent of any transfer of energy that may occur simultaneously by means of radiation. The amount of heat that is transferred per unit time across a unit area of any small plane located perpendicularly to the direction of the heat flow, divided by the temperature gradient, is called *thermal conductivity*.

From the point of view of kinetic theory, viscosity and heat conduction are closely related. In viscosity, the shearing stress, which represents the transport of momentum of flow in the direction of the velocity gradient, is proportional to the temperature gradient in the flowing gas. In heat conduction, the transport of thermal energy is proportional to the temperature gradient in the direction of the heat flow.

For monatomic gases, the thermal energy acquired by molecules in the hot region is exclusively translational. The thermal energy that polyatomic gases acquire in similar conditions includes, in addition, rotational and vibrational energies.

The rate of heat transfer due to thermal conductivity varies with pressure and, for certain conditions of the flow regime, is proportional to pressure. The flow regime can be characterized by Knudsen's number K_n:

$$K_n = \lambda/d \tag{4.8}$$

where λ is the mean free path of gas molecules and d is a characteristic dimension of the system accommodating the gas.

The heat transfer Q_n from a hot wire to the surrounding gas as a function of pressure P is illustrated in Fig. 4.5. In the range of low pressures where *molecular flow conditions* prevail, $K_n > 3$ intermolecular collisions are rare, compared with molecular collisions with the hot wire and the boundary walls. *The heat transfer is proportional to pressure.* As the number of molecules increases, the flow regime shifts from molecular to that of transition or viscous slip flow ($0.01 < K_n < 3$). The mean free path of gas molecules becomes significantly less than the distance separating the hot wire from the walls. The proportionality between heat transfer and pressure observed before is no longer maintained. At still higher pressures the flow regime of gas becomes *viscous* (*laminar* or *turbulent*) and is characterized by $K_n < 0.01$. *Thermal conductivity becomes independent of gas pressure,* provided that convective cooling (dotted line in Fig. 4.5) is avoided. Thermal conductivity is expressed in this region as

$$K = \tfrac{1}{4}(9c_p/c_v - 5)\eta c_v \tag{4.9}$$

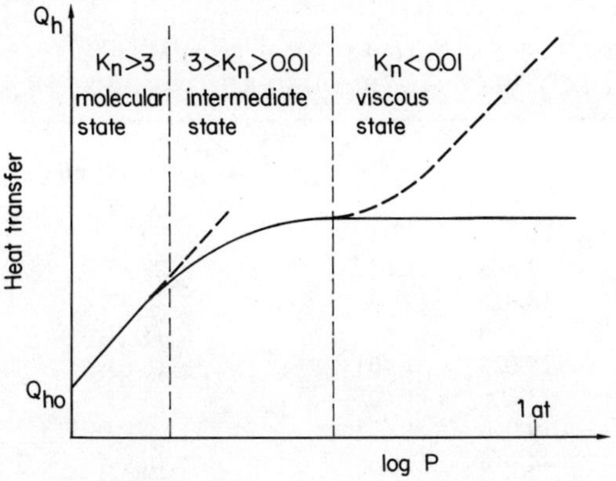

Fig. 4.5 Variation of heat transfer Q vs. pressure P. Reprinted with permission from Vacuum **23,** 307, W. Steckelmacher. Copyright 1963 Pergamon Press.

where c_p and c_v are the *heat capacity* of the gas at *constant pressure* and *volume,* respectively, and η is the coefficient of viscosity of the gas. The values of K for different gases are listed in Table 4.2.

The rate of energy transfer E_T *per unit area* from a wire of radius r_w at temperature T_s, stretched along the axis of a cylinder of radius r_c at temperature T_i through the surrounding gas ($K_n < 0.01$), is

$$E_T = \alpha_r \Lambda_T P(273/T_i)^{1/2}(T_s - T_i) \tag{4.10}$$

where Λ_T is the free-molecule thermal conductivity at 0°C and

$$\alpha_r = \alpha_t/1 - [(1 - \alpha_t)(r_2/r_1)] \tag{4.11}$$

Table 4.2
THERMAL CONDUCTIVITY K OF DIFFERENT GASES AT 273.16 K

Gas	K ($Wm^{-1}s^{-1}K^{-1}$)	Gas	K ($Wm^{-1}s^{-1}K^{-1}$)
Hydrogen	1.00×10^{-6}	Helium	8.20×10^{-7}
Nitrogen	1.36×10^{-7}	Neon	2.60×10^{-7}
Oxygen	1.39×10^{-7}	Argon	9.32×10^{-8}
Air	1.39×10^{-7}	Krypton	5.02×10^{-8}
Carbon monoxide	1.27×10^{-7}	Mercury	2.87×10^{-8}
Carbon dioxide	8.13×10^{-8}		

Table 4.3
VALUES OF FREE-MOLECULE THERMAL CONDUCTIVITY Λ_T FOR DIFFERENT GASES[a]

Gas	M	C_p/C_v	Λ_T ($\times 10^6$ W cm^{-2} K^{-1} mm Hg^{-3})
H_2	2.016	1.41	60.72
He	4.003	1.67	29.35
H_2O	18.016	1.30	26.49
Ne	20.180	1.67	13.07
N_2	28.020	1.40	16.63
O_2	32.000	1.40	15.57
Ar	39.940	1.67	9.29
CO_2	44.010	1.30	16.96
Hg	200.600	1.67	4.15

[a] Reprinted by permission of John Wiley and Sons, Inc. from: Scientific Foundation of Vacuum Technique p. 48 by Dushman, S., Lafferty, J. M., editor, Copyright 1962 by John Wiley and Sons, Inc.

The values of Λ_T for different gases are given in Table 4.3. It can be seen that the rate of energy transfer E_T per unit area is, among other parameters, a function of the nature of the gas.

The theoretical aspect of the energy transfer used as a means to measure pressure was treated by Von Ubish (1952, 1957).

Gauges measuring pressure in terms of energy transfer are the Pirani gauge (Section 4.3.6), the thermocouple gauge (Section 4.3.7), and the thermistor gauge (Section 4.3.8). These gauges are utilized on a very large scale in vacuum technology.

4.3.2 The Decrement Gauge

In the decrement gauge, a light body immersed in the gas whose pressure is to be measured is set to oscillate, and the time needed to decrease the maximum amplitude to one half of the initial value is determined. This time, referred to as *half-time,* is symbolized by $t_{0.5}$. The oscillating body can be a thin glass disk (D) suspended by a glass fiber (F) between two horizontal coplanar disks (C) a short distance apart (Fig. 4.6a), a glass bar suspended on a glass fiber (F) carrying a glass sphere (S) at the opposite extremities (Fig. 4.6b), or a quartz fiber (F) (Fig. 4.6c).

For the simplest gauge (quartz fiber), theoretical considerations lead to the relationship between the pressure P of gas at a given temperature and the time $t_{0.5}$

$$PM^{1/2} = (B/t_{0.5}) - C \tag{4.12}$$

where M is the molecular weight of gas and B and C are constants determined by the geometry of the fiber. When several gases are mixed the term $PM^{1/2}$ must be replaced by $p_iM^{1/2}$, where p is the partial pressure of each of the gases (i).

Oscillation is initiated by an external magnetic field acting on a tiny piece of metal glued to one end of the moving body. Vibration is observed through an adequately illuminated telescope. The gauge is calibrated by plotting the time $t_{0.5}$ decay as a function of pressure for each species of gas.

The lower pressure limit is set by frictional forces originating in the suspension and spurious harmonics of the moving body. These harmonics yield oscillations in planes different than the original one, thus rendering readings difficult. Harmonic disturbances can be avoided for the fiber-type gauge by using two fibers joined together at one end and set apart at the other end, as illustrated in Fig. 4.6d (Jones and Roberts, 1978).

4.3.3 The Resonance Gauge

In the resonance gauge, a thin body is set to oscillate at its resonant frequency by an adequate system. The amplitude of oscillations is held constant by continuously feeding electrical energy, at a controlled rate, into the system. The energy input is used to give a measure of pressure. Schwarz (1957) described a gauge in which an iron membrane could be set swinging to its resonant frequency by an electrostatic or electromagnetic driving system. The instrument was referred to as an *acoustic gauge* since the frequency of vibrations is in the audible range. Becker (1960) used a thin ribbon (R) made of highly flexible metal stretched across the magnetic

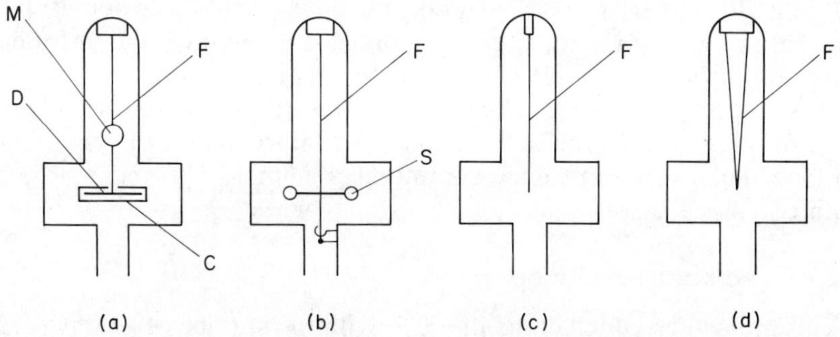

Fig. 4.6 Schematic diagram of different designs of decrement gauges (see text). C, coplanar disk; D, glass disk; F, glass fiber; M, mirror; S, glass sphere.

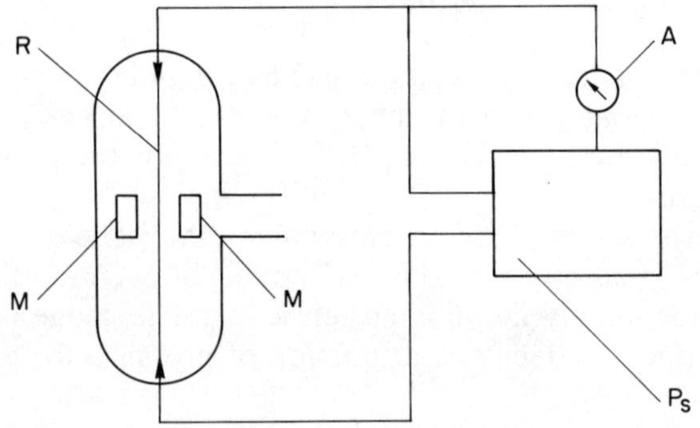

Fig. 4.7 Principle of the resonance gauge head. A, ammeter; M, permanent magnets; P_s, power supply; R, ribbon.

field (M) of two permanent magnets (Fig. 4.7). The vibration of the ribbon generates an ac current, the amplitude of which is compared with a reference in the power supply (P_s). The amplitude of the vibration is held constant with a current feedback to the ribbon. This feedback current is proportional to the damping of the oscillating ribbon, thus being a measure of pressure in the gauge head.

Different designs have been used for the construction of resonance gauges, such as piezoelectric crystals (Pacey, 1959/1960), diaphragms (Dimeff *et al.*, 1962), and vanes (Christian and Leck, 1966). Photoelectric, capacitance, and inductance transducers used together with advanced electronics have made it possible largely to overcome inaccurate readings such as those obtained with the use of simple optical techniques. However, this type of gauge has not yet left the laboratory stage.

The useful operating range of the resonance gauge is about 10^4 Pa (100 Torr) to 0.133 Pa (10^{-3} Torr). It has been reported that pressures below 10^{-2} Pa (10^{-4} Torr) were measured, but their accuracy is doubtful since the stiffness of the vibrating-element hinge is the limiting factor for determination of pressures below 10^{-1} Pa (10^{-3} Torr) (Schwarz, 1961).

Resonance gauge heads are not sensitive to vibrations provided that their frequency is not the same as the resonant frequency of the oscillating surface. Such gauges provide continuous readings. They cannot be used with corrosive gases.

4.3.4 The Rotating Gauge

Gas molecules which come in touch with the surface of a body rotating at high speed in a rarefied environment acquire momentum in the direction of the movement. This momentum can be transmitted, if conditions are conducive, to the adjacent surface of another body suspended a small distance away. The suspended body starts rotating until the force accounting for its movement is balanced by the torque of its suspension. If the rotating body is suspended in the rarefied environment by *magnetic levitation* (the body's weight is balanced by a magnetic force so that it remains suspended in space) and allowed to coast freely, the momentum imparted by gas molecules decelerates its movement. Either the torque in the suspension or the deceleration of the rotating body can be used as a means of measuring pressure since the rate of the transfer of momentum producing them is related to the number density of molecules.

Dushman (1915) designed a rotating gauge according to the suggestion of Langmuir (1913). This gauge uses the rate of momentum of molecules transmitted by a very fast rotating disk to a coaxially suspended disk a few millimeters away. The gauge called a *rotating disk* (Dushman, 1949, p. 291), shown in Fig. 4.8, consists of two coplanar disks D_1 and D_2 of aluminum and mica, respectively. The disk D_1 is rotated at constant speed by means of an external rotary magnetic field (R_f). The rotation of the disk D_2, produced by the momentum transferred by gas molecules, generates a torque in the quartz suspension fiber (F). A mirror (M), an illumination system, and an external scale permit reading the displacement of D_2 from its rest position. The angle of torque is

$$\alpha_t = (\mathrm{Ct} \times \tau_o^2 r^4 / I_i)(P\omega)(M/T)^{1/2} \tag{4.13}$$

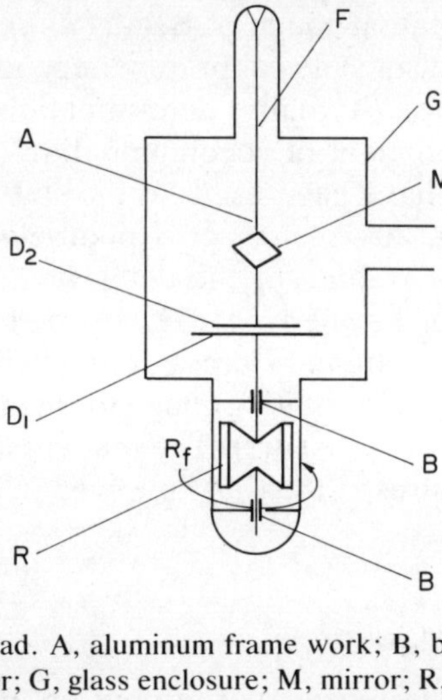

Fig. 4.8 Rotating gauge head. A, aluminum frame work; B, bearings; D_1, aluminum disk; D_2, mica disk; F, quartz fiber; G, glass enclosure; M, mirror; R, rotor; R_f, rotating magnetic field. Reprinted with permission from "Scientific Foundation of Vacuum Technique" S. Dushman. p. 291. Copyright 1949 John Wiley and Sons, Inc.

where the constant Ct depends on the nature of the gas and the accommodation coefficients, τ_o is the oscillation period, r the radius of D_2, I_i the moment of inertia of D_2, P the total pressure, ω the angular velocity of rotation, and M and T the molecular mass and the thermodynamic temperature of the gas. The proportionality between the torque in the suspension of D_2 and the quantity $P(M/T)^{1/2}$ is valid only in the range where the mean free paths of gas molecules is larger than the dimensions of the vessel. Pressure measurements in the range 4×10^{-5} Pa (3×10^{-7} Torr) are inaccurate because of both the eddy currents induced in the aluminum frame-work supporting the mirror and the pronounced tendency of D_2 to oscillate.

Briggs (1955) described an improved rotating gauge, but Beams *et al.* (1962) developed an instrument by which they overcame the difficulties arising in pressure measurement, mainly because of the suspension of the rotating body. The principle of the gauge relies on the drag effect produced by rarefied gas on a magnetically suspended rotor (steel ball) brought to high speed and subsequently allowed to spin with driving power off (Beams *et al.*, 1946). The approximate formula given by Beams *et al.* (1946) relates the deceleration of the rotor (N/N_0) to pressure P as

$$\log(N_r/N_{0r}) = (5P/r_B\rho_B)(m/2\pi kT)^{1/2}(t - t_0) \tag{4.14}$$

Here N_r and N_{0r} are the number of revolutions of the rotor at times t and t_0, respectively, P is the total pressure, and r_B and ρ_B the radius and the mass density of the rotor, respectively.

Theoretically, the rotating gauge of Beams *et al.* (1946) can be considered a primary standard since its calibration depends only on the diameter and the mass of the rotor. Actually, determination of pressure is a function of the thermal coefficient of accommodation of the gas.

Different techniques have been used for the stabilization of the rotor in the magnetic field, e.g., by means of a photoelectric servomechanism (Harbour and Lord, 1965), and for the determination of times t and t_0.

The research work of Fremerey (1972, 1973), Fremerey *et al.* (1974), Fremerey and Boden (1978), and Comsa *et al.* (1977, 1980) made possible the industrial realization of a rotating gauge for measuring pressures in the range 10^{-2}–10^{-5} Pa (10^{-4}–10^{-7} Torr). The main limitation to the measurement of very low pressures is the time involved in determining the deceleration of the rotor.

4.3.5 The Radiometer Gauge

Knudsen (1910a,b,c) proposed as the physical principle for a vacuum gauge the exploitation of the radiometric force, which produces the movement of the radiometer (a four-vaned instrument devised by Sir Williams Crook which rotates when a source of light or heat is brought near it and which was utilized for the detection of low pressures). Let us consider two parallel plates V and H a distance apart in a rarefied environment (Fig. 4.9a). The distance between the plates is less than the mean free path of gas molecules. The temperature T_1 of plate V is the same as that of the residual gas, whereas the temperature of plate H is $T_2 > T_1$. Assuming that all gas molecules impinging on the side of V away from H have the same velocity, they will rebound with no velocity change. Gas molecules impinging on the side of V which faces plate H transport a momentum greater than that of the former molecules. Hence, plate V is repelled from plate H. Knudsen measured the repulsion force, using the arrangement schematically shown in Fig. 4.9b. A rectangular vane, cut out in the center leaving two lateral strips (standing for plate V), is suspended by a fiber F between two rectangular strips H. The distance between the vane and the strips (which can be heated) is very small. The force F_r by which the vane is repelled yields a torque in fiber F, which is measured, like in a galvanometer, by the deflection of the mirror M. This force is proportional to the pressure P and *independent of the molecular weight of the gas* (Fig. 4.6). Lockenwitz (1938) showed that when a third plate is introduced parallel to V and H the repelling force on the vane V is independent of the vane temperature. However, Wu and Birendra (1972) demonstrated

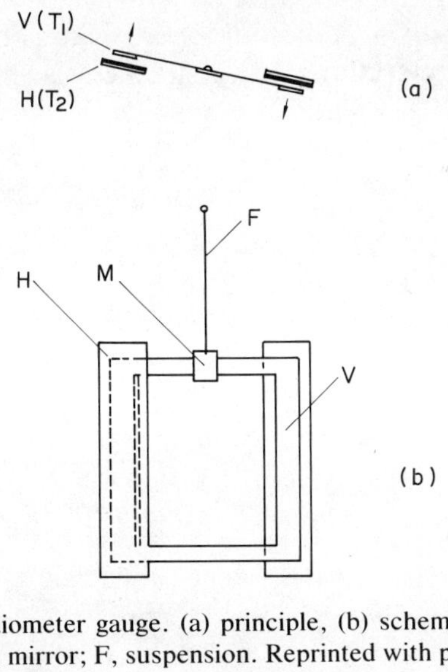

Fig. 4.9 Knudsen's radiometer gauge. (a) principle, (b) schematic of the instrument. V, light vane; H, heater; M, mirror; F, suspension. Reprinted with permission from "Scientific Foundation of Vacuum Technique." S. Dushman. p. 295. Copyright 1949 John Wiley and Sons, Inc.

that in the case of the modified Knudsen–Lockenwitz theory, the repelling force on the moving vane is dependent on its temperature.

The radiometric principle has been used in the design of instruments by a number of workers, notably Fryburg and Simons (1949), Klumb and Schwarz (1944), Kreisel (1976), Lockenwitz (1938), and Du Mond and Pickels (1935). Problems connected with the material used for the suspension of the moving surface have restricted the use of radiometer gauges to the laboratory. The pressure range spanned extends from 10^{-1} Pa (10^{-3} Torr) to 10^{-5} Pa (10^{-7} Torr).

In 1969, Evrard and Boutry (1969) designed a gauge, referred to as an absolute micromanometer, which works on the principle of the radiometric force and uses a frictionless, torque-free suspension of the moving surface. In principle, a light, thin graphite disk is supported with a nonuniform magnetic field and set spinning with stability around a diameter on a vertical axis (Fig. 4.10). The disk is suspended within a glass tube inside a cage built of molybdenum strips, such as illustrated in the upper part of Fig. 4.11. The glass tube is connected to the ultrahigh vacuum setup. The cage is set up in good thermal contact with a molybdenum rod, cooled in turn by liquid nitrogen, acetone, or ice water. The glass tube which accommodates the cage is at room temperature. A thermomolecular torque, proportional to pressure, due to the momentum transfer will act on the rotating disk. This torque will accelerate or decelerate the disk according to its initial direction of rotation. A narrow beam of light reflected by the disk on the cathode of a photomultiplier makes possible the automatic determination of deceleration. The instrument is small and is claimed to

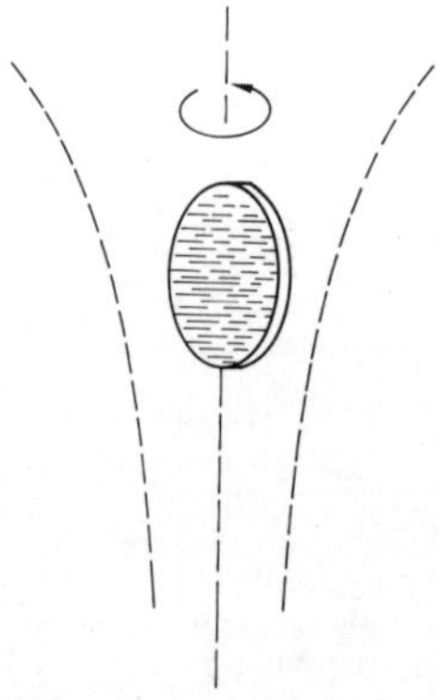

Fig. 4.10 Stable magnetic suspension of a graphite disk. (After Evrard and Boutry, 1969).

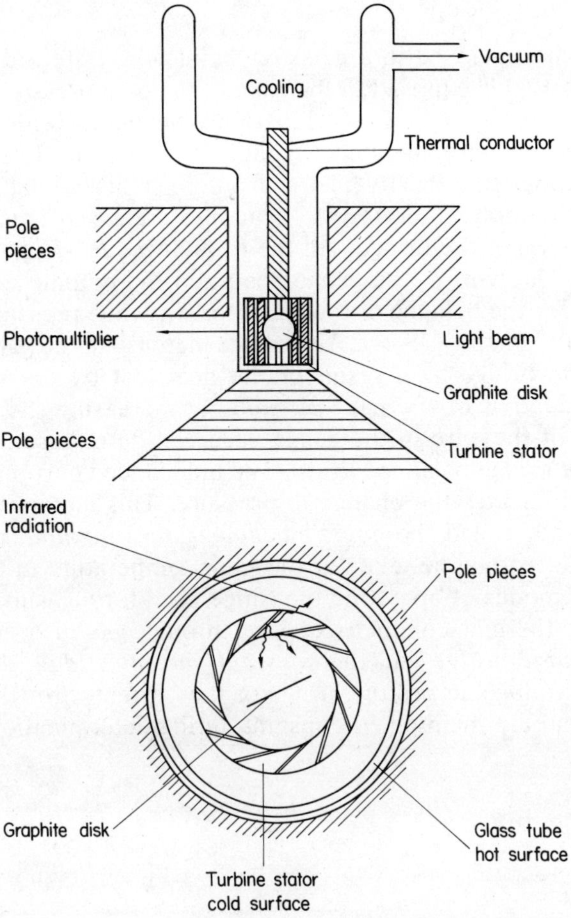

Fig. 4.11 Frontal and sectional view of an absolute micromanometer using magnetic levitation. (After Evrard and Boutry, 1969.)

be insensitive to shocks and vibrations transmitted from the vacuum setup. The range it spans when used as a pressure standard is 10^{-1}–10^{-5} Pa (10^{-3}–10^{-7} Torr). Rapid variation of pressure at the lower limit of the range cannot, however, be measured due to the long time interval (about 60 s) required for measurement. The instrument can be used as a viscosity gauge (exploiting the viscous drag of gas) when the temperature of the molybdenum cage is kept at the same temperature as that of the glass walls.

4.3.6 The Pirani Gauge

A metal wire enclosed in a glass or metal tube (referred to as *gauge head*) (Fig. 4.12a) is exposed to the gas of unknown pressure while being connected in one leg (R_1 in Fig. 4.12b) of a Wheatstone bridge (referred to as control unit). A wire, identical to that in the gauge head, sealed under a pressure of about 10^{-3} Pa (10^{-5} Torr) in a tube similar to that of the gauge head called "dummy," serving as vacuum reference, is connected in the adjacent leg (R_2) of the bridge. All the resistances R_1–R_4 have about the same value. The wire in the gauge head and the dummy are heated by adjusting R_5 for the balance of the ammeter (I), while pressure is suitably low (about 10^{-3} Pa, 10^{-5} Torr). A pressure increase in the gauge head will unbalance the bridge as a result of the heat lost by the wire through thermal conduction of the gas. By suitably increasing the voltage, the temperature of the wire in the gauge head can be restored to the initial value and the bridge balanced again. The unbalanced current through the ammeter (I) indicates the change in pressure. This method of operating the gauge head, called the *constant temperature method,* requires that each pressure reading be taken at the same temperature of the wire.

Two other modes of operating the gauge head for measuring pressures by means of thermal conductivity loss through gas also are used, the *constant voltage bridge* and the *constant current bridge,* depending on whether the voltage across the bridge or the current through the wire of the gauge head are maintained constant during determinations. The out-

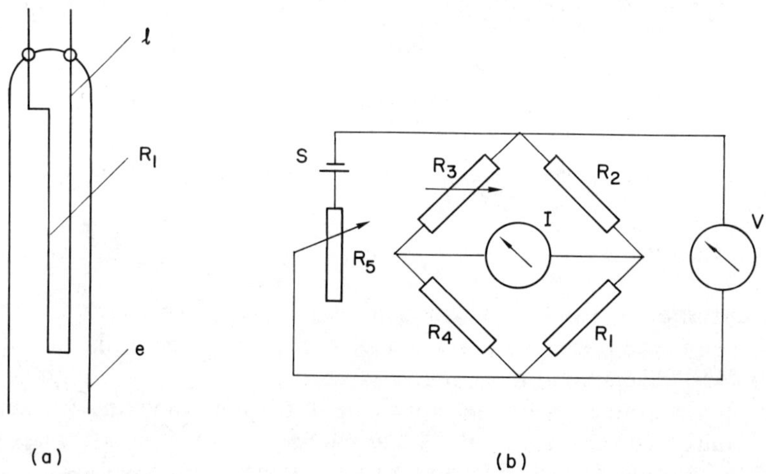

Fig. 4.12 (a) Pirani gauge head and (b) control unit. e, envelope; I, ammeter; l, lead wire; R_{1-5}, resistors; S, dc source; V, voltmeter.

of-balance current of the ammeter gives the value of the unknown pressure on a calibrated scale. *All the methods discussed measure indirectly the temperature of the wire by measuring the electrical resistance of the wire*. The most accurate method is that of *constant temperature* since the end losses of the hot wire through the leads are maintained constant over the range measured, and sensitivity is not diminished at high pressures.

The heated wire in the gauge head dissipates energy by radiation and thermal conductivity through both the suspending leads (l in Fig. 4.12) and gas. The temperature of the wire is determined by the balance between the energy supplied for heating and the thermal losses through the leads and through the gas to the surrounding walls. Thermal losses through the leads are proportional to the mass of the material of the leads as well as to the temperature difference between their ends inside and outside of the gauge head. Radiation losses are proportional to the emissivity, the temperature of the heated wire, and the surface of the wire. The functional dependence of radiation losses on the emissivity of a hot surface, complicated geometries of the gauge head, and surface conditions cannot be described precisely. Hence, errors arising from an erroneous application of Stefan–Boltzman's function are possible (Hunt, 1968).

Since losses both through thermal conductivity through the leads and through radiation are pressure independent, it is desirable to minimize them in order to make the temperature of the wire as pressure sensitive as possible. This suggests the use of materials with relatively low emissivity, such as tungsten or platinum, for the heated wire as well as short, thin leads for glass welding. Tungsten or platinum, which have low emissivities, offer the advantage of having a high temperature coefficient of resistance, thus enhancing the sensitivity of the electric control of the bridge. Gold-plated tungsten or noble alloy wires are used when pressure is measured in a contaminated atmosphere (Heinje and Vink, 1969). Platinum–iridium filaments or ribbons as well as rhodium-plated filaments are also used since they give increased protection of the hot wire in the presence of chemically active gases and have reduced emissivity. Usually the hot wire is in the form of helical coil or a hair pin.

Temperature variations of the gauge head walls due to ambient conditions constitute an important source of uncertainty in pressure measurement. To reduce their effects the dummy containing the reference wire is frequently placed in the same envelope with the gauge head which is connected to the vacuum system. One of the drawbacks of this technique is that the characteristics of the gauge head can change when measurements are performed in helium. Thus, if conditions are conductive, helium permeates through the walls of the dummy, hence changing the vacuum reference.

English *et al.* (1965) investigated a means of temperature compensation taking into account, besides the variations of ambient temperature, the heating of the gauge head tube due to the heat dissipated from the hot wire at high pressures. For compensation, they used a special resistor which was wound on the tube containing the heated wire. Electronic compensation of ambient temperature has also been reported (Górski *et al.*, 1977).

Calibration of the gauge head is performed with dry air or nitrogen. Since the rate of energy transfer E_T [Eq. (4.10)] is a function of the free-molecule heat conductivity Λ_T, which varies with the temperature of the gas, calibration of gases other than those mentioned or vapors is possible by using calibration curves such as those illustrated in Fig. 4.13. Calibration, however, cannot be retained for a long time in routine practice owing to changes in the emissivity of the hot wire produced by contaminants. In such cases the gauge indicates a higher pressure than the actual one. Aging effects on the gauge head result in both reduction of the cross section of the wire and structural modifications which alter the accommo-

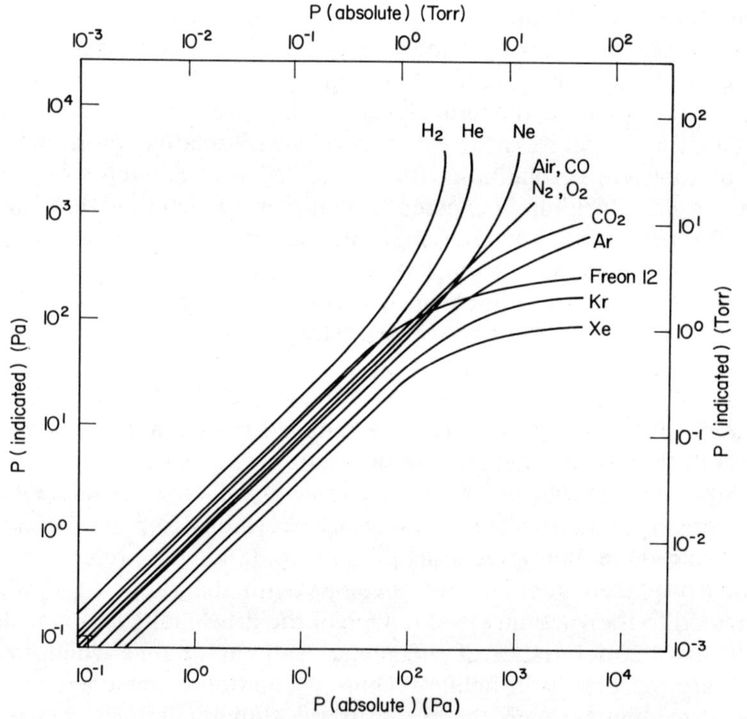

Fig. 4.13 Calibration curves for a commercial type of Pirani gauge head.

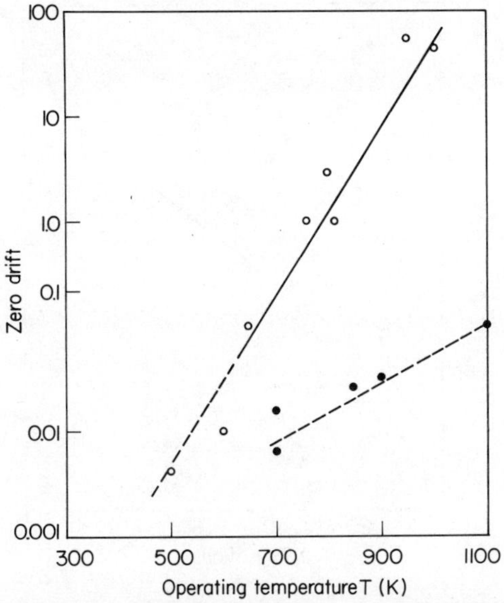

Fig. 4.14 Pirani gauge zero drift. The change in zero reading expressed as an apparant change in pressure, at zero pressure, caused by operating for 1 h at 10 mTorr (air) is plotted as a function of the wire temperature (temperature at zero pressure.) ○, tungsten; ●, platinum. Reprinted with permission from "Pressure Measurement in Vacuum Systems. J. H. Leck." p. 60. Copyright 1964 Chapman and Hall.

dation coefficients. Thus, after operating a gauge head (heated filament tungsten or platinum wire) in air at 1 Pa (10^{-2} Torr) in a constant voltage bridge, a zero drift was noticed (Leck, 1964, p. 60) as illustrated in Fig. 4.14. Zero drifts causing errors of $\pm 10\%$ in the gauge readings also were noticed at gauge heads (platinum wire) after a sudden change from a low (1 Pa, 10^{-2} Torr) to a higher pressure (100 Pa, 1 Torr) (Leck, 1964, p. 59).

The influence of contaminants can be greatly reduced by cleaning. Such procedures consist of filling the tube of the gauge head with a solvent, e.g., trichlor-ethylene or gasoline, shaking the gauge head, draining, refilling with solvent several times, then allowing to dry. For badly contaminated gauge heads, the wire is connected to a power supply, brought to a red hot color, and heated in air for about 10 s. This procedure is drastic and may damage the wire if repeated often. It also contributes to a rapid aging of the wire.

The time response (see Section 2.3.6) of the gauge head is rather long (see Table 2.4) owing to the temperature dependence of the heated wire on pressure in the gauge head. This temperature dependence also contributes to the nonlinearity of the output signal, as illustrated in Fig. 4.15.

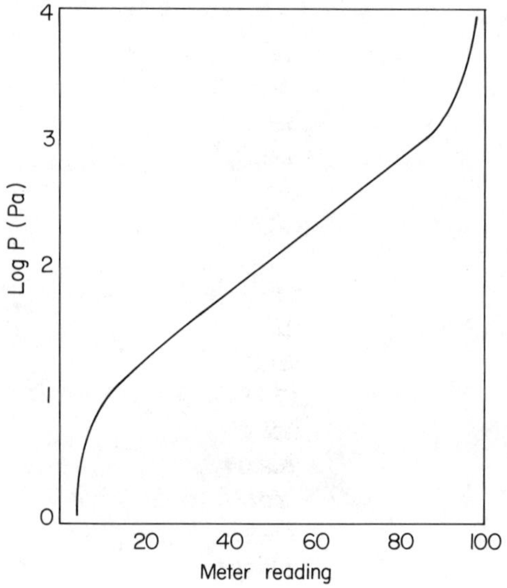

Fig. 4.15 Variation of the output signal of a Pirani gauge with pressure (for air).

The stability of Pirani gauges is impaired because the output signal usually drifts at the limits of the pressure scale, due to the electronic circuitry of the control unit. Such drifts can be avoided partially by operating the assembly gauge head control unit one hour before measurement.

The reproducibility of calibration of Pirani gauges, under *clean vacuum conditions,* over a period of six months is fairly good (Poulter *et al.*, 1980).

Pirani gauges measure pressures usually in the range 6×10^{-1}–60 Pa (5×10^{-3}–5×10^{-1} Torr). Different methods have been used for the extension of the upper pressure range. The method most used is natural or forced convection of the gas in the gauge head which is pressure dependent up to atmospheric pressure (Steckelmacher and Fletcher, 1972a; Steckelmacher, 1973).

The main application of Pirani gauges is in monitoring the forevacuum of pumping systems. For this purpose, they are integrated in an automatic control circuit which switches the system on or off when pressure has reached the preset limits.

4.3.7 The Thermocouple Gauge

The measuring principle of the thermocouple gauge is the same as that of the Pirani gauge. The difference between them is that in the case of the former the temperature of the hot element in contact with the gas is *directly measured* by means of a thermocouple.

A metal wire or ribbon (w in Fig. 4.16) is exposed to the unknown pressure while being heated under *constant current,* by an ac or dc supply. A thermocouple (t_h), spot welded at the mid-distance between the leads hooking the heated wire, operates the ammeter (I) whose pointer moves over a scale calibrated in pressure units. The heated wire is so designed that the temperature it assumes at pressures below 10^{-1} Pa (10^{-3} Torr) is less than 200°C. When pressure increases, the heat losses from the wire increase, hence the electromotive force (emf) from the thermocouple decreases. The output signal of the thermocouple is a nonlinear function of pressure, as illustrated in Fig. 4.17.

The emf of the thermocouple which measures the temperature of the hot wire is dependent on the cold junction temperature. Electronic simulation of the ice point is now used for reference. However, variations of the ambient temperature act on the gauge head, causing errors in pressure determination at the higher and lower limits of the measuring range.

The drawbacks resulting from spot welding of the thermocouple at the heated element are overcome by imbedding both the heated wire (ribbon) and the thermocouple in a thermoconducting head (Fig. 4.16b).

Variations of the basic design discussed include the use of two thermocouples in parallel instead of a heating wire. The thermocouples are connected to an ac power supply and the sum of their emf measured. Benson (1957), who used this arrangement, describes the influence of constructive

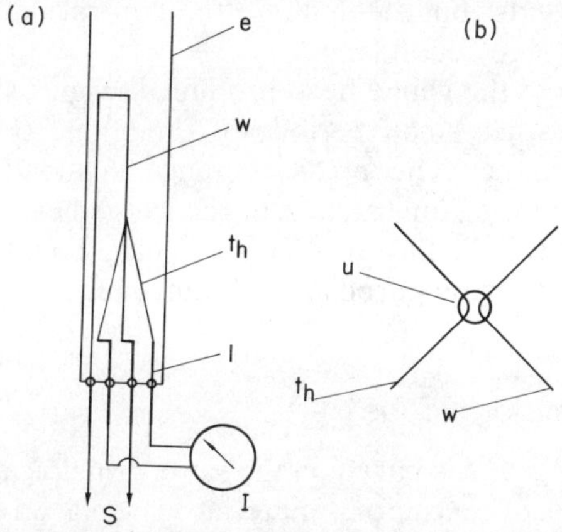

Fig. 4.16 (a) Thermocouple gauge head and (b) thermoconducting bead. e, envelope; I, ammeter; l, leads; S, dc or ac source; t_h, thermocouple; u, thermoconducting bead; w, heated wire.

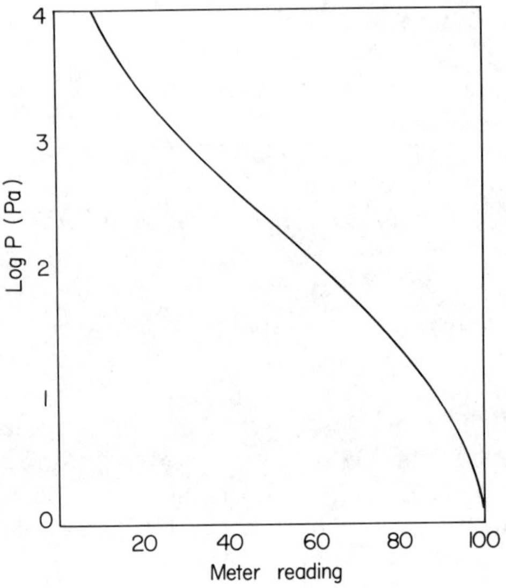

Fig. 4.17 Variation of the output signal of a thermocouple gauge with pressure (for air).

factors (e.g., the size of the thermocouple wires) and operating parameters (e.g., the temperature of the thermocouples or transient changes in ambient temperature) on the sensitivity of the gauge.

The time response of the gauge head (see Section 2.3.6) is usually of the order of milliseconds, but elements in the control unit, like relays, contribute to raising it to seconds.

Contamination of the gauge head produces serious shifts of the characteristics during rather small periods of time, due to alterations of the thermocouple surface. The pressure range spanned by thermocouple gauges depends on the construction of the gauge head and usually covers 2.6×10^{-3}–1.3×10^{-2} Pa (20–10^{-4} Torr). Thermocouple gauges are robust and are used for the same purposes as Pirani gauges, where high accuracy is not required.

4.3.8 The Thermistor Gauge

The thermistor gauge is based on the same principle as the Pirani gauge. The heating wire of refractory material is replaced by semiconductor oxides which have a negative coefficient of resistance about three times larger than that of tungsten or platinum (Varićak, 1962; Varićak and Stafic, 1959). Hence, in a Wheatstone bridge network, they produce a much larger out-of-balance current at a given pressure than refractory metallic wires. The control unit contains basically the same circuitry as that used for Pirani, the wire in the dummy being replaced by a thermistor, sealed under high vacuum. The ammeter current is set at zero for pressures of about 10^{-3} Pa (10^{-5} Torr) and the instrument inbalance observed as a measure of pressure.

Higher performances are obtained if the surface area of the thermistor is increased, thus permitting the extension of the usual high limit of the measuring range from about 10^{-3} Pa (10^{-5} Torr) to about 10^5 Pa (760 Torr) (Bretschi, 1976). One of the possible ways of enlarging the surface area consists of attaching to the thermistor a metallic foil. The technique is, however, complicated. Shioyama *et al.* (1978) used TaNi films instead of conventional materials for the construction of a thermistor gauge head. This enabled them to obtain small-sized transducers having large area and a rather large negative coefficient of resistance. Such films permit pressure measurements in the range of about 10^5 Pa (760 Torr) to 10^{-3} Pa (10^{-5} Torr). The use of special types of thermistors (type NTC Philips) together with a specially designed electronic circuit enabled Hidber and Süss (1976) to linearize the output signal of the gauge head as a function of pressure. The time response of the gauge head they designed is 20 ms over the whole pressure range from 1.3×10^{-2} up to 1.3×10^3 Pa (10^{-4} up to 10 Torr). It is interesting to note that the time response was measured by simulating pressure changes with a mechanically chopped He–Ne beam emitted by a laser focused on the thermistor.

In addition to conventional utilizations in laboratory and industry, thermistor gauges are used as such (Thiele and Beyers, 1967) or in combination with Pirani gauges (Israel, 1962) for upper-air pressure measurements in special rocket-born systems.

4.4 Gauges Measuring Phenomena Produced by Charged Particles Crossing the Gas

4.4.1 Principle

One of the methods utilized in vacuum technology to measure gas contents at different degrees of rarefaction consists of gauging the density of the gas. Any of the interaction phenomena of a charged particle with the gas under test offers possibilities of directly obtaining *density* data. If the temperature and nature of the gas are known, the data obtained may be used to infer pressure. Fluorescence, energy loss, scattering, and ionization are available for this purpose. The charged particles are provided either by radionuclides or electrons suitably accelerated in an electric field. Gauges based on these phenomena are used in vacuum technology as well as in other fields, such as chemistry, geophysics, aerodynamics, metallurgy, and space research. Except for the light-scattering gauge, which can be considered a pressure standard, all the other gauges based on the principles discussed must be calibrated to relate data measured to pressure.

(a) *Gas fluorescence.* Fast electrons which cross a volume of gas having a certain density lose a fraction of their energy to the gas molecules. For given conditions, an amount of this energy is irradiated by the gas as light quanta. The light yield ϕ_i (the ratio of the intensity of fluorescence light I_f to the fractional energy W_t lost by electrons to gas molecules, expressed as

$$\phi_i = I_f / W_f \tag{4.15}$$

is measured by a transducer. The output of the transducer is a function of pressure (Grün, 1958).

Fluorescent probes are used for measuring local densities in low-density wind tunnels or air density profiles of a high-flying aircraft in free flight (at night). The light output is not affected by the air flow (Schumacher, 1960).

Light quanta emitted from gas in the ultraviolet region as a result of electrons fired through its volume have been used by the construction of a vacuum gauge permitting pressure measurements of less than 10^{-11} Pa (10^{-13} Torr) (Alexeff, 1962) (see Section 4.4.2).

(b) *Energy loss of charged particles.* A narrow beam of highly energetic particles is fired through a gas, crossing the volume where the gas density is to be determined. In their passage through the rarefied gas the particles become scattered from the beam in a number of elastic and inelastic processes. The beam thus decays to a fraction of its initial intensity. A detector located along the geometrical axis of the beam receives the attenuated beam. The distance between the detector and the source emitting the charged particles, as well as the area of the detector, are selected such as to intercept only attenuated particles. The output I of the detector is

$$I = I_0 \exp(\alpha_a \rho l) \tag{4.16}$$

where I and I_0 are the final and initial current densities, respectively, of the particle beam (in amperes per square centimeter or in counting rate per unit area of detector), α_a is the attenuation coefficient, ρ the gas density, and l the distance between the source emitting the particles and the detector.

Accelerated electrons with energies ranging up to 100 keV and alpha or beta particles are used. Determinations are easier when particle counting techniques are used (Schumacher, 1960) instead of energy measurement.

Gauges built on this principle (see Section 4.4.3) are used for the measurement of air density in wind tunnels (Hurblut, 1959) or for the measurement of density profiles across shock waves (Venable and Kaplan, 1955). The energy loss principle was used by Helbing *et al.* (1962) for gauge calibration in the high vacuum range.

Inelastic scattering phenomena and limitations due to the detectors reduce the possibility of measuring pressures lower than about 1 Pa (10^{-2} Torr).

(c) *Scattering of charged particles.* Particles emitted from a source are shaped into a beam which crosses the gas whose density is being measured. In their passage through the gas some particles collide with gas particles (atoms or molecules) and, if conditions are conducive, are scattered *forward* from the beam. A detector conveniently located with respect to the axis of the beam receives the forward-scattered particles and converts their energy into a signal, decoded in pressure units by an electronic circuit.

In the absence of gas molecules or atoms, none of the particles in the main beam reaches the detectors. Except for spurious signals from either the detectors or the electronic circuit, there is no signal. The instant gas atoms or molecules are present, some electrons from the main beam are scattered forward and hit the detectors. Theoretical data indicate that the energy of particles acting in the scattering process, as well as the geometrical setup of the measuring assembly, are important parameters in the construction of forward-scattering gauges.

Electron guns or radionuclides (in the case of self-integrated gauge systems such as those used for measurements in the upper atmosphere) supply electrons or alpha or beta particles possessing the energies required for the measuring process. Excellent sources were obtained by careful selection of radionuclides emitting beta radiations, according to criteria such as reduced health hazard involved in their use, high decay energy, and relatively long half-life time. ^{147}Pm, for example, has moderate toxicity and high decay energy (Table 4.4). Its relatively short half-life accounts for its use in meteorological research (Ballard *et al.*, 1968, 1969, 1970; Seller and Morel, 1969; Sellers *et al.*, 1969).

Generally, the radionuclide used as an electron source is secured in a holder which, according to the way the gauge is utilized, is or is not part of a collimator. For gauging densities in unconfined volumes such as in space, collimators are unnecessary.

If the gauge is to be operated in a confined volume, a collimator is

Table 4.4

CHARACTERISTICS OF RADIONUCLIDES USED IN SCATTERING AND IONIZATION GAUGES[a]

Isotope	Relative toxicity[b]	Lifetime[c]	Mode of decay[c]	Decay energy[c] (MeV)	γ rays present	References[d]
^{3}H	Low	12.6 yr	β^-	0.01861	No	8, 9, 11, 12, 16, 17, 18, 19, 20, 23, 24, 25, 27, 31
^{63}Ni	Moderate	9.2 yr	β^-	0.067	No	4
^{147}Pm	Moderate	2.5 yr	β^-	0.225	No	15, 21, 26, 29, 32, 35, 37
^{210}Po	Very high	138.4 d	α	5.408	Yes	3, 10, 19, 23, 30
^{226}Ra	Very high	1,600 yr	α		Yes	1, 2, 4, 5, 6, 7, 11, 13
^{239}Pu	Very high	24,000 yr	α	5.234	Yes	14
^{240}Pu	Very high	6,580 yr	α	5.255	No	22, 33, 36
^{241}Am	Very high	458 yr	α	5.640	Yes	28, 34

[a] After Berman (1975).

[b] Relative toxicity established according to the radiodoses tolerated by the human body (Radiation Protection Procedures, 1973).

[c] CRC Handbook of Chemistry and Physics 54th ed. (1973–1974).

[d] References

1. Downing and Mellen (1946)
2. Mellen (1946)
3. Gimenez and Labeyrie (1951)
4. Sibley and Roehring (1953)
5. Sicinski *et al.* (1954)
6. Beynon and Nicholson (1956)
7. Vacca (1957)
8. Spencer and Bogges (1959)
9. Vanderschmidt (1959)
10. Akishin (1960)
11. El-Moslimany (1960)
12. Roehring and Vanderschmidt (1960)
13. Vitkus (1960)
14. Khavkin (1961)
15. Falkenberg and Schumacher (1962)
16. Blanc and Dagnac (1963)
17. Martin (1963)
18. Blanc and Dagnac (1964)
19. Cambou *et al.* (1964)
20. Dagnac (1964)
21. Kirsanov (1964)
22. Boziau (1965)
23. Cotin (1965)
24. Armas (1966)
25. Armas *et al.* (1966)
26. Schumacher *et al.* (1967)
27. Vanderschmidt and Simons (1967)
28. Andrieux and Iapteff (1968)
29. Ballard *et al.* (1968)
30. Howard *et al.* (1968)
31. Righby and Wright (1968)
32. Ballard *et al.* (1969)
33. British Patent (1965)
34. Sellers and Morel (1969)
36. U. S. Patent (1969)
37. Ballard *et al.* (1970)
38. Radiation Protection Procedures (1973)
39. CRC Handbook of Chemistry and Physics 54th ed. 1973–1974

necessary to prevent electrons emitted by the source from striking the walls of the vessel, which can scatter them back to the detectors. However, the drawback of the collimator is that it reduces the intensity of the particles emitted by the source. To ensure the accuracy of the measurement both electrons in the main beam and those scattered by the walls of the collimator must be prevented from reaching the detectors. Hence, a protection shield must be set between the main beam and the array of the detectors.

The detectors receive the scattered electrons and convert their energy into light or electrical pulses. Scintillator-type detectors made of glass or organic materials are used. Light pulses are conveyed to a photomultiplier via light pipes and transformed into electrical pulses which are fed into the electronic circuitry. Radiation detectors of the solid state or Geiger types are utilized to directly convert electron energy into electrical pulses.

In selecting the detector, besides its type (scintillator, solid state, etc.) characteristics such as geometrical dimensions, noise and background count, sensitivity to the impinging electron energy, signal-to-noise ratio, gain, and resolution have to be considered. The detector must have a surface capable of accepting a high electron flux, yet must be small enough so as to exhibit a low noise and background count and high signal-to-noise ratio.

Pulses from the detectors are fed into the electronic circuitry, amplified, and scaled. For special applications, such as measurements in space research, the pulses are modulated at convenient frequency, telemetered, and finally demodulated by a receiver. Both transmitter and receiver work within the range of special frequencies.

The response of the detector to scattered electron stimulation is shown in the characteristic curve in Fig. 4.18 which depicts in a log–log plot the dependence of the count rate on the density (solid curve). On this curve the density ranges can be arbitrarily delimited, i.e., range (B) corresponding to average densities, with a low-density (A) and high-density (C) range on either side. Within the range B, the curve is linear. With increasing density, double scattering and energy attenuation phenomena are more prevalent than single scattering. The count rate, which is no longer proportional to density, reaches a maximum strongly dependent upon the geometrical parameters of the measuring setup and then decreases. In region A, noise and background count in the detector grow higher than the count rate owing to single scattered electrons, and the characteristic curve again departs from linearity.

Spurious counts resulting from electrons generated under particular air circumstances can also alter the characteristic curve. For example, in applications where atmospheric densities are to be measured (Schuma-

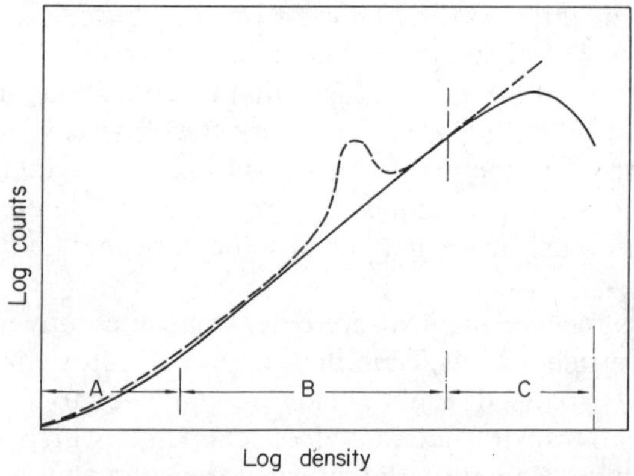

Fig. 4.18 Characteristic curve of count (or pulse, or current) response versus density.

cher *et al.*, 1967), electrons generated in shock waves or by electrical charges on the surface of the vehicle carrying the gauge may be scattered to the detectors and counted, thus giving inaccurate results. Strong or weak magnetic fields can also alter the path of the main beam. Scattering gauges (see Section 4.4.4) are designed for both general vacuum work and space sounding applications (Schumacher *et al.*, 1967; Ballard *et al.*, 1969). The range of densities of scattering gauges extends from 1.5×10^{-2} to 1.5×10^{-7} kg/m^3.

(d) *Light scattering.* Light of wavelength λ_0 which crosses a gas volume is scattered in any direction making an angle θ with the incident direction. The intensity of the scattered light is directly proportional to $(1 + \cos^2 \theta)$ and inversely proportional to λ_0^4 (Rayleigh scattering).

Marton (1971, 1974) proposed the use of *molecular counting,* relying on Rayleigh scattering of coherent visible light emitted from a laser, as the basis of a *primary pressure standard.*

The relationship between the total number N_s of scattered molecules and the intensities of scattered (I_s) and incident (I_0) light, expressed by rewriting the original Rayleigh equation as

$$I_s/I_0 = (8\pi/3)[(2\pi/\lambda_0)^4 N_s \alpha_g] \tag{4.17}$$

shows that light scattering can be used for molecular counting. In Eq. (4.17) α_g is the polarizability of gas molecules.

The pressure range of a molecule-counting instrument based on the principle described is about 15 orders of magnitude. The signal-to-noise ratio of such an instrument would be fairly good provided that spurious sources of light scattering, such as produced by particulate matter within the gas or the damage of the measuring system surface, are eliminated. Photodesorption of the instrument walls due to the light beam impact and the uncertainty introduced by the need to express the scattering volume exactly are the main difficulties of the method.

Marton's (1974) proposal for an instrument using light scattering for vacuum measurement is described in Section 4.4.5. No further theoretical progress or practical achievement has yet been reported.

(e) *Ionization.* Charged particles emitted by a suitable source within a gas volume collide in their passage with gas molecules and produce ionization, if conditions are conducive. The quantity of ionized particles depends upon both the kinetic energy of the ionizing particles and the ionization potential of the gas and provides information about molecular concentration. Ions created in such a process are accelerated by an electric field towards an electrode where they are collected, thus setting up an electric current in an outer circuit. This current is directly proportional to the gas density as long as no other phenomena interfere. If the temperature of the gas inside the ionization volume is known, density information can be used to calculate pressure. The output response can be considered instantaneous.

Conventional ionization gauges use electrons emitted either by a hot filament (thermionic gauges), a Si–C *p–n* junction, or an electrical cold discharge as the ionizing agent. In gauges using radioisotopes, also called radioactive gauges (referred to as radiological gauges), ionization is induced by a stream of alpha or beta particles emitted by a radioactive source.

4.4.2 The Fluorescence Gauge

Electrons emitted from a tungsten filament (electron energy about 100 eV) (1 in Fig. 4.19), shaped in a beam by a collimator (5), cross the gas and excite the gas atoms. The photons emitted, which have wavelengths in the ultraviolet region of the spectrum, pass through a lithium fluoride window (2) and enter a Geiger–Muller photon counter (3) where they are counted. A small glass bead (4) fused at the end of the central wire prevents electrical discharges from the point of the wire. The gauge is evacuated by an ion pump, the current drain of the pump being used as an indication of the pressure in the gauge. The lowest pressure measured in this way was 6.6×10^{-6} Pa (5×10^{-8} Torr). Alexeff (1960), who designed this gauge, measured the pressure of air, Ar, H_2, CO_2 and O_2.

The choice of a Geiger–Muller counter instead of a photomultiplier tube is justified by the fact that the Geiger–Muller counter is not sensitive to visible light, to the light coming from the electron source, or to thermal

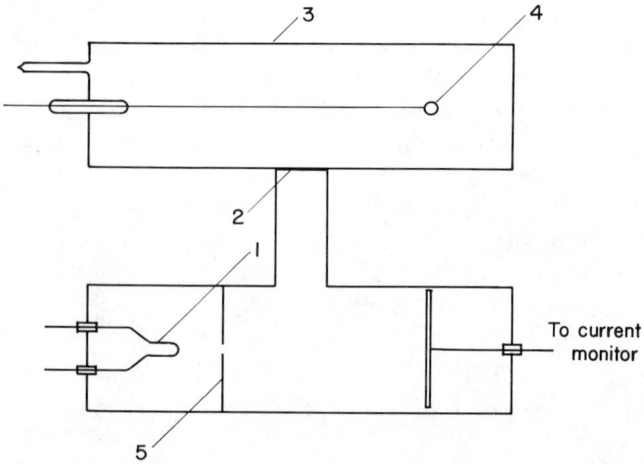

Fig. 4.19 Schematic diagram of an ultrahigh vacuum manometer. 1, tungsten filament; 2, lithium fluoride window; 3, photon counter; 4, glass bead; 5, collimator.

noise. The background counting rate which is observed when the electron beam is not present is due to natural radioactivity and cosmic rays.

The instrument which *theoretically* can measure degrees of gas rarefaction lower than 10^{-11} Pa (10^{-13} Torr) has also been tested as a leak detector.

The schematic diagram of a fluorescence gauge for pressure measurement in a wind tunnel is illustrated in Fig. 4.20. The fluorescence of gas is produced by high-energy electrons (60 keV, 200 μA). The light output of the gauge is not affected by the gas flow and is linear with pressure, at

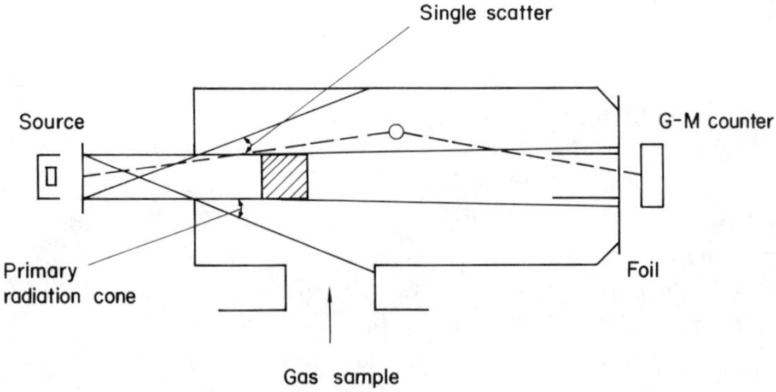

Fig. 4.20 Fluorescence gauge. Reprinted with permission from Nucleonics **18,** 106. B. W. Schumacher. Copyright 1960 McGraw-Hill.

pressures smaller than 133 Pa (1 Torr). The lower limit of measurement for air is about 10^{-3} Pa (10^{-5} Torr).

4.4.3 The Attenuation Gauge

Alpha particles emitted from a source (Fig. 4.21) enter an ionization chamber which accommodates the gas whose density must be determined. The energy of alpha particles which penetrate the ionization chamber varies according to the wedge thickness. As gas density in the ionization chamber is reduced, more alpha particles would cross the wedge in front of the Geiger–Muller counter. The gauge proposed by Schumacher (1960) was not built at the time it was suggested.

Sources emitting beta particles instead of alpha particles are also used (Fig. 4.22). Under suitable conditions, the exponential reduction in the counting rate of particles applies [Eq. (4.16)], and gas density can be continuously determined.

4.4.4 The Electron-Scattering Gauge

A beta-ray source (S in Fig. 4.23) emits electrons which, after traversing a thin window (T), are formed in a beam within a collimator channel (C). A beam stop (B) prevents electrons in the beam from directly reaching the detector (D) of the Geiger–Muller type. Electrons scattered from the beam reach the detector and are counted.

Instead of a beta source, an electron gun can be used to provide electrons with energies higher than that of electrons emitted from a radioac-

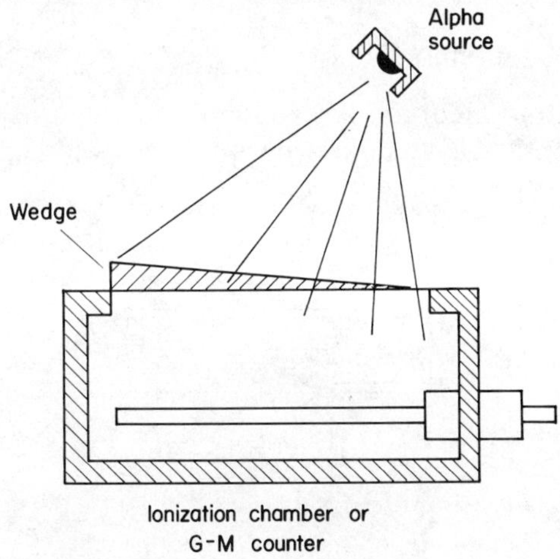

Fig. 4.21 Alpha-range gauge. Reprinted with permission from Nucleonics **18,** 100. B. W. Schumacher. Copyright 1960 McGraw-Hill.

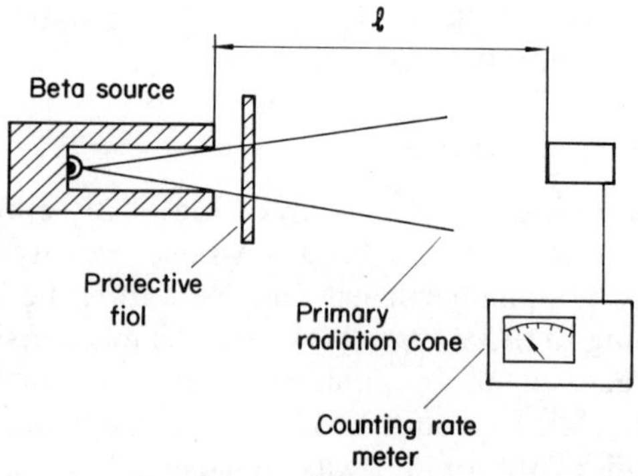

Fig. 4.22 Attenuation gauge. Reprinted with permission from Nucleonics **18,** 110. B. W. Schumacher. Copyright 1960 McGraw-Hill.

tive substance. Detection conditions become improved, but the probability of electrons of being scattered for a certain angle decreases.

The scatter count for constant primary current (10^{-12} A), when electron beams of 40, 60, or 80 keV were used, was a linear function of pressures between 10 and 10^{-3} Pa (10^{-1}–10^{-5} Torr). When electrons were emitted from a ^{147}Pm source (about 18.5 Bq, 500 mC) the useful range was from 20 to 10^{-1} Pa (0.2 to 10^{-3} Torr) (Schumacher, 1965).

Figure 4.24 illustrates the schematic diagram of a parachute-borne beta-ray densitometer for the determination of atmospheric densities in the 30–60-km altitude region. Electrons are provided by ^{147}Pm (1.12×10^8 Bq, 80.4 mC).

4.4.5 The Light Scattering Gauge

The radiation from an argon laser (20 W) operated in the CW mode is expanded by an optical system in order to avoid polarization of gas mole-

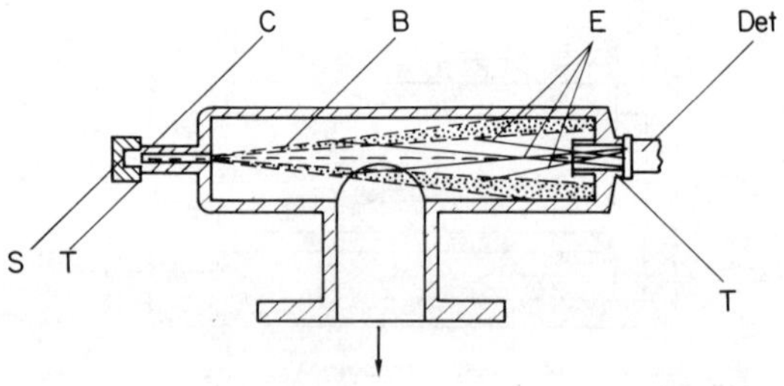

Fig. 4.23 Electron-scattering gauge. B, shield; C, collimator; Det, Geiger-type detector; E, scattered electrons; S, source; T, thin window. (After Schumacher, 1960.)

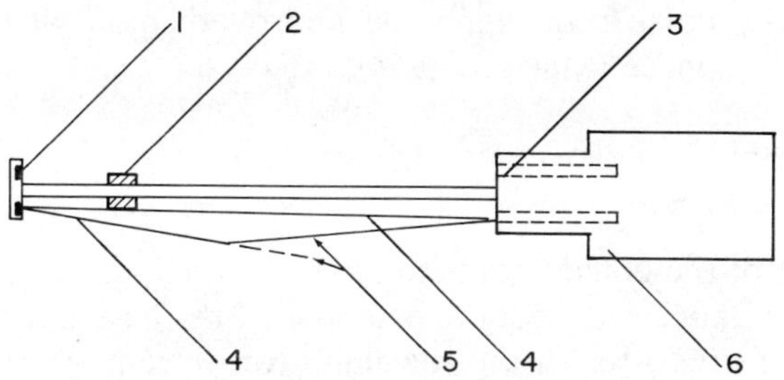

Fig. 4.24 Diagram of a parachute-borne beta-ray densitometer. 1, annular-shaped radioactive source; 2, shield; 3, annular array of Geiger–Muller counters; 4, typical scattered path; 5, scattering angle, 6, electronics. Reprinted with permission from *Int. J. Appl. Radiation and Isotopes,* **20,** 343. B. Sellers *et al.* Copyright 1969 Pergamon Press.

cules by the primary beam. The beam thus obtained passes through a window in a measuring cavity made of quartz (Fig. 4.25). The cavity is surrounded by an integrating sphere at which one or more highly sensitive photomultipliers are suitably located. A small fraction of the incident light is diverted by a beam splitter or another adequate device (not shown in Fig. 4.25) in order to measure the intensity of the beam I_0. The intensity I of the scattered light is measured with photomultipliers. Once the ratio I/I_0 is determined, the total number of scattering molecules N can be calculated [Eq. (4.17)]. After traversing the measuring cavity, the beam enters a light trap, possibly a Raman-type horn.

The output of the system varies linearly with gas density, which permits extrapolation of gas densities from atmospheric pressure down to extremely low degrees of rarefaction assuming that sources of error, such as the accuracy of the photometric determination of I and the reflectivity of the cavity walls and of the window through which the laser beam enters the cavity, are reduced to an insignificant level.

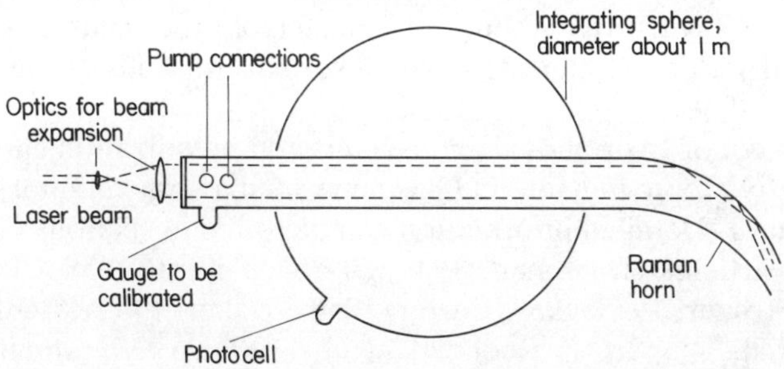

Fig. 4.25 Schematic diagram of the light-scattering gauge. (After Marton, 1974.)

The signal-to-noise ratio might be improved upon by using a pulsed laser as a light source (Marton, 1964).

4.4.6 General Considerations of Ionization Gauge Problems

As a result of the considerable number of studies that have been done, the ionization gauge has become established as the universal instrument for measuring pressures in the ultrahigh (or extreme ultrahigh) vacuum range. However, it is by no means the ideal instrument, since determinations depend on the nature of the gas and are affected by all kinds of inaccuracies.

Thus, during the operation of an ionization gauge head, the gas equilibrium and gas species are modified as a result of rather complicated processes. These include, according to the type of instrument (hot or cold cathode), electronic and chemical pumping and dissociation and desorption of the gas previously removed. The energy for the activation and maintenance of these processes is that which is dissipated in the electrical discharge that takes place within the gauge head. Most of the phenomena likely to be involved are primarily dependent on the chemical composition of the gas measured.

Much experimental work has been conducted by using inert gases since they are neither chemically nor physically sorbed at room temperature. Ionic combination of inert gases has been reported (Loeb, 1958), and modern techniques make use of inert gas fluorides in excimer lasers. Such combinations, however, are unlikely to occur at the energies dissipated in conventional gauge heads.

Chemically active gases complicate pressure measurements, rendering the analysis of the results difficult. Thus, hydrogen dissociates at hot tungsten cathodes, and the resulting atoms evaporate and settle on the surfaces of the instrument, producing an anomalous pumping speed and contaminant species such as CO, CH_4, and H_2O, (Hickmott 1960, 1965). Oxygen reacts with hot cathodes of tungsten or other materials containing carbon and yields CO and CO_2, which sorb at the walls of the gauge head (Singleton, 1966).

The subject of gas removal and re-emission in ionization gauge heads is an exhaustive topic in itself and has generated a large amount of research and literature. Detailed information can be found in sections of books and in review articles (Dushman, 1949, p. 704; Alpert, 1958, chap. 7; Von Gerhard Strozer, 1958, 1959; Carter, 1959a; Baker, 1962a; Redhead, 1975, Vol II, chap. 4, part 7; Redhead *et al.,* 1962, p. 323, 1968, chap. 6). Several simplified assumptions, not always experimentally justified, have

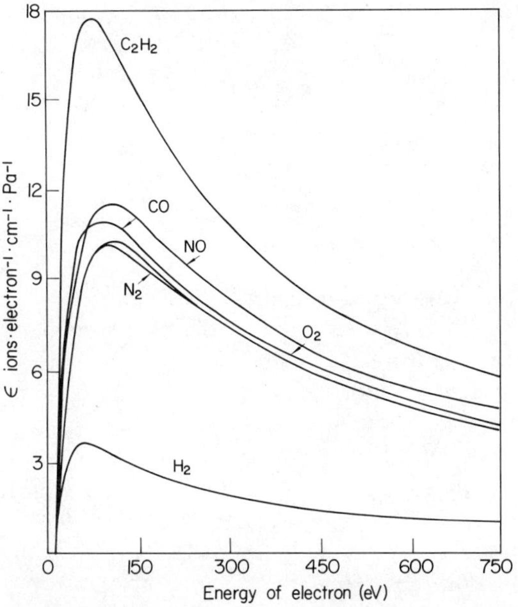

Fig. 4.26 The ionization probability for several gases. (After Tate and Smith, 1932.)

been advanced to explain the processes controlling the behavior of ionization gauges as both sinks and sources of gas. At present the effects associated with these processes are quite well known. However, a picture on an atomic scale of the physical phenomena underlying these processes is not yet available (Grant and Carter, 1965).

In addition to these problems, under certain circumstances the ionization gauge head generates currents which are not related to the pressure measured and may produce electronic oscillations leading to the instability and inaccuracy of readings. It is therefore useful to discuss the main attendant problems of these gauges in order to realize their reliability and accuracy.

(a) *Sensitivity.* If electrons are the ionizing agent, their energy must be greater than the ionizing potential of the gas. The number ε of positive charges produced per electron per centimeter path per pascal of gas pressure at 0°C (referred to as the ionization probability) depends on the gas species and the electron energy, as illustrated in Fig. 4.26.

For *hot-cathode gauges,* which use a *measurable, controllable, pressure-independent electron current* I^- (referred to as emission current) to ionize the gas, the ionization current I^+ and the gas pressure P are related as

$$I^+ = I^- l\varepsilon P \tag{4.18}$$

Here l is the length travelled by an electron to produce ionization. The *absolute sensitivity* s such as expressed by Eq. (2.16) results by noting that

$$s = l\varepsilon \tag{4.19}$$

and by substituting this value in Eq. (4.18).

The absolute sensitivity s depends on the electron energy, the gas species, and their temperature (through ε), as well as the gauge geometry (through l). The gauge geometry determines, in turn, the field distribution and ion collection efficiency.

For ionization gauges where the *ionizing current is not measurable and is pressure independent,* such as in *cold-cathode gauges,* sensitivity is expressed by Eq. (2.17).

Attempts to calculate the absolute sensitivity s from the cross section for ionizing collisions by electrons, for both hot- and cold-cathode gauges, have not proved useful because of a number of parameters, such as the gas species and temperature, the field distribution, and the ion collection efficiency (Nottingham, 1962; Schuetze and Stork, 1962b, Kornelsen, 1966; Lange and Eriksen, 1966; Groszkowsky, 1966b, 1968, 1969a,b; Pelz and Newton, 1967, Redhead *et al.*, 1968, p. 131; Redhead, 1969; Lafferty, 1972). Auger emission at the collector has also been found to be a reason for changes in the absolute gauge sensitivity (Gopalaraman *et al.*, 1970).

The value of the absolute sensitivity s can be measured by calibrating the gauge against a pressure standard (see Section 7.3). The importance of parameter s can be understood upon considering pressure measurement below 10^{-6} Pa (10^{-8} Torr) and above 10^{-1} Pa (10^{-3} Torr). At pressures of 10^{-6} Pa (10^{-8} Torr) or below, the degree of gas rarefaction is such that ionizing collisions occur only once in about 50,000. Pressure can be measured [Eq. (2.16)] by increasing either I^- or s. The increase of the emission current is, however, limited by the power dissipated in the cathode which in turn can contribute other constraints (see following paragraphs). An increase of the absolute sensitivity for a certain gas involves substantially lengthening the path l travelled by electrons before making a collision [see Eq. (4.19)]. This can be done either by using electrostatic fields or by combining such fields with magnetic fields, which will cause electrons to oscillate or spiral up within the ionizing volume. Both approaches are made possible by carefully designing the gauge head. At pressures above 10^{-1} Pa (10^{-3} Torr) the path length l of electrons decreases, becoming comparable with the dimensions of the gauge head. The number of ionizing collisions is reduced because electrons are scattered to other electrodes. Since ions may also be scattered to other electrodes, the ion current I^+ collected represents only a fraction of the total current produced. The upper pressure limit of conventional gauges (about 10^{-1} Pa, 10^{-3} Torr) can, however, be extended to approximately 100 Pa (1 Torr) by modifying the basic design of the gauge head.

The unit of the absolute sensitivity is 1/Pa (1/Torr). Sometimes the ratio of the ion current to the corresponding measured pressure (referred to as the *conversion constant*) is used as the unit of s (A/Pa, A/Torr).

(b) *Linearity.* When properly operated the output of the hot-cathode ionization gauges is linear with the pressure measured over their useful range. The output of cold-cathode gauges is proportional to pressure raised to a power approaching unity as long as erratic behavior does not develop. (For the determination of linearity, see Section 7.2.)

(c) *Factors affecting the characteristics of the gauge head.*

(i) *Sinks of gas.* In an ionization gauge head, gas particles activated by the adsorption of electrical or thermal energy strike the electrodes and the glass walls, and if conditions are conducive, the particles are driven to the interior of these surfaces. There the particles become trapped after being neutralized. The mechanisms which control the transport of gas to the surfaces are *electronic pumping with its two components, ionic and activated chemical pumping, chemical pumping,* and for hot-cathode gauge heads, *pumping at the incadescent cathode.*

We first consider *electronic pumping.* The available experimental evidence proves that ions produce the so-called ionic pumping. This is the mechanism by which inert gases are removed in ionization gauge heads and transported to the surface of the electrodes and glass walls as ions. Since the surface area of the electrodes is small compared to that of the walls, the largest part of the ionic current will be directed to the walls. The number of ions which goes to the walls is a function of several parameters, such as the region in which they are formed, the design of the electrodes, and the geometry of the gauge head. A mixture of inert and chemically active gases, or chemically active gases alone, is pumped by a combination of ionic and activated chemical pumping mechanisms. Instead of ions, metastable atoms or molecules have also been considered as the means of pumping inert gases (Carleton and Bills, 1955; Bloomer and Haine, 1953).

Various theories have been advanced to explain ionic pumping. They may conveniently be considered as belonging to two schools of thought, one regarding ionic pumping as a surface mechanism, the other considering that ions are embedded in the lattice at the surface of the material. Both theories admit the existence of discrete capture sites at which ions may adsorb. These sites, distributed all over the internal surface areas of a clean gauge head, exhibit various bond strengths (expressed in terms of energies of adsorption). Upon impinging on a vacant site, an ion can be captured or returned to the gas with or without retaining its activity. Upon impinging on a filled site, an ion can set free a neutral atom or molecule formerly trapped and take or not take its place.

The surface theory postulates the existence of discrete capture sites present over the entire internal surface of the gauge head (Carleton and Bills, 1955; Bills and Carleton, 1958).

The theory whereby gas is removed by being embedded in the lattice describes the penetration of singly or multiply charged ions to a depth of a few atomic layers. The probability that ions penetrate and are trapped as neutral atoms is a function of the interatomic forces between the incoming ions and the surface as well as of the ions energies. On entering the lattice, ions transfer momentum to the neighboring atoms, which results in vibrations, point defects, or ejection of atoms from the lattice (sputtering). The incoming ion then comes to rest and becomes trapped by forces which have been considered electrostatic (Redhead *et al.*, 1962, p. 323). Predictions of this mechanism as well as experimental evidence have been presented for both glass and metals (Young 1956; Carmichael and Knoll, 1959; Blodgett and Vanderslice, 1960; Leck and Carter, 1960; Carter and Leck, 1961a,b; Cobic *et al.*, 1961a; Kornelsen, 1964; Grant and Carter, 1966a). Varnerin and Carmichael (1955a,b, 1957) postulated that a metallic layer on the glass walls hinders trapped He atoms from being spontaneously released. However, Grant and Carter (1967b) proved experimentally that He is set free from glass almost as soon as it enters and that there is no reason why He ions should penetrate a thin layer of tungsten less easily than glass. Baker (1962a) postulated that the trapped particles are prevented from escaping from glass as a result of the lattice damage due to the bombardment by incoming ions. A slow annealing out of the point defects increases spontaneous gas desorption, resulting in a so called secondary cleanup. An alternative theory to that presented here is that ions (especially of He) which bombard glass targets become embedded in cracks or defects in the surface where they are trapped (Carter, 1959a).

However, no experimental evidence has been obtained to date to give full support to one mechanism or the other. Practically, it seems likely that each mechanism contributes a certain share according to various parameters, such as the nature of the gas and the energies involved in the electrical discharge.

The mechanism of ionic pumping, however, cannot explain the high pumping speed of nitrogen in an ionization gauge head. Carter *et al.* (1962b), Cobic and Carter (1962), and Carter (1963) attributed these large volume rates to a chemisorption mechanism. Jaeckel and Teloy (1962) considered that the activated particles impinging on the solid surfaces are molecules of nitrogen in a metastable state and not ions. In their experimental work they used a hot-cathode gauge head of special design, operated with accelerating potentials between 0 and 35 V.

The amount of gas pumped by the glass walls of the gauge due to the ionic mechanism is several times larger than that pumped by other glass parts or by the collector. The ionic current to the collector is negligible compared with that to the glass envelope (Young, 1956; Comsa and Musa, 1957a). Ishikawa (1965) estimated the pumping speed to the collector of a Bayard–Alpert gauge (BAG) head for Ar, He, and Kr as being two or three decades lower than the total pumping speed.

Ionic pumping is a complex function of various parameters such as the species of gas present and their partial pressures, the accelerating potential, the design of the gauge head, the materials used for the construction of the gauge head, the surface area of both the electrodes and the glass walls, and the temperature of the gas and enclosure.

The *activated chemical pumping* occurs as a result of the removal by chemisorption of the gas particles (particularily chemically active gases) excited to metastable states. According to Jaeckel and Teloy (1962) such particles are able to adsorb whereas neutral molecules cannot.

Chemical pumping is the mechanism by which gas is removed as the result of exchanges of electrons between reacting atoms that have only thermal energies (Bloomer, 1960). The bonding is usually stronger than that produced by van der Waals forces.

The chemisorption of the chemically active gases (e.g., H_2, N_2, and O_2) at the metallic electrodes produce pumping rates exceeding those which result from ionic pumping. Hobson (1961) found that a BAG head operated for the first time in ultrahigh vacuum exhibited a pumping speed of 2 dm^3/s (2 L/s) for N_2. This pumping speed originates at the grid and decreases nearly to zero after the removal of about 10^{15} molecules of gas.

The chemical pumping which continues even after switching off the emission current may greatly exceed the electronic pumping under certain conditions.

Pumping at the incandescent cathode may also control transport of gas. When chemically active gases such as hydrocarbons or water vapor are present within a hot-cathode ionization gauge head, their removal from active circulation may occur via chemical reactions with the impurities present in the cathode (these reactions result in solid compounds), dissociation into highly reactive fragments which, in turn, may either sorb (Alpert, 1962) or react with other gases and thus yield new gaseous products. This process usually alters the actual sensitivity of the gauge head (Werner and Leck, 1969).

Pumping in a mixture of gases, particularly containing hydrocarbons, is discussed in Section 2.3.2 as the Blears effect.

(ii) *Sources of gas.* Aside from leaks and permeation of gas (especially He) through the glass envelope (Alpert, 1953), the residual gas within the gauge head is modified as a result of the desorption of gas previously pumped electronically. The trapped particles can return to the gas phase in the following ways: by *spontaneous gas desorption,* a process involving both diffusion and desorption, by *induced gas desorption,* and by *chemical reactions*.

In *spontaneous gas desorption,* the trapped gas remains bound at the adsorption sites for a certain time, expressed by a parameter called *mean residence (sojourn) time*. This parameter describes the storage aspect of the pumped particles and depends on both the activated energy for desorption and the temperature of the surface at which gas is trapped. A typical plot of the sorption and spontaneous desorption of Ar in a BAG head is shown in Fig. 4.27. The spontaneous gas desorption is manifested

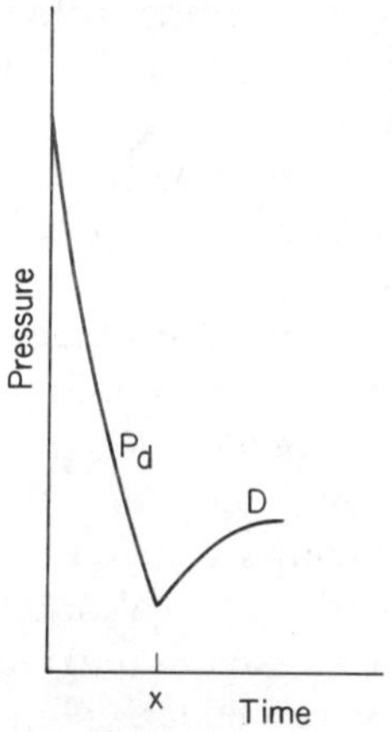

Fig. 4.27 The pump-down (P_d) and spontaneous desorption (D) curves of argon produced in a Bayard–Alpert gauge head in a sealed-off system. At time x the accelerating voltage is switched off. Reprinted with permission from *Brit. J. Appl. Phys.* **12,** 385. B. Cobic *et al.* Copyright 1961 The Institute of Physics.

by the increase in pressure with time which occurs after switching off the accelerating potential (X in Fig. 4.27). The gas re-emitted as a result of the spontaneous desorption starts affecting the performance of the ionic pumping before saturation in pumping is reached. Spontaneous gas desorption has been studied for both glass and metal surfaces (Carmichael and Trendelenburg, 1958; Robinson and Berz, 1959; Baker and Giorgi, 1960; Cobic *et al.*, 1961a; Blodgett and Vanderslice, 1962; Smeaton *et al.*, 1962; Cavaleru *et al.*, 1964).

Induced gas desorption is another mechanism whereby trapped particles can return to the gas phase. Under certain conditions, energetic particles and thermal radiation emitted from the incandescent cathode may release gas particles trapped on the glass walls or on metallic parts of the gauge head. The desorption of gas products induced by energetic particles results in both molecular and ionic products, which cause uncertainties in the measurement of the pressure in ultrahigh vacuum.

Electron impact desorption (EID, or electronic desorption) occurs in hot-cathode ionization gauge heads when energetic electrons, which impinge on the gas layer adsorbed on the electron collector, release neutrals, ions, atoms, and fragments (Redhead *et al.*, 1968, chap. 6; Redhead, 1970). The mechanism is initiated by the electron excitation or dissociation of the molecules at the surface of the layer. The gas species desorbed are CO, CO_2, H_2, O_2, water vapor, halogens, and hydrocarbons. The number of neutrals desorbed is large compared with that of ions. Redhead (1963) measured about 1% O^+ set free when a molybdenum grid covered with oxygen is bombarded with electrons. This mechanism contributes also an output signal to the ion collector.

In cold-cathode gauge heads, where the electron current to the anode is proportional to the pressure, the number of positive ions released from the collector is reduced. This positive current influences to a certain extent, however, the striking characteristics of the electrical discharge.

The pressure at which EID phenomena become troublesome can be predicted for hot-cathode gauge heads if their absolute sensitivities are known. The efficiency η_{EID} of EID is defined as the ratio of ions produced I^+_{EID} to the emission current I^-:

$$\eta_{EID} = I^+_{EID}/I^- \tag{4.20}$$

Equating the current I^+_{EID} with the current produced in the gas phase I^+ [Eq. (4.18)] gives

$$P = \eta_{EID}/s \tag{4.21}$$

The efficiency η_{EID} for the case of oxygen fully adsorbed on molybdenum is about 10^{-5} (Redhead, 1963). Thus, for a conventional BAG head with a sensitivity $s = 0.09$/Pa (12/Torr), the EID current will equal the ion current produced in the gas phase at $P = 1.1 \times 10^{-4}$ Pa (8.3×10^{-7} Torr).

The following techniques for minimizing the errors in pressure measurement with hot-cathode gauges are recommended (Redhead, 1975, p. 281): the use of large values of the emission current, the use of modulation techniques, the use of specially designed gauge heads such as the extractor gauge or the bent-beam gauge, the use of a mass spectrometer, and the proper selection of materials for the construction of the electron collector.

Ionic replacement (bombardment-induced release of particles) is the mechanism by which the positive ions of a gas which strike the glass envelope of a gauge head set free atoms trapped in a previous pumping operation. On investigating the pumping of Ar, H_2, and N_2 in a conventional triode, Schwarz (1944) found that the gas sorbed on the glass walls in an earlier operation was re-emitted into the gauge when a different gas was subsequently pumped. He attributed this effect to a replacement phenomenon caused by the energetic ions impinging on the envelope. After pumping N_2 until saturation was obtained, the gas was evacuated and replaced by Ar, and the gas discharge was again activated. The adsorbed nitrogen was released and replaced at the walls by argon. This replacement was found to be reversible and led Schwartz to conclude that the ionization gauge head is an unreliable instrument for measurement below certain pressure limits.

Several research workers explain the mechanism of the ionic replacement on glass surfaces as a result of the momentum transferred from the impinging ions (Brown and Leck, 1955; Carter *et al.*, 1962a; James and Carter, 1962a,b, 1963; Grant and Carter, 1967a). According to Carmichael and Trendelenburg (1958), positive ions which impinge on metal surfaces may release trapped atoms by the sputtering of either gas atoms or the target, or both. Despite these explanations the process by which the bombarding ions transfer energy to the trapped atoms so as to release them is not yet completely understood (Grant and Carter, 1966b). The ionic replacement in nickel and tungsten has been investigated by Carmichael and Waters (1962) and Kornelsen (1964).

The infrared (IR) radiation emitted from the hot cathode is adsorbed by both the glass and the metallic surfaces in the gauge head. The thermal energy developed (mainly in glass) promotes the desorption of the trapped gas (Edelmann, 1966). Elevated temperatures of the glass envelope assist both high rates of permeation and release of water vapor. Heated metallic electrodes or metallic layers contribute in a complex way to gas desorption and chemical reactions. Thermal effects of the type quoted become important in gauge heads where the hot cathode is assymetrically located (e.g., BAG heads).

Ultraviolet (UV) and visible radiations from the cathode may generate photoelectrons at the ion collector under certain circumstance (Redhead, 1961; Schwarz, 1961).

Chemical reaction at the hot cathode may also occur. This kind of reaction occurs only in the operation of hot-cathode gauge heads. They are produced either as a result of thermal dissociation of gas or vapor at the hot cathode and further recombination of the resulting fragments with impurities or by the combination of gas with the hot-cathode material. Gases and vapor usually present in measuring systems are oxygen, hydrogen, and water vapor.

Oxygen molecules in contact with the hot tungsten cathode ($T > 2000$ K) dissociate to atomic oxygen (Singleton, 1966). Atomic oxygen reacts with the carbon impurities of the cathode to yield CO and CO_2 (Rhodin and Rowner, 1961). CO is continuously generated as a result of carbon at the surface of the cathode diffusing from the bulk. The parameters which control this replacement are the temperature of the cathode, the partial pressure of oxygen in the residual gas ($<1.3 \times 10^{-4}$ Pa, $<10^{-6}$ Torr), and the reaction time (10–60 h) (Becker *et al.,* 1961). Assuming that the carbon impurity in the cathode is consumed due to the reaction discussed, atomic hydrogen still combines with carbon impurities present on the walls of the gauge head, yielding CO.

Oxygen atoms interact with the tungsten cathode and produce WO_2 and WO_3 as surface species, which, depending on temperature, interact with each other to yield W_2O_6 and W_3O_9 (Schissel and Trulson, 1965).

Hydrogen molecules dissociate at high rate to atomic hydrogen in the presence of a tungsten cathode ($T > 1000$ K). Atomic hydrogen reacts very quickly with glass or metal surfaces, yielding CO, H_2O, and CH_4 (Hickmott, 1960). After adsorbing a monolayer of atomic hydrogen, the work function of the tungsten cathode increases, and the emission current decreases (Abey, 1969). Operation of a tungsten cathode at low emission current (hence low temperature) as suggested by Hickmott (1965) has the drawback of contributing to the EID mechanism (Singleton, 1967).

The use of lanthanum boride or thorium oxide cathodes substantially reduces the chemical reactions arising from the presence of hydrogen in the residual gas. Lanthanum boride cathodes, however, are sensitive to oxygen poisoning especially at temperatures in excess of 1300 K (Gallagher, 1969; Avidenco and Malev, 1977).

Water vapor dissociates at the hot tungsten cathode to oxygen and atomic hydrogen initiating a vapor cycle (Dushman, 1949, p. 679). Oxygen reacts with tungsten, the tungsten oxide yield being deposited on the glass walls of the gauge head in a thin film. Atomic hydrogen reduces the film, thus setting free water vapor, which continues the cycle. This results in an increase of the tungsten layer on the glass walls and a subsequent reduction of the tungsten cathode thickness. The cathode lifetime is dramatically reduced. The phenomenon in vacuum is known as *water vapor cycle* (Dushman, 1949). Garbe *et al.* (1960) who studied the pressure of water in the residual gas in electronic tubes, found that CH_4, CO, and CO_2 are also produced as a result of the chemical reaction of water with hot cathodes.

Contamination effects due to chemical reactions of the type discussed can be minimized by enclosing the hot cathode within a metallic sheath which does not permit the combination of atomic species previously generated, reducing the carbon impurity in the cathode by heating it in hydrogen (partial pressure about 10^{-5} Pa, 10^{-7} Torr) at about 2000 K for 20 h, and the utilization of low work function emitters (see Section 4.4.6).

(iii) *Residual currents.* Residual currents (referred to as *background currents*) are those which would exist at the collector electrode if the molecular density within the gauge head were zero. Such currents occur in gauge heads where the *electron current* which produces ionization *is independent of gas pressure* (hot-cathode gauge heads) and are produced by the photoemission of soft x rays (forward and reverse x-ray effect) and EID.

The *forward x-ray effect* is considered first. The lowest pressure measured with conventional gauge heads of the triode type before 1950 was about 10^{-6} Pa (10^{-8} Torr), although the thermionic and photoelectric characteristics of electronic devices which were evacuated indicated that much lower pressures were attained. Nottingham (1947) suggested that the current in the gas phase to the ion collector, proportional to pressure, is obscured by a residual current also collected at the ion collector. This *pressure-independent current,* is produced as follows.

Electrons emitted from the cathode, which strike the grid (electron collector), generate soft x-ray photons. A *small fraction* of these x-ray photons collide at the ion collector and in turn release photoelectrons which are accelerated to the grid. *The resulting current at the ion collector, of positive sign* is superimposed on the ion current produced in the gas phase, masking it at pressures lower than 10^{-6} Pa (10^{-8} Torr). The two currents are indistinguishable from each other in the outer circuit. Bayard and Alpert (1950) overcame the pressure limitation due to the forward x-ray effect by modifying the electrode geometry. The Bayard–Alpert gauge head can measure pressures two orders of magnitude lower than the conventional triode.

In the *reverse x-ray effect,* only a small fraction of the soft x-rays produced at the grid is intercepted at the ion collector. The remaining fraction is intercepted by the envelope of the gauge head. Two situations are then possible (Hayward *et al.,* 1963).

First, the potential of the envelope (usually glass) is near that of the cathode, and the ion collector is operated at 30–45 V negative with respect to the cathode. Photoelectrons emitted from the envelope do not have enough energy to reach the ion collector. They are collected either at the grid or recollected at the envelope.

Second, the envelope (glass covered with a thin electrically conductive layer or made entirely of metal) is at ground potential and the ion collector is operated at the same ground potential. Photoelectrons emitted from the envelope possess enough energy to reach the ion collector. The resulting current at the ion collector is negative and is comparable with or larger than that yielded by the forward x-ray effect. The representative path of photoelectrons in a gauge head with grounded metal envelope is illustrated in Fig. 4.28. A bias of about 10 V between the ion collector and the envelope may prevent the collection of the reverse x-ray current.

The pressure at which the ability of the gauge head to measure the ion current in the gas phase is limited by x rays is referred to as the *x-ray limit* of the gauge. (For x-ray limit determination, see Section 7.4.)

In *electron impact desorption* (EID), in addition to gas desorption from the electron collector, positive ions released from this electrode which reach the ion collector contribute a signal not related to gas pressure. Redhead (1962) found that the x-ray limit of the BAG head is increased by as much as a factor of 400 owing to the gas species chemisorbed at the

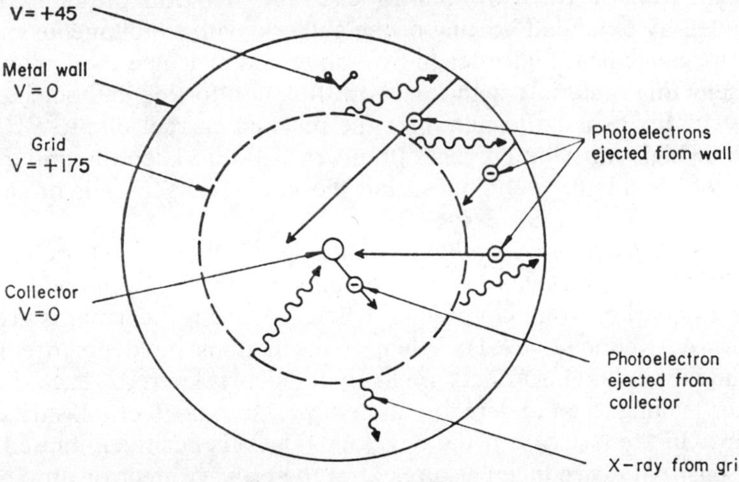

Fig. 4.28 Representative paths of photoelectrons in a gauge head with a grounded metal envelope. (After Hayward *et al.*, 1963.)

electron collector. The electron bombardment of an adsorbed layer of oxygen on the electron collector resulted in an anomalous residual current at the ion collector but had little or no influence on the measurement when the modulation technique was used (Redhead, 1963). By using mass spectrometric techniques, Hartman (1963) found that the anomalous current to the ion collector of a BAG head is produced by oxygen ions released by the EID mechanism. At pressures lower than about 10^{-8} Pa (10^{-10} Torr), the residual current to the ion collector due to EID is frequently larger than the x-ray current, thus becoming the limiting factor of low pressure measurement (Redhead and Hobson, 1965).

Singleton (1967) found errors produced by hydrogen desorption due to EID and pointed out that operation of the gauge head at 10-mA emission current appreciably reduces the errors produced by the residual current. For accurate results, Singleton recommended the use of modulation techniques. As a result of a theoretical analysis Lawson (1967) calculated that the material of the grid accounts for the anomalous currents yielded by EID in BAG heads. His experimental results are consistent with his theoretical conclusions, and he claims that the residual current due to EID can be eliminated by a correct choice of material for the grid, e.g., platinum-clad molybdenum wire.

Sodium or potassium, as well as other alkali metal impurities, coming from the hot tungsten cathode can adsorb at the electron collector and subsequently be released by EID. These ions contribute, in turn, an anomalous current to the ion collector (Denison *et al.*, 1963). The difficulties arising from the release of impurities evaporated from the cathode can be avoided by extended heating of the cathode with simultaneous pumping of the gauge head, in order to evaporate and evacuate the impurities, or by selecting materials with low work function for the cathode.

The techniques used to minimize the residual current due to EID are the use of high emission current (10 mA in BAG) to keep the grid clean, the use of modulator techniques, and the use of gauge heads of special design.

(iv) *High frequency oscillations.* Radio frequency (rf) oscillations of the *Barkhausen–Kurtz* type develop between the electron and ion collectors, under almost all conditions, when the electron current is greater than 1 mA (Redhead, 1961). When rf oscillations build up to a large amplitude the following effects are likely to occur: electrons reach the ion collector leading to an underestimation of pressure or electrons strike the envelope. In the last case if the envelope (glass) is positively biased, the gauge sensitivity is reduced as a result of the primary electrons travelling directly towards the walls rather than oscillating, their effective ionizing path being thus reduced, or secondary electrons which contribute to the measured electron current not initiating ionizations due to their low energies (Carter and Leck, 1959).

It is also possible that any rf output might be rectified in the electronic circuitry of the control unit, resulting in erratic readings.

The effects of rf oscillations can be minimized or avoided by controlling the potential of the gauge head envelope by covering it with a thin $SnCl_2$ film (Gomer, 1953; Redhead, 1961), housing the electrode structure within a cylindrical grid screen (Nottingham, 1955), decoupling the grid and the cathode with a capacitance (Pierre, 1961), lowering the potential of the walls from −30 to −90 V according to the type of the gauge head, or by locating asymmetrically the assembly grid cathode with respect to the ion collector and the envelope (Szwemin and Pitkowski, 1977).

Radio frequency oscillations also develop in all types of cold-cathode ionization gauges. They lead to mode changes in the discharge with discontinuities in the sensitivity and erratic jumps of the reading as the main consequence. The occurrence of rf oscillations is attributed to plasma (Conn and Daglish, 1953) for Penning gauges or to a rotating space charge in the cases of a cold-cathode discharge with magnetron geometry (Redhead, 1965) and a trigger gauge head of the Penning type (Poulter, 1972).

Radio frequency oscillations bear no relationship to pressure and are unaffected by changes in the circuitry of the control unit (Redhead, 1960).

(v) *The envelope of the gauge head.* The envelope which accommodates the electrode structure controls the pumping speed of the gauge head (see Section 7.5.1, the reverse x-ray effect) and to some extent the sensitivity. The main parameters contributing to this control are *the electric potential* and *temperature of the envelope* and *the gas permeation through the envelope.*

The electric potential of the envelope. An insulating material, bare or covered with a thin conducting layer, subjected to the simultaneous bombardment of ions and/or electrons, acquires an electric potential. The magnitude and polarity of this potential depends on various parameters, such as the competition between the buildup of both positive and negative charges, the secondary emission of electrons from the material, the pressure and composition of the gas in which the electrical charges are produced, the electron current, and the energy of the electrons which contribute to ionization. Such a situation is present in a hot-cathode ionization gauge head where the envelope constitutes an insulating target to energetically charged particles. Depending upon the specific conditions, the glass may stabilize to one of two potentials, one near that of the cathode potential, the other approximately that of the anode. The resulting bistability influences the rate of bombardment of the impinging ions and consequently the sensitivity and pumping speed of the instrument.

Schwarz (1940) noticed that the glass walls of his experimental tube stabilized to either negative or positive potentials, causing different pumping speeds. He attributed the dual stabilization of the potential of the walls to both ion and electron bombardment and to the diffusion of occluded gas on the surface of the glass. Bloomer and Haine (1953) noticed that no measurable ionic pumping occurred when the glass walls of their experimental gauge head were coated with a thin layer of Aquadag. It seemed likely that this layer acquired a positive potential, which inhibited the transfer of positive ions to the solid surface. Their experiment was confirmed by Young (1956) who applied the potential of the ion collector to the Aquadag layer. Bruining (1954) ascribed the duality of electrical potentials acquired by the glass walls of electronic valves to the emission of secondary electrons from these walls. The potential is negative or positive depending on whether it is the cathode or the anode which is switched on first. This behavior, known by the electronic valve manufacturers of the early forties as the S-effect (Schalt-Effect in German literature), occurred particularily when evaporated getter films on the glass walls favored the conduction of leakage currents. Comsa and Musa (1957b) calculated that the glass walls of the diode pump they developed acquire a potential of about 5 V negative with respect to the cathode. In all their experimental work, the cathode was switched on before activating the accelerating grid. They explained the electric potential assumed by the walls of the didode as a result of: the equilibrium between the current of ions (produced as a result of the thermal energy emitted from the cathode) and leakage currents on the glass; and the electron current.

Carter (1959b), on performing experiments with a diode pump similar to that of Comsa and Musa, proved that the instrument possesses two distinct pumping speeds, hence the glass walls possess different potentials. He concluded that stabilization at a higher potential is possible when the glass walls contain much occluded gas. In this case the coefficient of the secondary emission from the glass exceeds unity, thus permitting equilibrium with the incoming electrons. When the amount of occluded gas is reduced as a result of outgassing, the value of the secondary emission becomes less than unity, and stabilization at a higher potential is not possible. It therefore seems likely that Comsa and Musa (1957b) did not notice the stabilization of the glass walls at higher potentials due to the experimental techniques used in which the outgassing procedures were very stringent. The most comprehensive study concerning the two distinct and stable modes of a BAG head as a result of the electrical potential acquired by the glass envelope was performed by the *Liverpool school* (Carter and Leck, 1959; Cobic *et al.,* 1961a; James and Carter, 1962c). When the gauge is operated in the usual way the values of the potentials of the glass envelope and the cathode are close. The electron current to the walls in this case equals the positive current. The higher the pressure, the closer become the potentials of the cathode and walls. The other mode of pumping is characterized by the potential of the envelope approaching that of the electron collector. Such a situation is possible when the two electron currents, to and from the walls (secondary emission), balance. This is an infrequent occurrence and may take place under special operating conditions.

The experimental evidence of the two modes of pumping due to stabilization of the potential of the glass envelope was presented by Leck and Carter (1960). They noticed two different sensitivities of the gauge head during the measurement of adsorption characteristics of gaseous ions on glass and metals. They used a BAG head in which the glass walls were shielded from the electron structure by a nickel cylinder (Fig. 4.29a). The instrument was operated without being baked, and the variation of the electron current arriving at the cylinder was plotted as a function of the electrical potential of this cylinder. The plot yielded a curve (1 in Fig. 4.29b) exhibiting the dynatorn characteristics of conventional radio vacuum valves. After an hour of continuous operation, curve 2 (Fig. 4.29b) was obtained. Curve 1 could not be reproduced when reoperating the instrument after it had been switched off for 24 h. Upon isolating the cylinder its potential was measured and found to be 110 V above that of

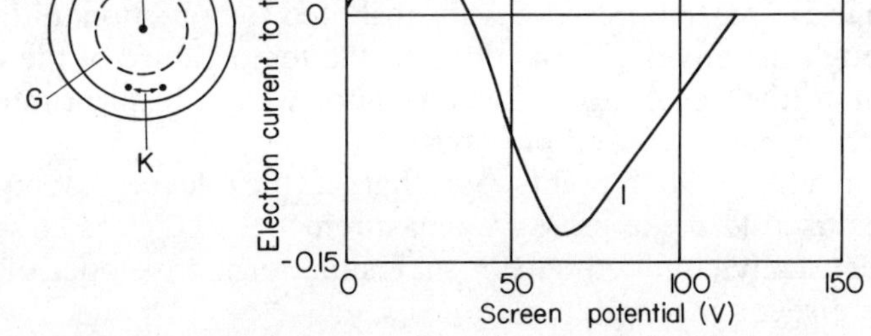

Fig. 4.29 (a) Section through a modified BAG head. C, ion collector; G, grid, K, cathode; S, cylindrical nickel shield; W, glass wall. (b) The dynatron characteristics of the BAG head immediately after switching on the electrical potentials (curve 1) and after 1 h (curve 2). Reprinted with permission from *Brit. J. Appl. Phys.*, **10**, 365. G. Carter and J. H. Leck. Copyright 1959 The Institute of Physics.

the filament. After running the instrument for approximately 30 min, the potential of the cylinder suddenly shifted to a value close to the voltage of the filament. This shift was reproducible only by increasing the emission current by a factor of two or three. There was no need to switch off the gauge head for 24 h.

The problem of bistable behavior of the BAG head as a function of the internal geometry of electrodes and operating conditions (the gas pressure, electric potentials) has also been thoroughly studied (Jahrreiss and Eckertová, 1967; Konjevic *et al.*, 1969; Edelmann and Engelman, 1974). The potential of the glass walls also influences the residual current produced as a result of the reverse x-ray effect.

The temperature of the envelope. The temperature of the gauge head envelope can seriously impair the accuracy of pressure measurement, mainly in the ultrahigh vacuum range, due to changes in the permeation, adsorption, and desorption rates as well as because of thermal transpiration.

The envelope temperature changes as a result of the high energy dissipated in the cathode, which produces the electronic emission required, as well as the difference between the temperatures of the gas in the system and the gauge head, usually at ambient, or variations of the ambient temperature such as are encountered in space research.

Redhead (1961) found that for usual operating conditions of a BAG head at about 10^{-6} Pa (10^{-8} Torr), the envelope attained a temperature of about 50°C. On cooling the envelope with an air blast jet, the pressure in the measuring system changed by a factor of two, and the composition of the gas, analyzed with a mass spectrometer, radically changed. Thus, the partial pressures of the main components He and H_2 varied from 1.73×10^{-8} Pa (1.3×10^{-10} Torr) and 1.46×10^{-8} Pa (1.1×10^{-10} Torr), respectively, when the envelope was at about 50°C to 1.19×10^{-8} Pa (9×10^{-11} Torr) and 2.66×10^{-9} Pa (2×10^{-11} Torr) after cooling the envelope. The reduction in He partial pressure was attributed to reduction in the permeation through the envelope. In addition, the temperature of the envelope may promote the desorption of water vapor which, in turn, initiates the *water vapor cycle* discussed previously.

Steckelmacher (1962) points out that at the elevated temperatures sometimes used to make pressure measurements, the glass envelope becomes electrically conducting. For such situations, envelopes of metal or alumina are recommended.

Thermal transpiration occurs when the temperature of the gas in the vacuum system substantially differs from that of the gauge head. Brombacher (1967, p. 27) advised that if the resulting error is considered significant, equalizing the pressures between the system and the gauge head is a better approach than correcting the gauge readings for thermal transpiration.

Choumoff and Iapteff (1972) analyzed the effect of the envelope temperature on pressure measurement at the x-ray limit of the gauge head. A hot-magnetron gauge head with a metallic envelope was used for the experiments. Two thermal models were built, one considering the situation where thermal equilibrium between ambient temperature and envelope temperature can be attained, the other where such equilibrium is never reached. The two models correspond to the situation met in laboratory and space research. The parameters of the control unit used were maintained constant for experiments with both models. Choumoff and Iapteff found that the ionic current produced in the gas desorbed from the envelope has an important share in the residual current, and the selection of a reference temperature is extremely important to the measurement.

Gas permeation through the envelope. Pressure measurement in *ultrahigh vacuum* may be impaired by the presence of He diffused through the glass envelope of the gauge head. He from atmosphere *diffuses through noncrystalline materials* at significant rates *and desorbs* after crossing the thickness of the wall. The process of permeation, composed of diffusion and desorption of gas, depends on the material and temperature of the gauge head. The permeation rate of He through quartz is very high and diminishes for other glass compositions such as pyrex and soft glass (soda lime) (Norton, 1962; Altemose, 1961).

The simplest way to prevent atmospheric He from penetrating the envelope is to make the envelope of metal. The permeation rate can be reduced by using a ceramic–titanium envelope (Lafferty, 1961a).

(vi) *Gauge head history*. One of the factors critically influencing the accuracy of pressure determinations in ultrahigh vacuum is the gauge head history. Permanent changes in the gauge head sensitivity and stability often occur because of one or both of the following situations: gases or vapors adsorbed or deposited on the electrodes or on the tubulation of the gauge head in a previous use are set free in the actual operation by electron impact desorption (EID) or Blears effect, respectively; and the emission characteristics of the cathode or the electrode geometry have been altered as compared with their initial status, during previous operations of the gauge head.

Electron impact desorption produces desorption of gases chemisorbed or adsorbed at the ion collector in former utilizations. Thus, Redhead (1962) reported an increase of as much as a factor of 400 in the x-ray limit of BAG heads due to chemically active gases (O_2, CO), water vapor, and carbon monoxide chemisorbed on the electrodes. A detailed examination of the effect of oxygen on pressure measurement with BAG heads enabled

Redhead (1963) to observe pressure indications of as much as 75 times the true pressure. Singleton (1967) used a BAG head with a thoria-coated cathode to measure hydrogen in the presence of small amounts of carbon monoxide. The pressure indicated by the gauge head was about 10% higher than the pressure when the operating current of the cathode was 10 mA and 10 times higher when the current was 0.1 mA. The sensitivity of BAG heads supplied by different manufacturers changed by up to a factor of four after being operated for one day in a hydrocarbon atmosphere (Young, 1973). Young attributed the reduction in sensitivity to the formation of a high electrical resistance layer on the ion collector and showed that sensitivity can be restored by electron bombardment of the ion collector. Systematic investigations carried out at PTB (Physicalisch-Technische Bundesanatalt) Institut, Berlin, have shown that changes in the sensitivity of hot-cathode ionization gauge heads can be related to changes in the ion collector surface properties (Becker and Messer, 1980).

The Blears effect, as well as sorption of oil on the electron collector, leads to a reduction in tubulated gauge head readings below the actual pressure in the system. Such a process continues indefinitely.

Changes in the emission characteristics of the cathode occur particularly at oxide cathodes where the coating peels off the filament after long periods of use.

Changes in the geometry of electrodes resulting from previous mishandling, incorrect outgassing procedures (e.g., deformation of the grid pitch due to sagging), or repeated thermal cycling affect the path length of electrons which produce ionization, hence sensitivity. Apart from the effects which can influence the accuracy of pressure measurements due to changes in the emission characteristics of the cathode or to changes in the electrode geometry, the gauge head history can be dealt with by evacuating the gauge head one decade (for calibration purposes, two decades) below the pressure point to be measured (calibrated), before pressure determinations or by allowing enough time (several minutes) to reach equilibrium between gas and gauge head.

Good practice consists in repeating pressure point determinations, each time using a different treatment of the gauge head. This may cast some light on the magnitude of possible effects.

(vii) *Magnetic fields.* As long as no external factors affect the length l travelled by an electron before producing ionization, sensitivity [see Eq. (4.19)] for a gas of certain chemical composition, temperature, and pressure depend only on the geometric configuration of the electrodes and their electrical potential. A magnetic field may modify sensitivity due to its effect on l.

Normand (1962) calibrated hot-cathode ionization gauge heads (conventional triode, BAG) in the presence of *static homogeneous* magnetic fields. He experimentally demonstrated the variation of the gauge head sensitivity with orientation in the magnetic field. The intensity of the magnetic field in his experiments was 100 times higher than that of earth.

Martin (1962) investigated the response of conventional triode, BAG, and Schultz–Phelps gauge heads as a function of their orientation in *homogeneous static* and *pulsed* magnetic fields. The intensity of these fields was over 500 times that of the earth's. Equations were obtained by disregarding the influence of space charge, nonuniformity of potential distribution due to finite length of electrodes, recombination phenomena, and collection efficiency. However, good agreement was obtained between the qualitative theoretical predictions and experimental results.

In conclusion, the location and orientation of hot-cathode ionization gauge heads, operated in the presence of magnetic fields, is essential for the accuracy of pressure measurement.

The variation of the collected ion current efficiency (Penning, cold magnetron) or of sensitivity (hot magnetron) as a function of the relative position of the gauge head with a magnetic field was also studied (Bernardet and Gourdon, 1963; Nichiporovici and Khanina, 1969; Webb, 1974).

Investigation of the influence of the earth's magnetic field on the accuracy of pressure measurement with hot-cathode gauge heads have not been made yet.

(d) *Ion collection.* Theoretical and experimental work with regard to x-ray control on the accuracy of pressure determination with hot-cathode ionization gauge heads has proved that the *electron energy* and the *gauge head geometry* are first-order parameters, whereas the *material of the electrodes* and the *coverage of electrodes with gas* are second-order parameters (Schutze and Stork, 1962a,b; Kornelsen, 1964; Comsa, 1966a; Groszkowski, 1968; Steckelmacher, 1969; Lafferty, 1972; Weston, 1979). The various techniques based on this work, aimed at increasing the ratio of the ion current in the gas phase to the x-ray photocurrent, were developed according to the following approaches: (i) the modification of the electrode geometry, (ii) the use of an electric field to either retard or suppress photoelectrons emitted by the ion collector, and (iii) the increase of the gauge head sensitivity.

(i) *Modification of the electrode geometry.*

One modification is *reduction of the solid angle subtended by the ion collector to the x rays from the grid.* The geometry of the early triode gauge head was modified by Bayard and Alpert (1950) in that the cylindrical collector (I_c in Fig. 4.30a) was replaced by a thin wire (I_c in Fig. 4.30b). The diameter of the ion collector wire was gradually tapered by an elec-

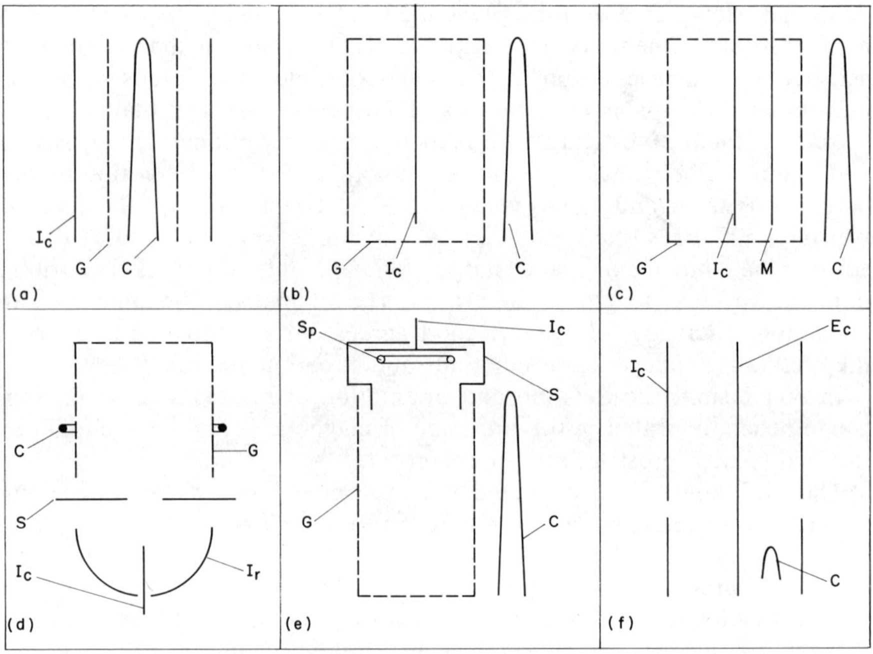

Fig. 4.30 Hot-cathode ionization gauge heads. (a) Conventional triode; (b) Bayard–Alpert; (c) modulated Bayard–Alpert; (d) extractor; (e) suppressor; (f) orbitron. C, cathode; E_c and G, electron collector; I_c, ion collector; I_r, ion reflector; M, modulator; S, shield; S_p, suppressor.

trolytic etching process to further minimize the wire size. The solid angle subtended by this ion collector to the x rays from the grid is several hundred times smaller than that of the cylindrical ion collector, but the sensitivity of the gauge head was low. In another constructive variant, both ends of the grid structure were closed by grid-like end caps to prevent ions from axially escaping (Nottingham, 1955). The sensitivity of the gauge head having these features was improved over the original design, and the x-ray limit was 10^{-8} Pa (10^{-10} Torr).

Theoretical and experimental work (Groszkowski, 1965a,b,c; Comsa, 1966a,b, 1967) proved that sensitivity is influenced to a high degree by the diameter and length of the ion collector in the case of open grid structures, is slightly influenced by changing the position of the ion collector wire (wire length equal to grid length), and is independent of the cathode length and location.

Venema and Bandringa (1958) and Van Oostrom (1962) further reduced the diameter of the ion collector wire to 0.025 mm and 0.004 mm, respectively. The x-ray limit of gauges having such thin ion collectors was about 10^{-10} Pa (10^{-12} Torr).

In addition to the technological problems raised by suspending such delicate wires, BAG heads with fine wire collector (BAG FC) lose sensitivity as a result of decreased ion collection efficiency. Thus, at low pressures ions oscillate many times about the ion collector, and their probability of escaping through the grid is reduced. At high pressures the collection efficiency is reduced because of the space charge in the vicinity of the ion collector. The ion collection efficiencies in a BAG head with the usual collector (about 0.150 mm) and in a BAG FC head as a function of the radius r_1 at which ions are formed to the radius r_g of the grid are illustrated in Fig. 4.31. The initial kinetic energy of electrons is indicated for each plot.

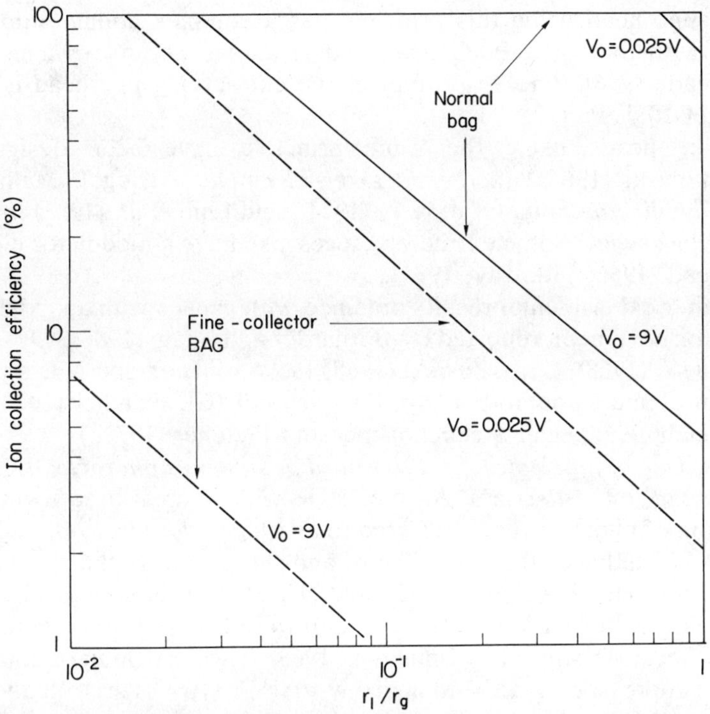

Fig. 4.31 Ion collection efficiency as a function of radii at which ions are formed (r_1) to that of the grid γ_g, V_o is the initial energy of electrons. For the normal BAG, the collector diameter is 0.200 mm and the grid diameter is 25 mm. For the fine collector BAG the collector diameter is 0.004 mm and the grid diameter is 20 mm. Reprinted with permission from "The Physical Basis of Ultrahigh Vacuum." P. A. Redhead *et al.* p. 311. Copyright 1968 Chapman and Hall.

Redhead (1960) suggested adding a wire electrode entering the grid volume parallel to the ion collector (referred to as *modulator electrode*) to help in measuring the residual current (M in Fig. 4.30c). This electrode can be operated at the potential of either the grid or the ground and permits discriminating the residual current from the ion current in the gas phase without greatly altering the residual current. The gauge head is known as the *modulated Bayard–Alpert gauge (MBAG)*.

Another technique for reducing the solid angle subtended by the ion collector to the x rays from the grid consists in *withdrawing the ion collector from the grid*. Gauge heads designed according to this technique are referred to as *gauges with external ion collector*. Ions produced in the gas phase are extracted through an aperture in a shield (S in Fig. 4.30d) and directed by an ion reflector I_r to the collector I_c. The cathode C is wound in a single turn, concentrically, around the closed-end grid cage G. The gauge head using this principle has a complex configuration compared with that of a BAG head and is known as an *extractor gauge* (Redhead, 1966). The x-ray limit of the extractor gauge head is on the order of 10^{-11} Pa (10^{-13} Torr).

Gauge heads using the same principle have been designed by Groszkowski (1966a) (*screened external collector* gauge), Melfi (1969) (*buried collector* gauge), Pittaway (1974), and Fujii *et al.* (1983) (*modified extractor* gauge). Some extractor gauges also have a modulator electrode (Redhead, 1966; Pittaway, 1974).

Further experimental results obtained with gauges with an external ion collector have been reported by Bernardet and Shaw (1968), Cleaver and Zakrewski (1968), Groszkowski (1968, 1969a,b), Barz and Kocian (1970), Bernardet and Choumoff (1970), Pittaway (1970a), Fletcher (1970), Beck and Reich (1972), and Steckelmacher and Fletcher (1972b).

(ii) *Use of an electric field to retard or suppress photoelectrons emitted by the ion collector.* An electrode with a small area to minimize emission of photocurrents (referred to as *suppressor electrode*, S_p in Fig. 4.30e), located near the ion collector and shielded from the x-ray source, produces a negative electrostatic field. This field is large enough to force any electron emission from the collector to return to it, thus reducing the x-ray effect (Metson, 1951; Shueman, 1963). The x-ray limit of Shueman's (1963) suppressor gauge head is below 10^{-10} Pa (10^{-12} Torr). Redhead and Hobson (1965) and Helmer and Hayward (1966), using the same principle, designed a modulator suppressor gauge head and a bent-beam gauge head, respectively.

(iii) *Increase of the gauge head sensitivity.* If the path of electrons is increased before being collected, their collection probability (hence, the sensitivity of the gauge head) is enhanced with no increase in x-rays photoemission (Herb *et al.*, 1963). In a gauge head built according to this principle (Meyer and Herb, 1953), electrons from the cathode (C in Fig. 4.30f) are ejected into the electrostatic field between two cylinders, the one with the large diameter being the ion collector (I_c) and the other, a thin wire, being the electron collector (E_c). The energy and angular momenta of electrons are such that they travel long paths before producing ionization. The x-ray limit of the orbitron gauge is 1.19×10^{-9} Pa (9×10^{-12} Torr).

(e) *Cathodes.* Electrons producing gas ionization are supplied according to the gauge head type by hot cathodes, cold cathodes, and radionuclides.

(i) *Hot cathodes* are based on thermionic phenomena (the emission of charged particles from the surface of a conducting material or compound at high temperature). Hot cathodes are also referred to as thermionic cathodes.

The emission of charged particles depends on *the work function* (the energy, usually measured in electron volts, required to move an electron from Fermi level outside the surface) of the material used.

Hot cathodes are required to: supply the necessary electron emission with minimum energy input (the electron emission must remain stable for constant operating conditions); have a reduced chemical reaction with the rarefied environment being measured; have a rather reduced evaporation rate at the operating temperature, hence a long service life; and have a vapor pressure as low as possible. As a rule, the pressure resulting from the evaporation of the cathode must be at least one order of magnitude lower than the lowest pressure which has to be measured.

The consequences of high energy input, hence high temperature, of the cathode have been examined [see (c)] in this section. Both the magnitude of the energy input and the chemical composition of the cathode contribute also to the emission of neutrals and charged particles from the cathode. Thus, alkali metals (Na, K, Rb, Cs) usually contained in tungsten and platinum wires are brought to the surface of the cathode, whereupon they evaporate as positive ions. *Neutrals,* which enter into the volume of the helical grid of the electron collector, limit the lowest measureable pressure (Alpert and Buritz, 1954). *Positive ions* are emitted by pure metal cathodes (W, Pt) and by the alkali metal compounds. The production rate of positive ions from tungsten cathodes is enhanced after bombardment with positive ions of noble gas followed by heating (Riddoch and Leck, 1958). The emission of positive ions from tungsten cathodes may appear either continuous or in bursts (Winters *et al.*, 1963). Positive ions contribute erroneous pressure measurements.

The evaporation of neutrals and positive ions can be reduced to a minimum by heating the cathodes (usually pure metal cathodes) at high temperature for prolonged periods of time while pumping the gauge head.

Cathodes used in ionization gauge heads can be grouped into the following classes: *pure metals, monolayer emitters, oxide cathodes,* and *refractory compounds*.

Among *pure metals, tungsten* is used on a large scale in ionization gauge heads operated in the medium and high vacuum range. The operating temperature of tungsten cathodes is between 2200 and 2500 K. The work function at the lowest operating temperature is rather high (see Table 4.5). The advantage of running the cathode at such temperatures is that contaminating electronegative gases are rapidly evaporated, hence their influence in increasing the work function and reducing the emission level is diminished. The drawback is that the radiated power causes the temperature increase of the envelope and of the electrodes with all the consequences discussed. Chemical reactions at the hot cathode, as well as the high vapor pressure of about 10^{-6} Pa (10^{-8} Torr) at 2200 K (Honig and Kramer, 1969), prohibit the use of pure tungsten cathodes in gauge heads used for measurements in ultrahigh vacuum. The service life of tungsten cathodes is 10–20 h at 1.0–0.1 Pa (10^{-2}–10^{-3} Torr) and about 1000 h at about 10^{-4} Pa (10^{-6} Torr) (Mueller, 1962). Air inrush into the gauge head while under reduced pressure and mechanical mishandling damage tungsten cathodes.

Rhenium is more resistant to the water cycle and active gases than tungsten. Rhenium's higher work function (4.7 eV) (Levi and Espersen,

Table 4.5

COMPARISON OF CATHODES USED IN UHV INSTRUMENTS[a]

Cathode	ϕ^b (V)	$T_e{}^c$ (K)	W^d (g/cm² s)	η^e (A/W)	Reference
W	4.53	2180	6.4×10^{-12}	1.3×10^{-4}	Jones and Langmuir (1927)
LaB_6/C	2.66	1370	$\sim 10^{-15}$	$\sim 10^{-3}$	Lafferty (1951)
LaB_6/Rh	2.70	1370	$\sim 10^{-10}$	$\sim 10^{-3}$	Buckingham (1965)
ThO_2/W	2.96	1500	$\sim 2 \times 10^{-16}$	$\sim 10^{-3}$	Shapiro (1952) Hanley (1948)
[BaSr]O	1.5	750	$\sim 10^{-17}$	$\sim 10^{-2}$	Wagener (1951)

[a] After Redhead *et al.* (1968, p. 300).

[b] ϕ, average work function.

[c] T_e, temperature for emission density of 10^{-2} A/cm².

[d] W, rate of evaporation at $T = T_e$.

[e] η, emission efficiency, amperes of emission current per watt of heating power at $T = T_e$.

1950) and evaporation rate (about 150 times that of tungsten) restrict its use as uncoated cathode material.

The second class of cathodes to be considered is *monolayer emitters*. The work function of pure metals is substantially reduced by surface adsorption of atoms with a low ionization energy, such as alkali or alkaline earth atoms. Thus, a monolayer of thorium on tungsten reduces the work function of the pure metal from 4.53 to 2.7 eV. Thorium oxide, added in small amounts (0.5–1.5%) to tungsten is reduced to thorium when the cathode is flashed at about 2700 K in a reducing atmosphere. Poisoning of thoriated tungsten by CO_2, H_2O, or O_2 from the residual gas is avoided by the *carburizing* technique. For that purpose, the cathode is heated in a mixture of hydrogen (or a hydrogen–nitrogen mixture referred to as *forming gas*) and xylene (or another organic vapor) to 2200 K. The free carbon dissociated as a result of the thermal energy supplied diffuses in tungsten and combines with it to yield tungsten carbide.

Uncarburized thoria tungsten cathodes may be repeatedly reactivated by high temperature flash or, better, by sputtering with ~300 eV argon ions (Edwards, 1978).

Thoriated iridium cathodes are used in gauge heads working in the medium vacuum range (Schultz and Phelps, 1957) since they are not affected by the presence of oxygen at the partial pressure corresponding to that range.

Oxide cathodes are the third class used in ionization gauge heads. The basic element which produces thermionic emission is a *n*-type semiconductor plated on the metal, usually tungsten or iridium. Oxide cathodes may be classified into three classes according to the chemical nature of the oxide utilized: thorium oxide coatings, rare earth oxide coatings, and alkali oxide coatings. *Thorium oxide* cataphoretically coated onto tungsten or iridium produce low work function cathodes (2.7–2.96 eV). Weinreich (1951) gave data on both iridium and rhodium filaments (thoria coated) and indicated the methods of preparing thoria oxide cathodes. Thoria-coated iridium filaments are very resistant to poisoning, and oxidation filaments of this type do not burn out when being exposed for a short time during operation to a sudden air inrush. Thus, a gauge head with a thoria-coated iridium cathode was exposed 20 times at atmospheric pressure for a short time, and no changes were noticed in its characteristics (Pencko, 1961). A thoria–iridium-coated cathode was used by Futch (1961) who measured the vapor pressure of water at 66.6 Pa (5×10^{-1} Torr) (see Section 6.2).

Thoria oxide cathodes are used in gauge heads for measurements in the ultrahigh vacuum region. They must have small diameters in order to minimize the heat flow to the leads, which results in reduced outgassing effects. Redhead *et al.* (1968, p. 300) recommend avoiding the use of ac high-voltage supply for degassing gauge heads with thoria-coated iridium filaments in order to avoid damage.

Cathodes with *rare earth oxide and alkali earth oxide coatings* have also been studied. Cleaver (1968) reported the use of a gauge head with a thoria-oxide-coated platinum–rhodium substrate for the measurement of pressures in the low vacuum range up to 100 Pa (1 Torr). Alkali earth oxides (Ba, Ca, Sr) have the lowest work function but have two main drawbacks. They emit negative ions in appreciable amounts and are easily poisoned by chemically active gases (CO, H_2O, O_2). Excepting special utilizations, such as Ba or Sr oxide cathodes for the omegatron gauge head, cathodes in this class are not reported as being utilized.

Refractory compounds make up the fourth class of cathodes. Among the borides, carbides, and nitrides which can be used as refractory compounds, lanthanum boride has the best properties for the construction of hot cathodes. Its work function is 2.66 eV (Lafferty, 1951). Cathodes are prepared by using LaB_6 in the form of a sintered rod, supported on carbon or on a refractory metal. Tungsten, rhenium, and tantalum are used as host materials. Before coating, the surface of the metal is carburized (see the discussion of monolayer emitters above) in order to reduce boron diffusion through the bulk.

Tantalum wire LaB_6-coated cathodes neither react with oxygen at 1000°C (Young, 1959) nor dissociate hydrogen (Hickmott, 1960).

Poisoned LaB_6 cathodes are reactivated by heating at a temperature higher than the operating one. Hayward and Taylor (1966) described a cataphoretic technique for coating cathodes in the gauge head *in situ*.

The maximum electron emission densities as a function of the cathode temperature for the cathodes just discussed is illustrated in Fig. 4.32. The approximate pressure indication of a BAG caused by evaporated cathode materials as a function of electron emission density is shown in Fig. 4.33. The plot relies on the assumption that the molecules evaporated from the cathode pass only once through the ionization volume before being adsorbed at a surface (Redhead *et al.*, 1968, p. 300).

Electron emission can also be obtained by using a solid-state device such as a reverse biased SiC *p–n* junction. Dobrott and Oman (1970) built a BAG head with such a source, which requires an input power of only 0.25 W for emitting a 1×10^{-9}-A current. The current in the gas phase in their gauge is *independent* of pressure below 100 Pa (1 Torr). It is claimed that the interaction of the electron-emitting source with gas in the gauge head is reduced.

(ii) *Cold cathodes.* The choice of materials for the construction of electrodes is not critical. Nickel is preferred because of the relative ease with which the parts can be shaped through plastic deformation.

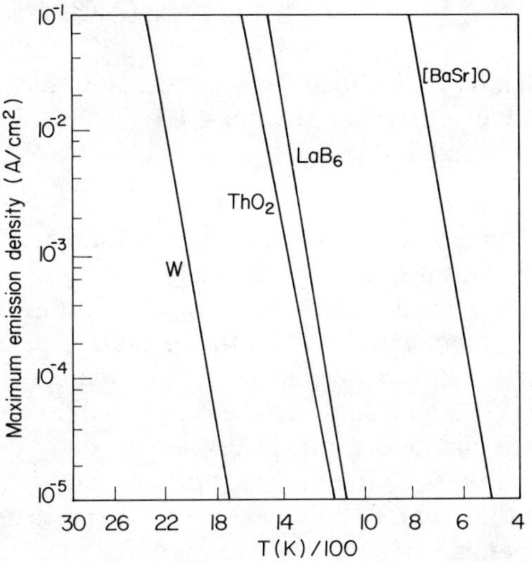

Fig. 4.32 Maximum electron emission density as a function of the cathode type. Reprinted with permission from "The Physical Basis of Ultrahigh Vacuum." P. A. Redhead *et al.* p. 301. Copyright 1968, Chapman and Hall.

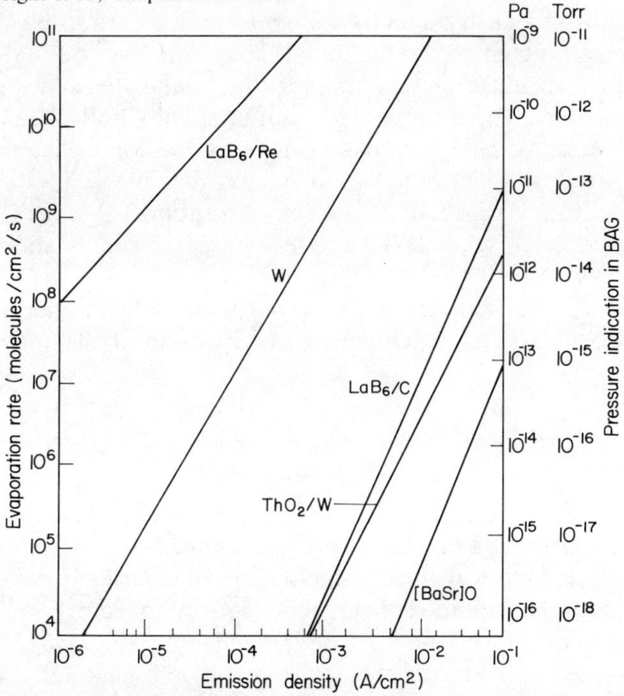

Fig. 4.33 The approximate pressure indication of a BAG, caused by evaporated cathode materials as a function of electron emission density. Reprinted with permission from "The Physical Basis of Ultrahigh Vacuum." P. A. Redhead *et al.* p. 302. Copyright 1968 Chapman and Hall.

(iii) *Radionuclides.* Radioactive substances used in the construction of gauge heads are discussed in Section 4.4.1. Table 4.4 lists the radionuclides most often utilized.

(f) *The basic functions of measuring circuits.* Measuring circuits in the control unit energize the gauge head, control and measure the currents and bias voltages, and display (or telemeter) the results of the measurement. For ionization gauge heads, the control unit must provide a means to outgass the gauge head and protect the cathode in case the pressure in the gauge head rises above a certain value. Remote control and the possibility of recording the measured data are optional features for both hot- and cold-cathode ionization gauge heads.

The electrical circuits of the control unit are required to regulate and maintain constant the *emission current* within a specified *drift* (stability); regulate and maintain constant the *bias voltage* of electrodes within a specified *drift;* and *measure* the magnitude of the *ion current, maintain* it within a specified drift, and *control the stability* of the display.

Regulation of the drift of the emission current is usually achieved by a system which senses the electron current and controls, by a feedback loop, the bias voltage of the cathode. Close and Yarwood (1972) developed an integrated differential amplifier which regulates and stabilizes the electron emission. The amplifier sets automatically both the correct current of the cathode and the bias voltage of the ion collector. Herbert (1976) designed an electronic circuit which regulates and stabilizes the collected electron current. The electron current drift is 10%/°C, the regulation is about 0.1%/°C, and the anode-to-cathode supply stability is better than 0.1%/°C.

Generally, the ion current is measured with conventional methods or with a feedback electrometer which maintains the collector input very close to ground.

4.4.7 Hot-Cathode Nonmagnetic Ionization Gauge Heads

(a) *The conventional triode.* Designed by Buckley (1916) and developed by Dushman and Found (1921), this gauge head resembles a triode vacuum tube having a central cathode (c) concentrically stretched in a cylindrical loosely wound grid (g) (the electron collector) and a metallic cylinder (a) referred to as anode (the ion collector) (Fig. 4.34). The gauge head is operated in the electrical circuit shown in Fig. (4.35). The cathode (c), a tungsten filament typically biased at 30 V dc above ground potential, emits electrons which are accelerated to the grid (g) by the potential difference between cathode and grid, typically 150–200 V. Most of the

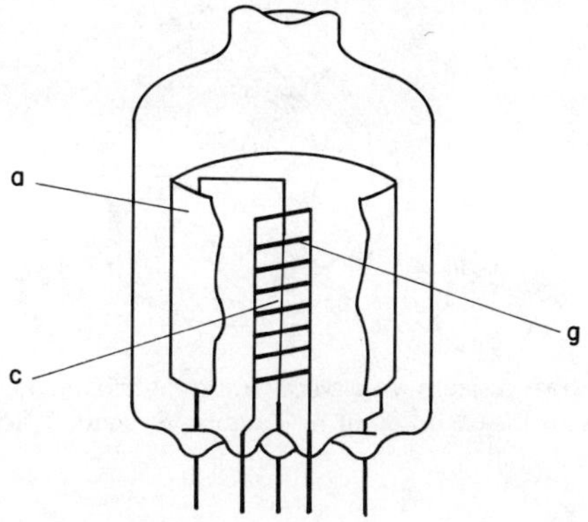

Fig. 4.34 Conventional triode gauge head. a, anode; c, cathode; g, grid.

accelerated electrons pass through the grid where their direction of motion is reversed because of the oppositely directed field of the ion collector (Fig. 4.36). The electron cloud distributed throughout the shaded area illustrated in Fig. 4.36 cycles back and forth until it strikes g where it is collected. Some of the electrons collide along their path with residual gas molecules and, if they possess enough energy, produce an ion electron pair. Owing to the distribution of potential in the ionizing region, electrons are decelerated so that they are inefficient for producing ionization over a large portion of their trajectory. The positively charged ions pro-

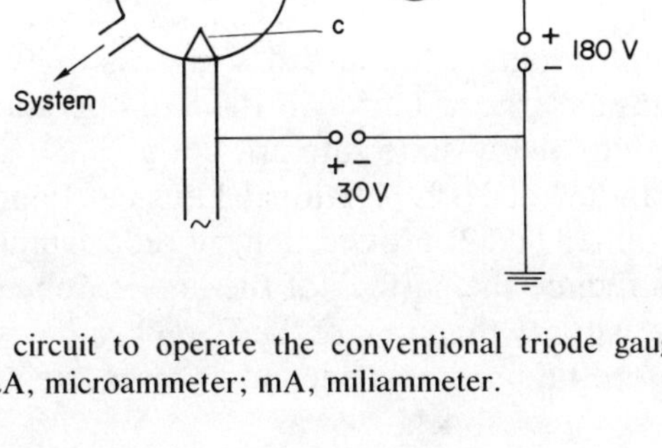

Fig. 4.35 Electrical circuit to operate the conventional triode gauge head. a, anode; c, cathode; g, grid; μA, microammeter; mA, miliammeter.

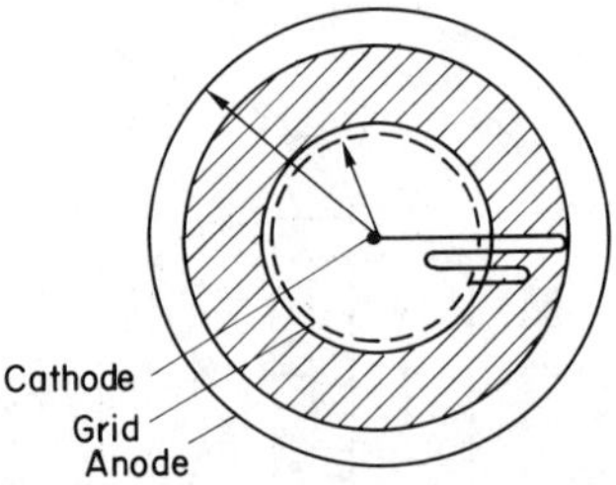

Fig. 4.36 Typical electron trajectory in a conventional triode gauge head. Reprinted with permission from "Pressure Measurement in Vacuum Systems." J. H. Leck. p. 79. Copyright 1964, Chapman and Hall.

duced are repelled by both the positively biased cathode and the electron collector, and they are collected at the ion collector, usually maintained at ground. Electrons resulting from the ionization process are collected at the grid together with the electrons emitted from the cathode. If the emission current (typical values ranging from 0.1 to 10 mA) is maintained constant, the ion current is proportional to gas pressure, as shown by Eq. (4.18). The nitrogen sensitivity of the gauge head measured with an accuracy of about 10% is on the order of 0.075 Pa^{-1} (10 $Torr^{-1}$). For the typical values of emission currents mentioned, ion currents of 10^{-10} to 10^{-9} A are obtained at equivalent x-ray pressures of 10^{-6} Pa (10^{-8} Torr).

The linear relationship between ion current and pressure predicted by Eq. (4.18) falls below 10^{-6} Pa (10^{-8} Torr) and above approximately 10^{-1} Pa (10^{-3} Torr). In the lower range of pressure this is due to the residual currents (x ray, EID) to the collector. In the upper pressure range electrons of *low energy* are created owing to the high frequency of electron and residual gas molecule collisions. These electrons, which cannot yield ionization, are collected at the grid and measured as part of the emission current. This reduces the value of the gauge head sensitivity.

Even in the linear range of the gauge head, *sensitivity slowly drifts with time* causing difficulties in both reproducibility of measurements in the long term and establishment of realistic time intervals between gauge head calibrations.

Poulter and Sutton (1981) found drifts in sensitivity for conventional triode gauge heads as large as 0.45% in 100 h of operation, but qualitative explanations of that steady-state rate are not given.

More recent studies at NPL (National Physical Laboratory, England) by Sutton and Poulter (1982) proved that by redesigning the triode gauge head, such as to reduce the surface of the anode (ion collector), a better stability of the sensitivity than that of the BAG head was obtained. Such a triode can therefore be used as a secondary pressure standard (see Section 3.6) for pressure measurements in the medium and high vacuum range. In the redesigned construction, which retains the cylindrical symmetry shown in Fig. 4.30a, the anode is a short cylinder made of high-optical-transparency mesh (about 90%) (a in Fig. 4.37). The cathode–grid–anode assembly is surrounded by a cylindrical plain screen (sc) positively biased. Both the ion collector and the screen are closed by grid-like mesh end caps to improve ion collection. Positive ions formed anywhere between the screen and the ion collector are trapped in the potential well of these two electrodes and repelled towards the ion collector.

Triode gauge heads built with this configuration were calibrated at about 3.75×10^{-3} Pa (5×10^{-1} Torr) in nitrogen. The high calibration pressure was selected such that the ion current in the gas phase should be much greater than the residual current. Sensitivity measurements for five gauges indicated a *mean* value $s = 0.047$ Pa^{-1} (6.27 Torr^{-1}) after 100 h of operation. The standard deviation of measurements was 5%. No significant change in the sensitivity was found when the gauge heads were operated in carbon monoxide or methane at 10^{-2} Pa (10^{-4} Torr) for more than 18 h. It is worth noting that BAG heads operated in similar conditions in N_2 showed large changes in sensitivity.

A new version of a triode gauge head designed by Sutton and Poulter (1982), in which the collector area was reduced from that of a cylinder to that of a fine ring, showed a sensitivity $s = 0.073$ Pa^{-1} (9.73 Torr^{-1}). The sensitivity of the new gauge head remains constant to within 1%/100 h at pressures up to 0.3 Pa (2.25×10^{-3} Torr), and the residual current equivalent pressure is less than 10^{-7} Pa (10^{-9} Torr).

Triode gauge heads constructed at the Electrotechnical Laboratory, Japan by Hirata *et al.* (1982) also retain the cylindrical symmetry of electrodes of the conventional triode. Nitrogen sensitivities measured on a

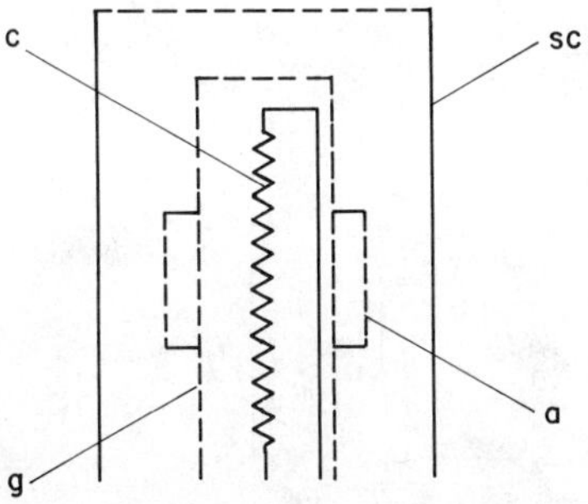

Fig. 4.37 New design of the conventional triode gauge head. a, anode; c, cathode; g, grid; sc, screen.

batch of 258 gauge heads showed an average sensitivity $s = 0.137\ \mathrm{Pa}^{-1}$ ($18.26\ \mathrm{Torr}^{-1}$) in the pressure range from 0.01 to 0.3 Pa (7.5×10^{-5}–2.25×10^{-3} Torr). The accuracy of sensitivity determinations is ±3%. Measurements of the variation in sensitivity as a result of changes in the electrode geometry show that a change in sensitivity of about 2% results when the filament (double helical tungsten wire) is shifted from its initial position by 0.5 mm.

Work performed at NBS (National Bureau of Standards, Washington D.C.) shows that triode gauge heads are more stable and uniform than BAG heads (Tilford, 1983).

(b) *The Bayard–Alpert gauge (BAG).* After Nottingham (1947) showed the existence of a residual current to the ion collector of the conventional triode gauge head, a number of investigators (Lander, 1950; Bayard and Alpert, 1950; Metson, 1951) attempted to reduce the current by reducing the solid angle, subtended by the ion collector, to the x rays from the grid, as discussed.

In the original design of the BAG head shown in Fig. 4.38 the electrodes are the same as in the triode gauge head but inverted from the conventional arrangement. The ion collector, a very fine metallic wire, is suspended inside the grid along its geometrical axis, and two cathodes (instead of one) are located outside the grid. The volume enclosed by the grid is comparable to that between the grid and cylindrical anode of the conventional triode gauge head. Both the electrical bias of the electrodes and the emission current of the new gauge head were similar to the parent triode at the time of development. The second cathode could be used (other than as a spare) in connection with the *flash filament technique* (Apker, 1948) for estimating low partial pressures of residual gases.

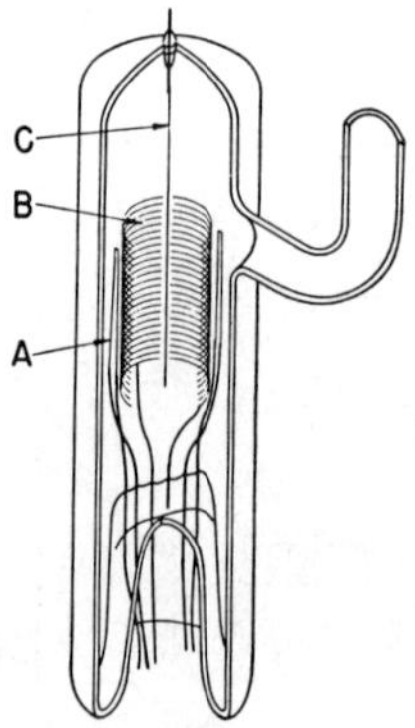

Fig. 4.38 First design of the Bayard–Alpert gauge head. (After Bayard and Alpert, 1950.) A, filament; B, grid; C, ion collector.

Upon reducing the dimensions of the ion collector from that of a cylinder to that of a thin wire, the x-ray cross section of the collector was greatly reduced so that the limiting pressure of the gauge head was 10^{-8} Pa (10^{-10} Torr) instead of 10^{-6} Pa (10^{-8} Torr). Bayard and Alpert (1950) brought original evidence of the residual current, produced by soft x rays, by measuring the collected ion current I_c as a function of the grid-to-cathode voltage V_g. Measurements (Fig. 4.39) were made at constant pressure in both a conventional triode gauge head (RCA type 1949) and the gauge head they designed. The variation of I_c versus V_g at $P = 1 \times 10^{-3}$ Pa (1×10^{-5} mm Hg) for the conventional triode gauge head shows a rapid rise with V_g up to 200 V followed by a slow variation of I_c from this value. The shape of the curve is similar to that illustrated in Fig. 4.26 for the ionization probability. When pressure in the triode gauge head is lowered below $P = 10^{-6}$ Pa (10^{-8} mm Hg) the variation of I_c versus V_g

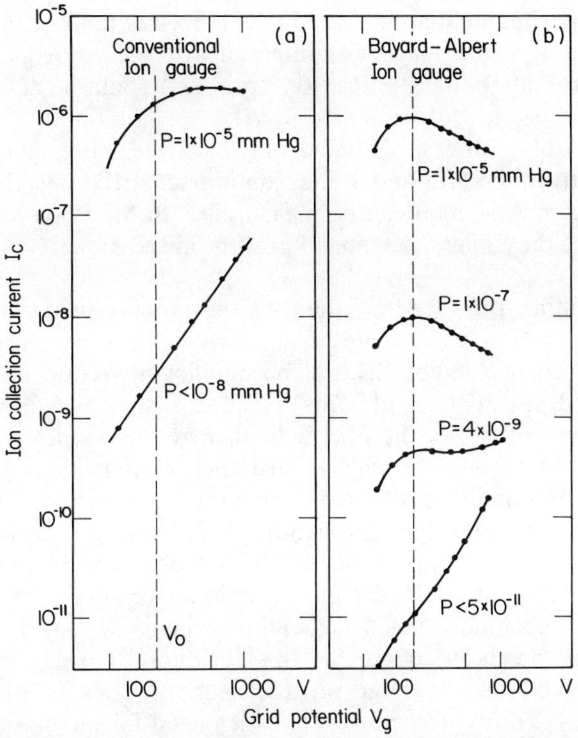

Fig. 4.39 Variation of $I_c = f(V_g)$ for (a) conventional gauge head (RCA type 1949) and (b) Bayard-Alpert gauge head. V_g is the operating voltage (150 V). Reprinted with permission from "Handbuch der Physik Vol XII." D. Alpert. p. 613. Copyright 1958 Springer Verlag.

yields a straight line (lowest curve in Fig. 4.39), typical for the x-ray current. Alpert (1953) pointed out that for this situation, I_c is proportional to V_g^a where the exponent a takes values between 1.5 and 2. Since then, several values for a have been reported by other investigators. These values range from 1.1 to 2, but no physical significance can be attached to them.

The variation of I_c versus V_g measured for the BAG head (Fig. 4.39b) indicates that at $P = 5.3 \times 10^{-7}$ Pa (4×10^{-9} Torr) the collector current still retains the characteristic shape of gas ionization probability, superimposed on a monotonously rising curve. At $P < 6.6 \times 10^{-9}$ Pa (5×10^{-11} Torr) the shape of the curve shows that the ion current in the gas phase is still present (initial portion of the lowest curve in Fig. 4.39) at operating voltages below 200 V. However, the fact that the upper portion of the curve can no longer be distinguished from a straight line (see the lowest curve in Fig. 4.39) indicates that pressure cannot be derived with any accuracy from measurements at such degrees of gas rarefaction.

These experiments demonstrated that the x-ray limit of the BAG head was reduced by two decades as compared with the triode gauge head and that pressures of about 10^{-8} Pa (10^{-10} mm Hg) could be accurately measured.

Unfortunately, other processes, of which the most important is ion desorption from the grid under electron impact (EID), also limit the lowest pressure range which can be measured with the BAG head, and it is possible that they might have been present in the experiments of Bayard and Alpert.

The limitation produced by the x-ray photocurrent can be measured (see Section 7.4) and eliminated from low pressure measurements. The residual current yielded by EID can be significantly reduced by driving off the adsorbed gas at the grid. This is accomplished by *outgassing techniques*. In one of these, the *grid is heated by the ohmic effect* resulting from passing a current through its structure. Distortion and sagging can occur if the temperature attained during outgassing is too high. Modern gauge heads have nonsag molybdenum or tungsten grids, which in some cases are carburized, thus increasing their thermal resistance. Typically, powers used for outgassing by this technique are about 7.5 W/cm^2.

The second technique consists in *heating the grid by electron bombardment*. Gauge heads adequate for this technique have a fine wire grid (usually molybdenum) welded on thick support rods to withstand high temperatures. The typical power densities used for outgassing are about 15 W/cm^2, giving rise to temperatures of about 1800 K.

In the modern construction of a BAG head the ion collector (0.125-mm diameter wire) is placed outide a capped grid. A single thoria-coated iridium filament ("burnout proof") or two tungsten filaments located either on opposite sides of the grid or next to each other on one side of the grid are the thermionic emitters of electrons. In the case of tubulated (glass) gauge heads, the portion of the glass walls surrounding the electrode assembly is covered with a transparent conducting layer, which helps in avoiding or reducing the influence of the electric potential of the gauge head envelope. In the case of nude gauge heads the metallic chamber walls where the gauge is located provide the necessary screening. Figure 4.40 illustrates the modern construction of a BAG head.

The potential distribution between the cylindrical grid and the coaxial ion collector is logarithmic, with the radius within the grid. Hence, the electric field between these electrodes is fairly uniform across most of the grid volume except within the close vicinity of the ion collector. The trajectories of most electrons penetrating into the grid volume will therefore suffer little deviation, their kinetic energy being close to that corresponding to the maximum ionizing probability of gas molecules with which they collide. Electrons escaping the grid are decelerated in the electrical field between the grid and the screen and are repelled from the grid. Figure 4.41 illustrates the electron trajectories for three different angles of incidence at the grid boundary in a cross section of a BAG head, as computed by Pittaway (1970b). According to Pittaway's calculations, the angle at which electrons approach the grid boundary should be close to normal so that their paths would be very long. The screen potential greatly influences the electron path length, hence sensitivity.

More recent studies of sensitivities of BAGs made under *ideal standard laboratory conditions* on nude and tubulated heads having only several (or a few hundred) operating hours show the following.

(i) Changes as large as 25% in sensitivity occurred after exposing the gauge head to air (the gauge heads were stored 15 months in a laboratory cupboard). Calibrations were performed at 6.6×10^{-1} Pa (5×10^{-3} Torr) in N_2 (Sutton and Poulter, 1982).

(ii) Variations from 70 to 110% of the specified nitrogen sensitivity occurred at the calibration of dual filament UHV gauge heads in the high vacuum range (McCulloch and Tilford, 1981).

(iii) Variations as low as 52% of the specified sensitivity (0.06 Pa^{-1}, 8 $Torr^{-1}$) were found at gauge heads calibrated in N_2 at 10^{-4} and 10^{-1} Pa (10^{-6} and 10^{-3} Torr) (Tilford and McCulloch, 1982). Besides N_2, Ar, h_2 and D_2 were also used for sensitivity determinations.

(iv) Unpredictably, drifts in N_2 sensitivity occurred, sometimes changing sharply by several percentage points (Sutton and Poulter, 1982).

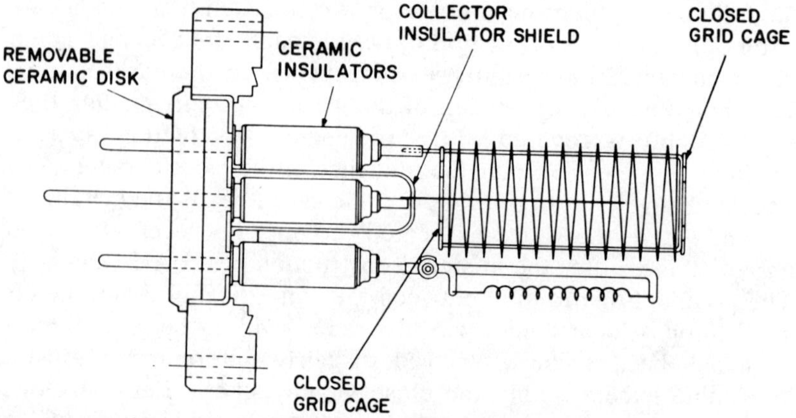

a

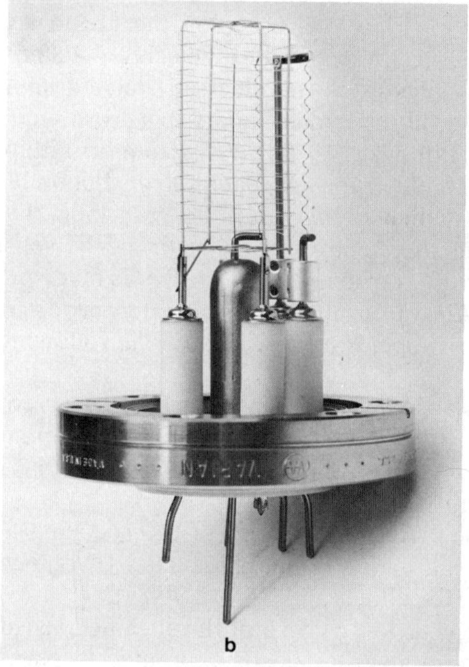

b

Fig. 4.40 Modern construction of a Bayard–Alpert gauge head. Schematic (a) and photograph (b) of the UHV, 24-gauge head for demountable use. (Courtesy of Varian Associates.)

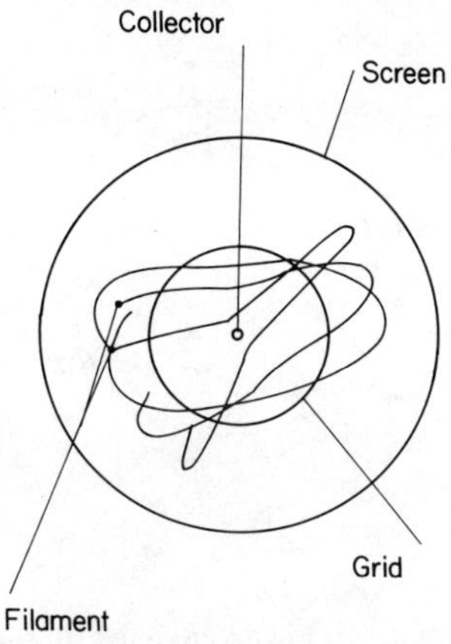

Fig. 4.41 Electron trajectories for three different angles of incidence at the grid boundary in a cross section of a BAG head. (After Pittaway, 1970b.)

At pressures higher than 10^{-1} Pa (10^{-3} Torr) the BAG head shows strong nonlinearity of the collected current I_c as a function of pressure P. The departure from linearity is considered to result from the following causes.

(i) Ions are scattered to other electrodes before reaching the ion collector and are collected at the cathode, together with ions produced outside the grid volume. The ion current to the cathode becomes comparable to the electron current maintained constant by the emission control circuit. The actual electron emission, effective for ionizing gas molecules, is thus reduced.

(ii) Ions reaching the insulated envelope (glass) accomodating the electrode structure of the gauge head give rise to *sudden* changes of the wall potential, altering the potential distribution around the cathode (Carter and Leck, 1959). Ion collection is thus enhanced.

(iii) Secondary electrons produced in ionizing collisions contribute a reduced ionization yield due to their low energy.

Suchannek and Sheridan (1974) noticed that the pressure readings of the BAG head in a glass envelope they used in atomic collision experiments were distorted in the range 10^{-2}–10^{-1} Pa (10^{-4}–10^{-3} Torr). This was attributed to the buildup of electric charges on the glass envelope of the gauge head. To prevent the influence of these charges, an open cylindrical screen grid (4 in Fig. 4.42) surrounding the electrode structure shielded the electrodes from the envelope. The screen was connected to the cath-

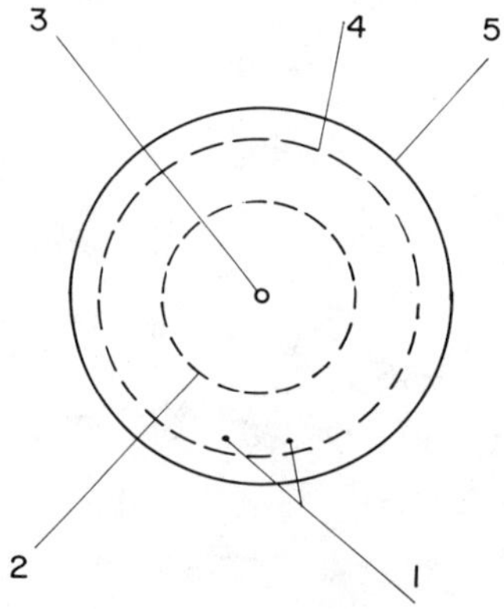

Fig. 4.42 Schematic diagram of a BAG head modified by surrounding the electrode structure with a screen grid. 1, cathode; 2, accelerating grid; 3, ion collector; 4, screen grid; 5, glass envelope.

ode. The collector current was a linear function of pressure between 10^{-1} and 10^{-2} Pa (10^{-3}–10^{-4} Torr) in the BAG head modified as discussed.

Studies aimed at the extension of the low pressure limit of the BAG head have led to the design of a gauge head capable of measuring pressures from UHV to 10 Pa (10^{-1} Torr) (Ohsako, 1982). The approach adopted for linearizing I_c versus P consists of minimizing the ion current to the cathode in the range 10^{-2}–10 Pa (10^{-4}–10^{-1} Torr), thus preventing the emission current from being reduced. It was demonstrated that a metallic electrode negatively biased (a stick or a semicylindrical plate), located close to the cathode, is effective for this purpose (Fig. 4.43).

Linearity in the high pressure range is obtained by using emission currents of 10 or 20 μA. In the UHV range, emission currents of several milliamperes are necessary in order to avoid desorption of ions from the grid by EID.

(c) *The modulated Bayard–Alpert gauge (MBAG).* Redhead (1960) developed a new method for measuring the true current in the gas phase at pressures where the value of this current is lower than that corresponding to the residual current. This method enables pressure determination by measuring the ion current at two different grid–ion-collector bias voltages. For that purpose, the structure of the BAG head was modified by the addition of a separate fine wire electrode within the grid on the side opposite to the cathode (Fig. 4.44). This electrode, referred to as the *modulator,* electrically isolated from the other electrodes, can be operated at different bias voltages relative to the cathode. The modulator can thus be switched from the grid voltage to the ion collector potential (normally grounded). With the modulator at grid potential, the sensitivity of

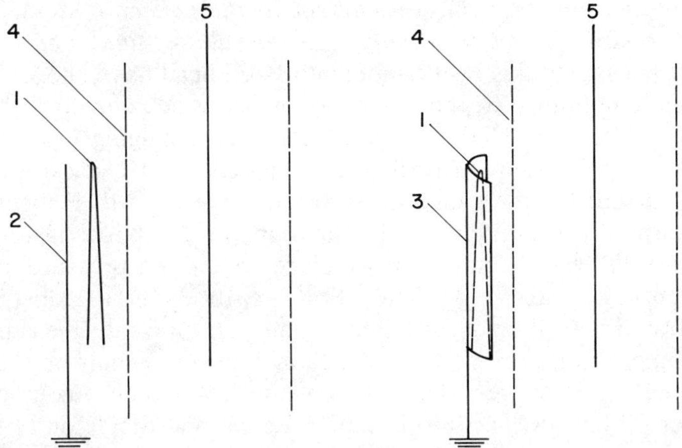

Fig. 4.43 Schematic diagram of a BAG head from UHV to 10 Pa (10^{-1} Torr), such as designed by Ohsako (1982). 1, cathode; 2, auxiliary electrode (stick); 3, auxiliary electrode (semicylindrical plate); 4, electron collector; 5, ion collector.

the gauge head is that of the unmodified BAG head. If the modulator is biased to the collector potential, a fraction of the ion current in the gas phase will be collected at this electrode, without affecting the residual current. α is referred to as the *modulation factor*. The difference between the two ion current measurements at the collector, corresponding to the different voltage biases of the modulator, is therefore proportional to gas density. *The method* (referred to as *Redhead Mode I* modulation) *is valid*

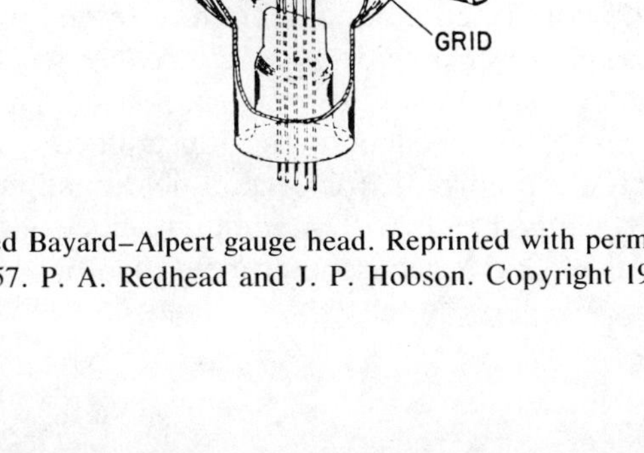

Fig. 4.44 Modulated Bayard–Alpert gauge head. Reprinted with permission from *Brit. J. Appl. Phys.*, **16,** 1557. P. A. Redhead and J. P. Hobson. Copyright 1965 The Institute of Physics.

on the assumption that the ion current to the collector, modulated by about 30 to 40%, *leaves the residual current almost unaltered.*

Nevertheless, studies of the modulate BAG head have shown that under certain conditions, depending on the gauge head geometry, the residual current could be slightly modulated when changing the modulator potential (Appelt, 1962). Appelt (1962) quoted a 15% variation of the residual current for the gauge head that he investigated. Hobson (1964) also confirmed residual current modulations up to about 2.5% in gauge heads with different geometries, which restricted the pressure measurement accuracy to above 4×10^{-9} Pa (4×10^{-11} Torr). Hobson (1964) demonstrated a further decade improvement in pressure measurements upon compensating for the dependence of spurious signals on the modulated potential. However, the techniques used for such a compensation are rather difficult to put into practice. Lange and Singleton (1966) discussed the effect produced by the alteration in electron trajectories (caused by the switch in modulator potential) of the sorption–desorption balance of the measuring system. Thus, with the modulator at grid potential, the ionically pumped gas is released because of electron bombardment. A modified modulation technique (referred to as *Mode II of Lange and Singleton modulation*) consists of switching the modulator from grid potential to some 20–30 V below grid potential. This technique applies to open-ended grid constructions. Redhead (1967) repeated Lange and Singleton's experiments on MBAG heads with closed-end grids and found a way to achieve modulation using a different switching method (referred to as *Redhead Mode III* modulation). Other modulation modes were also proposed by Tuzi *et al.* (1968) and Poulter (1970) and are referred to as Mode IV and Mode V modulation techniques, respectively.

Upon using the end caps of the grid instead of a modulator electrode, Van Oostrom (1967) achieved a modulation factor of 90% in a BAG head with a fine wire ion collector. Szwemin (1972) demonstrated that the modulation method proposed by Groszkowski (1970), who used a cage surrounding the ion collector for modulation, permitted the elimination of the adsorption–desorption process, which is the drawback of the conventional modulation method. The BAG head with cage modulator (referred to as GCM) performs well in both noble and active gases.

Pittaway (1973) confirmed a phenomenon observed by other investigators: the *ion current increases instead of decreases* when a MBAG head, contaminated following exposure to oxygen, is used. Desorbed ions from the modulator can be focused onto the ion collector, giving rise to an increased collected ion current, for certain field configurations.

Bayard–Alpert gauge heads with modulating electrodes similar to Redhead's (1960) MBAG have been constructed also by Appelt (1962), Lee

(1963), and Reich (1966). A three-grid MBAG was studied by Kanaji *et al.* (1977), who added two other external grids concentrically located with respect to the grid of the conventional BAG head. The authors claim that gas desorption produced by EID in their gauge head was not detectable. Hua *et al.* (1980) designed a MBAG head in which the ion collector and the modulator lie on the grid axis with their lead-throughs opposite each other. This constructive solution, referred to as pin-to-pin (PTP) construction, considerably reduces the solid angle subtended by the modulator with respect to the ion collector. The modulation mode is such that the modulator bias voltage always varies positively with respect to the cathode, with a low frequency ac voltage superimposed on a given dc potential. The gauge head can measure pressures down to 10^{-10} Pa (10^{-12} Torr). The rapidity in biasing the modulator between the values selected for the operation mode is an important parameter since the residual current may change in extremely short times in the presence of chemically active gases. In the past several years *automatic modulation methods* have been proposed by Groszkowski *et al.* (1977), Kanaji and Urano (1980), and Edwards and Lami (1980) to replace the manual method. Two automatic modulation methods were compared by Mizuno and Horikoshi (1977).

The modulation method permits the measurement of residual currents, as discussed in Section 7.4.2.

(d) *The suppressor gauge (SG).* The basic principle of the suppressor gauge head is the return to the ion collector of photoelectrons emitted from this electrode when bombarded with x rays. Based on the early work of Metson (1951) and Dalke and Schultze (1960), Schuemann (1962, 1963) developed a gauge head using the general arrangement illustrated in Fig. 4.45a. Positive ions are produced in a region of the gauge head and collected in another region on a plate electrode located at the end of the grid structure. An additional ring electrode located in the vicinity of the ion collector, but shielded from the ionization region, prevents any photoemission caused by x rays from leaving the ion collector. The shield accelerates ions produced in the grid region, focuses them at the ion collector, and prevents x rays generated at the grid from entering the ionization region. In addition, the shield protects the ion collector against the influence of the wall-potential variations. The gauge head, using the suppression principle, is illustrated in Fig. 4.45b.

The nitrogen sensitivity of the suppressor gauge head is 0.15 Pa^{-1} (20 $Torr^{-1}$). Schuemann (1963) measured currents corresponding to pressures as low as 2.6×10^{-10} Pa (2×10^{-12} Torr), by partially submerging the gauge head in liquid nitrogen. Redhead and Hobson (1965) pointed out that the reflectance of metals to x rays amounts to a few tenths for most metals, hence it is possible that reflected x rays can reach the suppressor

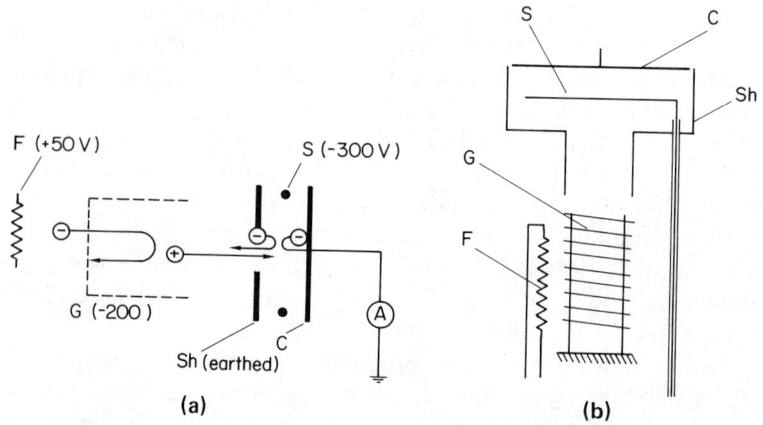

Fig. 4.45 Schematic diagram of (a) the photocurrent suppressor method and (b) the suppressor gauge head designed according to this method. C, cathode; F, filament; G, grid; S, suppressor ring; Sh, shield.

electrode. Photoelectrons emitted from the suppressor may go to the ion collector and there yield a negative residual current. Redhead and Hobson developed a gauge head retaining Schuemann's principle for the suppression of photoelectron emission. In their design, they reduced the area of the ion collector and added a modulator electrode protruding into the grid structure. The x-ray limit of their gauge head is about 2.0×10^{-11} Pa (1.5×10^{-13} Torr) without modulation and about 10^{-12} Pa (10^{-14} Torr) with modulation. The gauge head has about the same sensitivity as that of the BAG head.

Another gauge head based on the photoelectron suppression principle is the *bent-beam* gauge head reported by Helmer and Hayward (1966) (Fig. 4.46). In this gauge head, ions produced in the grid region are extracted through a slit and bent by means of a cylindrical electrostatic field. X-rays emitted from the grid can reach the ion collector plate only by reflection through the deflector. An 80% transparent grid is inserted between the slit at the deflector exit and the ion collector. Residual currents at the ion collector of 1.5×10^{-14} A and -1.5×10^{-15} A were measured when the bias voltages of the suppressor grid were 0 and -500 V, respectively. Pressure measurements were performed in a stainless steel system evacuated by an ion-sputtering titanium sublimation pump. The residual currents correspond to pressures of 2.0×10^{-11} Pa (1.5×10^{-13} Torr) and 2.0×10^{-12} Pa (1.5×10^{-14} Torr). The nitrogen sensitivity was 0.12 Pa^{-1} (16 $Torr^{-1}$). Helmer and Hayward pointed out that the use of an electron multiplier would increase sensitivity. The ion beam may be 100% modu-

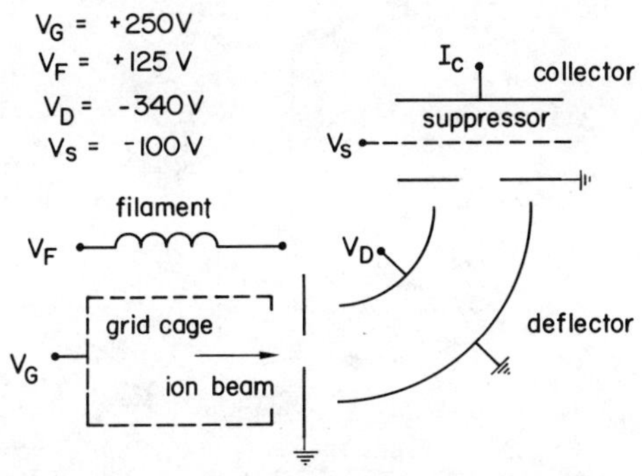

Fig. 4.46 Bent-beam gauge head. (After Helmer and Hayward, 1966.)

lated by turning off the bending field, thereby permitting measurement of pressures as low as 1×10^{-12} Pa (1×10^{-14} Torr).

Urbanek (1967) stressed the fact that in a bent-beam gauge head the amounts of oxygen set free from the grid by EID are 30 times smaller than in a conventional BAG head. The reduced desorption is attributed to the external position of the ion collector. The large amounts of metal used at the gauge head construction render outgassing procedures rather difficult.

Benvenuti and Hauer (1980) improved the pressure limitation (about 3.3 $\times 10^{-10}$ Pa, 2.5×10^{-12} Torr) of several commercial Helmer gauge heads which they used at the ISR at CERN. They reduced the tungsten sublimation from the cathode by coating it with thoria. Both the geometrical dimensions of the electrode structure and the optical transparency of the suppressor grid were modified. The improved construction of the gauge head permits measurements of pressure as low as 10^{-12} Pa (10^{-14} Torr). The nitrogen sensitivity is 0.30 Pa^{-1} (40 $Torr^{-1}$).

(e) *The extractor gauge (EG)*. In this gauge head, the influence of residual currents on the ion collector is minimized by reducing both the cross-sectional area and the length of the electrode and by removing it from the ionization volume to a suitable position outside. Ions from the gas phase are drawn to the collector region designed so as to minimize the incidence of incoming x rays on the ion collector. Ions desorbed from the grid by EID escape through the grid spacings or are collected on it.

In the EG head designed by Redhead (1966), the ion collector, a short fine wire, is mounted at the end of the grid cylinder on the axis (Fig. 4.47). A screen, separated from the grid, located at the end adjacent to the ion collector, shields this electrode from the grid. Ions produced in the gas phase are attracted towards the negatively biased shield, wherefrom most of them pass through a circular aperture. A positively biased hemispheri-

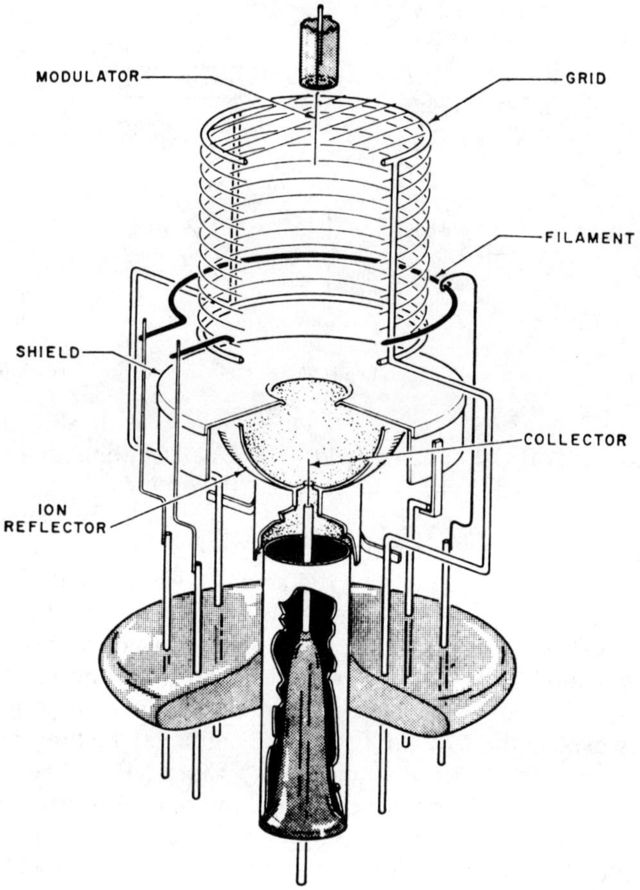

Fig. 4.47 Extractor gauge. (After Redhead, 1966.)

cal metallic reflector at grid potential focuses the ions on the collector. The operation biases are filament, envelope, and shield, +200 V; grid-to-filament voltage 105 V; ion collector at ground potential and emission current, maximum 2 mA. The corresponding nitrogen sensitivity is 0.0975 Pa^{-1} (13 $Torr^{-1}$), but it depends on the emission current and electrode potentials. Redhead (1966) incorporated a modulator at the far end of the grid cage to permit the assessment of the x-ray limit. Calculation of the x-ray limit upon considering geometric factors shows that the currents due to these rays are lower by a factor of 130 than those in the BAG head (4×10^{-11} Pa, 3×10^{-13} Torr).

Tests with chemically active gases (particularly oxygen), indicating that the gauge head readings are less affected by ions yielded in EID, led Redhead to assume that this lower current level is due to reduced efficiency in the collection of desorbed ions. The potential distributions and ion trajectories in Redhead's (1966) EG head computed by Pittaway (1974) confirm the reduced efficiency in the collection of the desorbed ions. The potential distribution and ion trajectories plotted by Pittaway (Fig. 4.48) indicate that ions formed at a radius greater than the aperture in the shield in the vicinity of the end cap are less likely to be extracted.

Groszkowski (1966a) designed a gauge head (referred to as JG) in which the ion collector, a fine short wire inside a glass sleeve, is located inside the anode opposite an aperture in one of the end caps. The JG gauge head is normally operated with a cathode (V-shaped tungsten filament) at 210 V, anode 310 V, electron current 4 mA, and ion collector to ground potential. The potential distribution and ion trajectories in Groskowski's gauge head are plotted in Fig. 4.49. It can be seen that almost all trajectories of desorbed ions pass through the aperture in the glass sleeve and are focused on the ion collector. The nitrogen sensitivity as reported by Groszkowski (1968) is about 0.38 Pa^{-1} (50 $Torr^{-1}$).

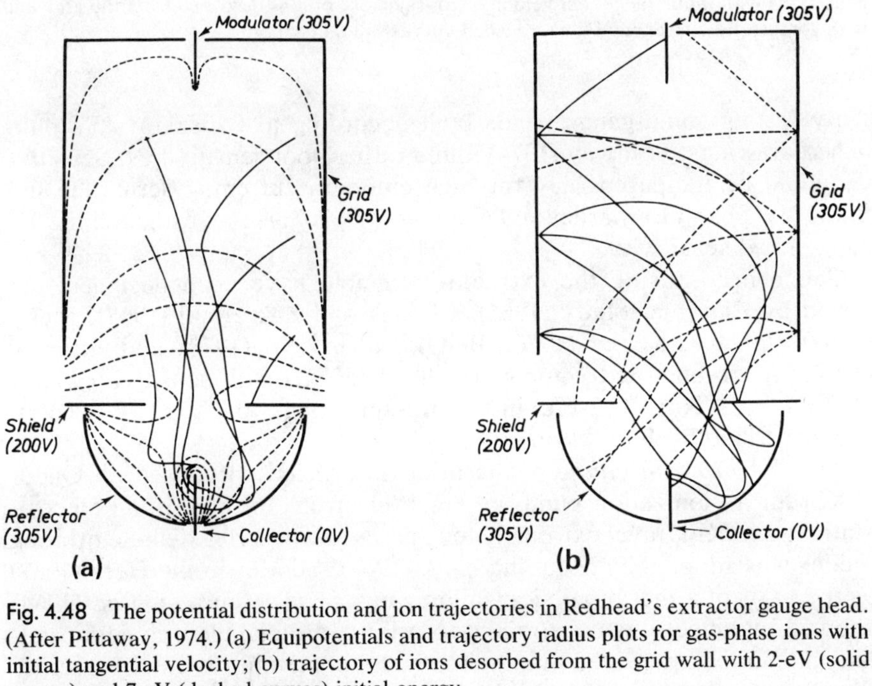

Fig. 4.48 The potential distribution and ion trajectories in Redhead's extractor gauge head. (After Pittaway, 1974.) (a) Equipotentials and trajectory radius plots for gas-phase ions with initial tangential velocity; (b) trajectory of ions desorbed from the grid wall with 2-eV (solid curves) and 7-eV (dashed curves) initial energy.

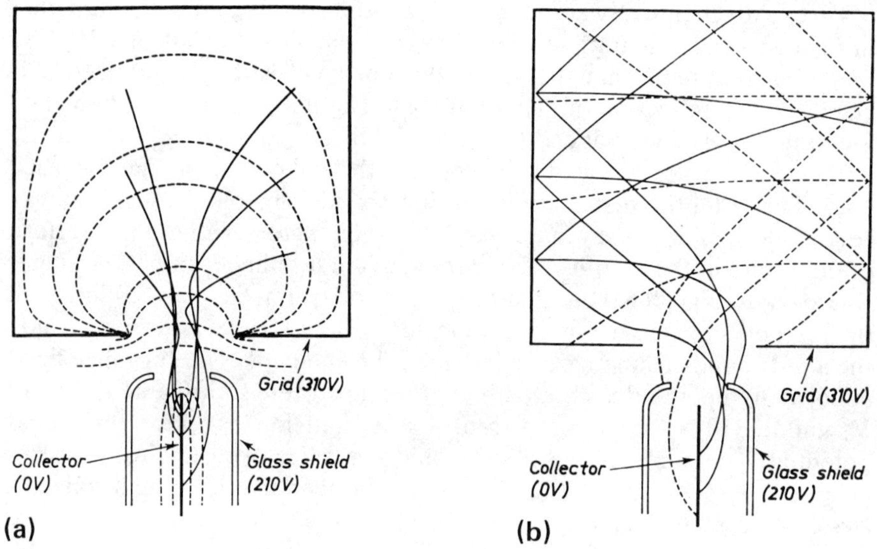

Fig. 4.49 The potential distribution and ion trajectories in Groszkowski's hidden-collector gauge head. (After Pittaway, 1974.) (a) Equipotentials and trajectory radius plots for gas-phase ions with initial tangential velocity; (b) trajectory of ions desorbed from the grid wall with 2-eV (solid curves) and 7-eV (dashed curves) initial energy.

By testing some gauge heads built according to Groszkowski's published drawings, Pittaway (1974) found rather poor sensitivities. Sensitivity could be increased only for high emission currents. Bernardet and Shaw (1968) and Bernardet and Choumoff (1970) also investigated the JG gauge head sensitivity.

Gauge heads using the extractor principle have been designed and tested by Clay and Melfy (1966), Cleaver and Zakrewski (1968), Melfy (1969), Barz and Kocian (1970), Beitel and Gosselin (1970), and Pittaway (1974). A modified extractor gauge head (MEG) with sensitivity 8.3×10^{-2} Pa^{-1} (11 Torr^{-1}), useful in the pressure range 10^{-10} Pa (10^{-12} Torr), was described by Fujii *et al.* (1983).

(f) *The orbitron gauge.* The orbitron principle proposed by Gabor (1962) for an ionization pump in which electrons injected in an electrostatic radial field travel extremely long paths and describe definite orbiting modes was adopted by Mouraud *et al.* (1964) and Meyer and Herb (1967) as the basis of a thermionic ionization gauge head. Mouraud *et al.* (1964) showed that electrons constrained to orbit in a radial electric field oscillate indefinitely, provided that their angular momentum at entry is greater than a certain critical value. In the gauge head they developed, an electrostatic field is created between an outer cylinder (ion collector) at 0 V and a fine quartz conductive wire (electron collector) at +500 V (Fig. 4.50). Electrons emitted from a short tungsten filament are prevented from following a direct path to the electron collector by a small reflector post around the quartz fiber. The reflector is operated at filament potential or at a small bias with respect to the filament. Ions formed in collisions between gas atoms and electrons orbiting on extended paths in the anular space are collected at the grounded collector, thus constituting a measure of gas density. The angular momentum necessary for electrons to orbit is obtained by varying the potentials of the electron source. Hooverman (1963) deduced the equations of motion of particles in the omegatron configuration and integrated them with a digital computer. The representative orbits he found as a result of his study are illustrated in Fig. 4.51.

Turner (1964), Gammon (1967), and Walker and Pacey (1968) also studied the pattern of orbits in the orbitron gauge head. In addition to the

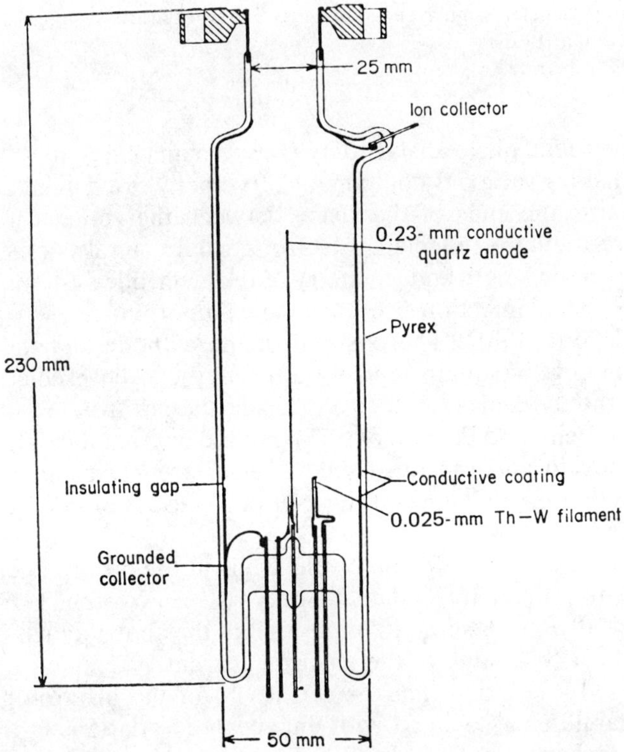

Fig. 4.50 Orbitron gauge head. (After Meyer and Herb, 1967.)

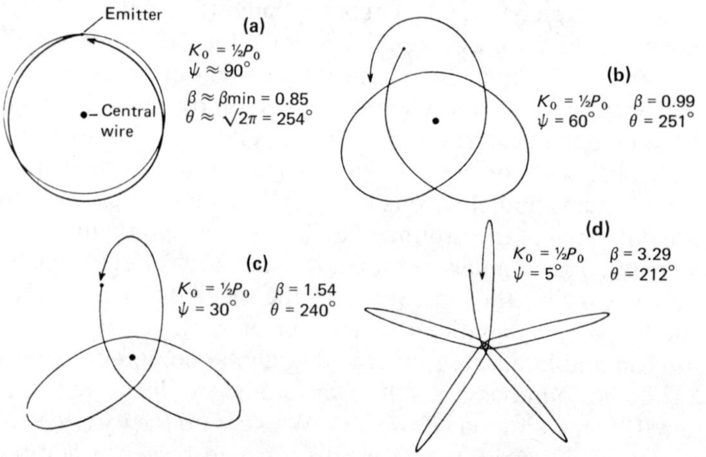

Fig. 4.51 Examples of orbits of particles in a logarithmic potential. K_0, kinetic energy; P_0, the strength of the potential; Ψ, angle of emission with respect to the radius vector; β, single dimensionless parameter, angular separation between two successive apogees (or perigees). (After Hooverman, 1963.)

orbits in the radial plane analyzed by Hooverman (1963), it was found that electrons having an axial component of velocity are forced along spiral paths towards the ends of the space between the concentric cylinders, from where they are reflected. Electrons would spiral back along paths depending on the length and geometry of the electrode system. The conditions of electron injection are extremely important for ion production (Feidt and Petit, 1981). The position of the cathode and the energy of electrons giving maximum ionizing efficiency can be predicted by laws governing the dynamics of electrons inside the orbitron (Feidt and Paulmier, 1972). Feidt and Paulmier (1978) pointed out that the optimum ionizing efficiency in an orbitron device is also strongly dependent on end effects and that the influence of these effects varies considerably with the injected electron current.

Path lengths of about 10^5 cm for electrons in the orbitron geometry are obtainable by running the cathode at only 10^{-6} A (Weston, 1979). A sensitivity of 7×10^{-2} Pa^{-1}, some 10^4 times that of the conventional BAG head, is obtained for this value of the electron current.

Gammon (1967) found that the sensitivity of the orbitron gauge head shows instabilities associated with unexpected variations in the ion current characteristics, as illustrated in Fig. 4.52. The explanation of these instabilities is in terms of the orbital motion of electrons.

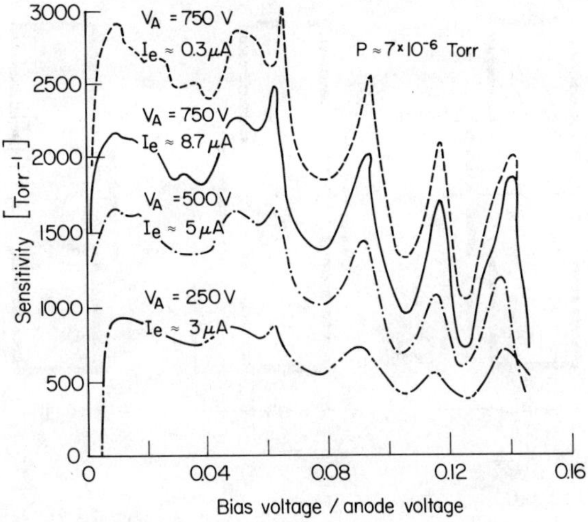

Fig. 4.52 Variation of the sensitivity of the orbitron gauge head with anode voltage and emission current. Reprinted with permission from *Vacuum* **17**, 330. R. B. Gamon. Copyright 1967 Pergamon Press.

Meyer and Herb (1967) suggested that oscillations are produced as a result of high ionization efficiencies, whereas Fitch *et al.* (1969) claimed that they are associated with potential disturbances near the filament. An approach to minimizing these instabilities is to inject electrons from a cathode located outside the ion collector (Fitch and Rushton, 1970). Operating the cathode at reduced values of the electron current also helps in minimizing these disturbances which result because of the behavior of the cathode as a sink and/or source of gas (see Section 4.4.6).

The emission of positive ions from the cathode may cause serious errors in pressure measurement. These errors can be overcome by avoiding the use of pure metals for the cathode and separating the ion collector into two cylinders by means of an insulating gap (Meyer and Herb, 1967). The cylinder in the vicinity of the cathode, grounded, would act as a guard ring for the measurement of the collected current on the upper cylinder.

The x-ray limit of the orbitron gauge head is about 10^{-9} Pa (10^{-11} Torr). The power required to operate the gauge head is ~90 mW. Blechschmidt (1974) discussed the low pressure limitations of this gauge.

(g) *The space-charge gauge.* In this gauge, proposed by Atkinson (1963) and discussed by Atkinson and Banks (1966), a dense beam of electrons emitted from a thermionic cathode is fired into a field-free space where it produces gas ionization (Fig. 4.53). Positive ions thus formed are collected in the negative potential well resulting from the electrostatic

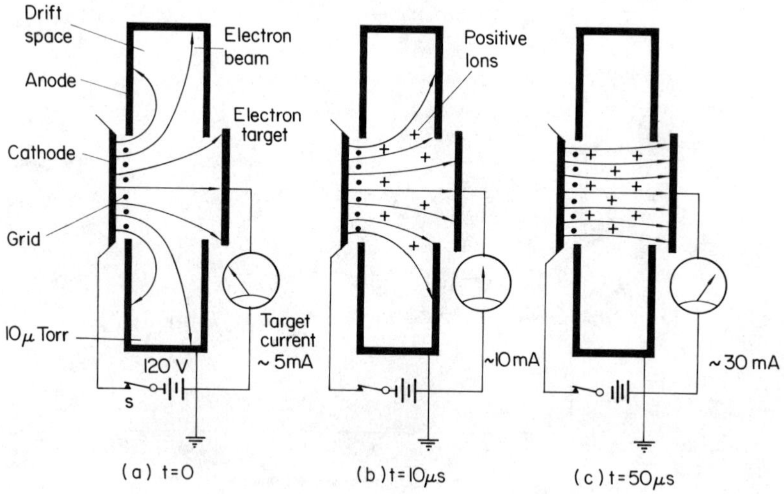

Fig. 4.53 Principle of the space-charge (ion focussing) gauge. Approximate trajectories of electrons are shown at different times (a)–(c) as positive ions are trapped in the electron beam. Reprinted with permission from *J. Sci. Instrum.* **43,** 512. H. H. Atkinson and P. H. T. Banks. Copyright 1966 The Institute of Physics.

repulsion between electrons in the drift space. Trapped ions modify the space charge forces, thus producing a decrease in the *transverse* spread of the electron beam. As a result of the change in electron trajectories, more of the beam is permitted to pass outside the drift space, where it is collected at a target. The electron current to the target rises with time and finally levels off. The rate of rise of the current to the target, measured with an oscilloscope, is proportional to the pressure of gas in the gauge head.

The gauge head (referred to as an *ion focusing* or a *time-of-rise* gauge head) can measure pressures from 10^{-2} Pa (10^{-4} Torr) to less than 10^{-6} Pa (10^{-8} Torr). The time rise corresponding to the pressure ranges mentioned is 1×10^{-6} and 1×10^{-5} s, respectively.

The main advantage of the space-charge gauge head is that it avoids the ion current collection, hence all the problems associated with the collection of positive currents. Atkinson and Banks (1966) reported that when reading pressures of about 10^{-3} Pa (10^{-5} Torr) with a BAG head and a space-charge gauge head in the presence of a sparking coil (5 m away from the measuring location), the BAG head read negative pressure whereas the space-charge gauge head readings remained steady.

The high electron current (about 30 mA) required for running the gauge head appears to be a serious drawback to its use in UHV.

(h) *High-pressure gauges.* Following the research of Schultz (1957) with respect to the high pressure limit of the BAG head, Schultz and Phelps (1957) designed a gauge head at which the collected current of positive ions remains linear with pressure in the range 10^{-2} Pa (10^{-4} Torr) to 133.3 Pa (1 Torr). They showed that the variation of I_c versus P can be linear above 10^{-1} Pa (10^{-3} Torr) provided that

(i) the distribution of the electric field between the electrodes permits the movement of electrons along straight lines between the cathode and the ion collector,
(ii) the sensitivity of the gauge head is small compared with the reciprocal of the highest pressure to be measured, and
(iii) the collection efficiency does not change with pressure.

Two types of gauge heads, with cylindrical and plane-parallel geometry, were designed to meet the requirements quoted above. The electric characteristics of gauge heads with either geometry are identical within experimental error.

The gauge head with plane-parallel geometry consists of a thoriated iridium filament stretched between two parallel plates, the electron and ion collectors, spaced about 3 mm apart (Fig. 4.54). Electrons emitted

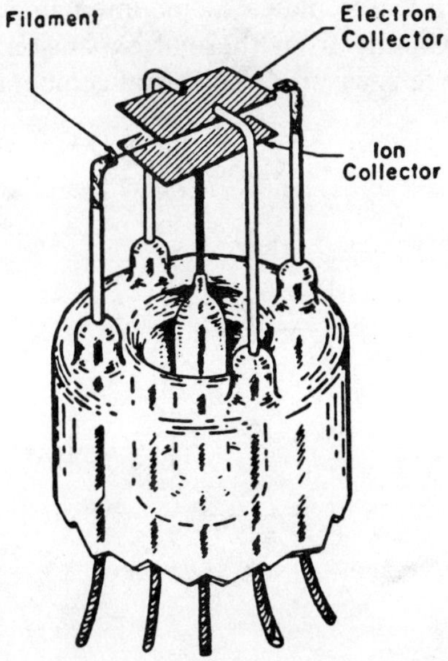

Fig. 4.54 High-pressure ionization gauge. (After Schultz and Phelps, 1957.)

from the cathode travel along a short path length to the electron collector with little chance of oscillating, thereby being involved in multiple collisions with gas atoms. The gauge sensitivity is therefore reduced ($s \simeq 80$ $Pa^{-1} \simeq 0.6$ $Torr^{-1}$ for N_2).

The large area of the ion collector permits a rather large collection efficiency.

The ratio of the collected current I_c to the emission current I_e varies linearly with pressure P from 133.3 Pa (1 Torr) to 10^{-3} Pa (10^{-5} Torr) for nitrogen. The x-ray limit of the gauge head is 10^{-3} Pa (10^{-5} Torr).

The gauge head designed by Schultz and Phelps (1957) has two drawbacks resulting from its construction:

(i) Some ions produced in electron–gas collisions may be collected at the cathode; this results in both a loss of sensitivity and possible damage of the cathode because of the ion bombardment.

(ii) Small changes in spacing of the electrodes produce changes in the gauge head sensitivity.

Cleaver (1967) developed a high-pressure gauge overcoming the drawbacks discussed. In his gauge head, two parallel plates having small area (the electron collector) are mounted at right angles to two other parallel plates with a large area (the ion collector) (Fig. 4.55). The cathode (a platinum–10%-rhodium wire, yttrium oxide coated) running axially down the center of the collector plates is mounted under spring tension to prevent buckling resulting from thermal expansion. Thus, dimensional stability of the cathode is assured. This arrangement avoids the collection

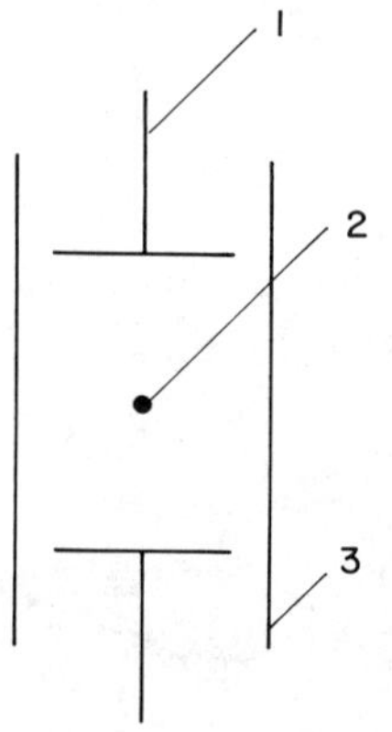

Fig. 4.55 Electrode arrangement in Cleaver's high-pressure gauge head. 1, electron collector; 2, cathode; 3, ion collector.

of positive ions at the cathode and its damage by ionic bombardment. The gauge head has a rugged construction. Sensitivity (14.6 Pa^{-1}, 0.11 $Torr^{-1}$) is constant within the range 133.3 Pa (1 Torr) down to 10^{-3} Pa (10^{-5} Torr) provided that the electron collector bias is no greater than 42 V. Cleaver (1968) measured air pressures within the range 133.3–5×10^{-3} Pa (1–10^{-5} Torr) with a sensitivity of 6.66 Pa^{-1} (5×10^{-2} $Torr^{-1}$) by slightly modifying the gauge head construction.

Weinman (1966) designed a gauge which can be operated at pressures between 3×10^{-1} Pa (2×10^{-3} Torr) and 533 Pa (4 Torr) corresponding to densities encountered in atmosphere between 35 and 85 km altitude. An additional electrode permits the operation of the gauge head under space controlled conditions up to this upper value of pressure. The positive ion current increases monotonically with pressure.

4.4.8 Hot-Cathode Ionization Gauge Heads with Magnetic Field

(a) *The hot-cathode magnetron gauge.* Conn and Daglish (1954a) described a hot-cathode gauge head where the path length travelled by electrons is increased by using a magnetic field, as in a magnetron, operating under cutoff conditions (*cutoff* designates the smallest theoretical value of a steady magnetic flux density which would prevent an electron emitted from the cathode at zero velocity from reaching the anode held at constant voltage). In this gauge head, electrons emitted from a filament into a cylindrical anode volume describe circular paths along an axial magnetic field established by two solenoid coils producing up to 1989 A/m (250 Oe). Electrons move to and fro between the end plates, which limit the length of the anode, perpendicular to the axis of the cylinder. The negatively biased plates are used both to prevent electrons from escaping out of the anode volume and to collect positive ions. If the magnetic flux density exceeds the cutoff value, most electrons fail to reach the anode and return to the cathode (or to a virtual cathode depending on the space-charge conditions). The electron density within the anode is therefore greatly increased, and so is the path length travelled by electrons compared with the conventional triode. Magnetic flux densities in excess of the cutoff value lead to an increase in temperature as a result of the higher energy of electrons and lead to the bombardment of the filament by positive ions. As a consequence the gauge head must be operated with magnetic flux densities greater than, but near, the cutoff. Since a sharp cutoff is not necessary, neither magnetic fields of great uniformity nor accurate alignment conditions are required. The cutoff value and the bias voltage of the anode are closely interrelated.

The sensitivity of the gauge head, increased as a result of the long paths of electrons, is several hundred times greater than that of the BAG head. Pressures measurements of 3×10^{-4} Pa (2×10^{-6} mm Hg) or greater have been reported.

Houston (1956) developed a gauge head using the magnetron principle for lengthening the paths of electrons, thus increasing the sensitivity of the gauge head. A uniform magnetic flux density of 13,528 A/m (1700 Oe) permitted the achievement of a sensitivity of 0.075 A Pa^{-1} (10 mm Hg^{-1}) for an emission current of 1 mA and a 1-kV potential at the anode. It is worth mentioning that the conventional BAG head operated with an electron emission of 10 mA has a sensitivity of 7.5×10^{-4} A Pa^{-1} (0.1 A $Torr^{-1}$). The estimated lowest pressure limit of the gauge head is 10^{-13} Pa (10^{-15} mm Hg).

For both gauge heads discussed, neither the x-ray limit nor the pumping speed was measured. In addition, an external solenoid, which provided the magnetic field, made use of the gauge head rather cumbersome.

The advantage offered by the use of the magnetron principle was fully exploited by Lafferty (1961a,b) who designed a UHV gauge which was simple to operate and had stable electrical characteristics. In the Lafferty gauge head, a hair-pin filament is mounted within a cylindrical anode with end plates (Fig. 4.56). Electrons emitted from the filament are forced to spiral along the magnetic field lines produced by a permanent external cylindrical magnet surrounding the gauge head envelope. One of the end plates, negatively biased, serves as an ion collector, the other as a shield preventing electrons from escaping the anode volume. Stable operation is obtained with electron emission currents in the range 10^{-7}–10^{-9} A. Higher values of electron emission lead to instabilities and nonlinearities of I_c versus P. An electron emission current of 4×10^{-7} A provides, for example, the same ion current as a conventional BAG head operated at 8 mA.

The influence of the magnetic field on the current of positive ions collected at the collector electrode is illustrated in Fig. 4.57 where the electron and collected ion current are plotted as a function of the magnetic flux density. The positive ion current to the collector is almost negligible for magnetic flux density values below the cutoff. After reaching the cutoff value (772 A/m, 97 Oe. for the gauge head illustrated in Fig. 4.57), the electron and ion current sharply decrease and increase, respectively. The decrease of the electron current is due to the electrons which miss the anode, whereas the ion current increase is the result of ionization produced by electrons travelling along extended paths. At a certain value of the magnetic flux density exceeding the cutoff, electrons are confined in a small volume, thereby ionizing fewer gas atoms (kinks on both electron

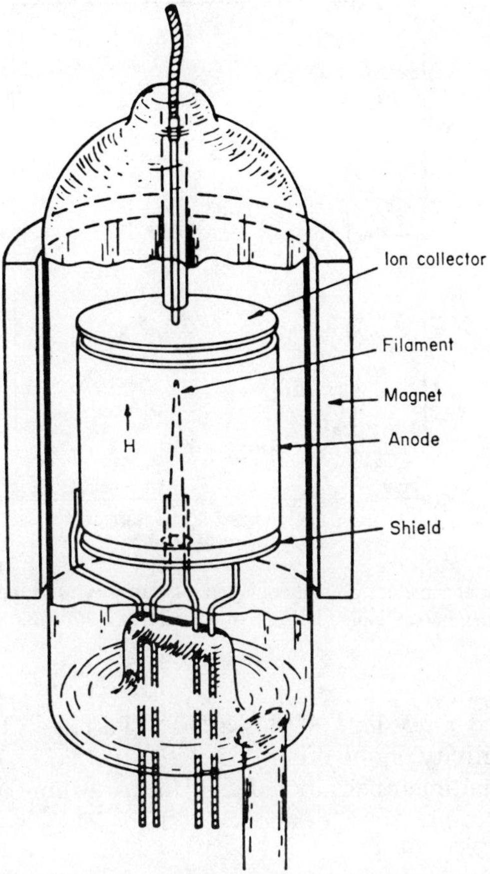

Fig. 4.56 Hot-cathode magnetron gauge head. (After Lafferty, 1961a.)

and collected current curves in Fig. 4.57). Beyond this value, the collected ion current drops slowly with increasing magnetic flux density.

The gauge head operation at higher pressures (about 10^{-4} Pa, 10^{-6} Torr) becomes space-charge limited. The electron current is then pressure dependent and cannot be continuously regulated as in the BAG head. Below about 10^{-6} Pa (10^{-8} Torr) the electron current is pressure independent, and therefore the collected ion current varies linearly with pressure.

The presence of the magnetic field suppresses photoelectrons from the ion collector over a rather large area. The ratio of the collected ion current to the x-ray photocurrent when the magnetic field is on is about 10^6 times larger than when the magnetic field is off.

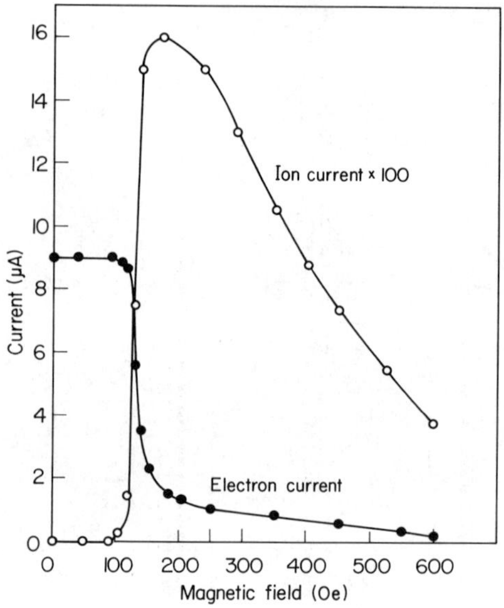

Fig. 4.57 Variation of the electron and collected ion currents as a function of the magnetic flux density in the magnetron gauge head. (After Lafferty, 1961a.)

The calculated x-ray limit settles at 10^{-12} Pa (10^{-14} Torr). The gauge head has a sensitivity on the order of 5×10^{-4} Pa^{-1} (375 $Torr^{-1}$). The use of an electron multiplier permits reducing the x-ray limit and getting linear response down to pressures of about 4×10^{-15} Pa (3×10^{-17} Torr) (Lafferty 1962a,b, 1963, 1968).

A hot-cathode magnetron with photocurrent suppressor was reported by Visser (1967). The lowest pressure the gauge can measure accurately is 1.0×10^{-13} Pa (7.7×10^{-16} Torr).

Discussions with regard to both electrical characteristics and performance of the magnetron gauge in ultrahigh vacuum can be found in the papers by Pelz and Newton (1967) and Davis (1968).

(b) *The Klopfer gauge.* Klopfer (1962) suggested the use of a gauge head with an electrode structure similar to that of the omegatron which makes no use of an rf field. The interesting features of this gauge head are the following.

(i) The ion collector is removed from the vicinity of the electron collector to a location where it can "see" neither x-rays emitted from the electron collector nor photons emitted from the cathode.

(ii) Electrons are formed in a ribbon-shaped beam which passes through the ionization volume without impinging on the entrance or exit slits in the accelerating electrodes.

Electrons emitted from a thermionic cathode K, are collimated by the electrostatic field of the electrodes which it crosses and are focused by a strong axial magnetic field in a ribbon-shaped beam (Fig. 4.58). After passing the ionization chamber where they produce ionization, electrons are collected at the electron collector T (referred to as the electron trap). The ionization chamber is the volume bordered by the electrodes G_3, A, G_4, H, and another two electrodes not shown in Fig. 4.58. Positive ions are drawn to the ion collector J under the influence of the electrostatic field of the ionization chamber. The same field prevents ions from escaping in a direction parallel to the magnetic field.

Owing to the location of the ion collector not many x-rays yielded by the electron beam can reach the ion collector directly. Photoelectrons released from the ion collector are returned to this electrode by the magnetic field. Thus, the proper alignment of the magnetic field becomes essential to the gauge head performance.

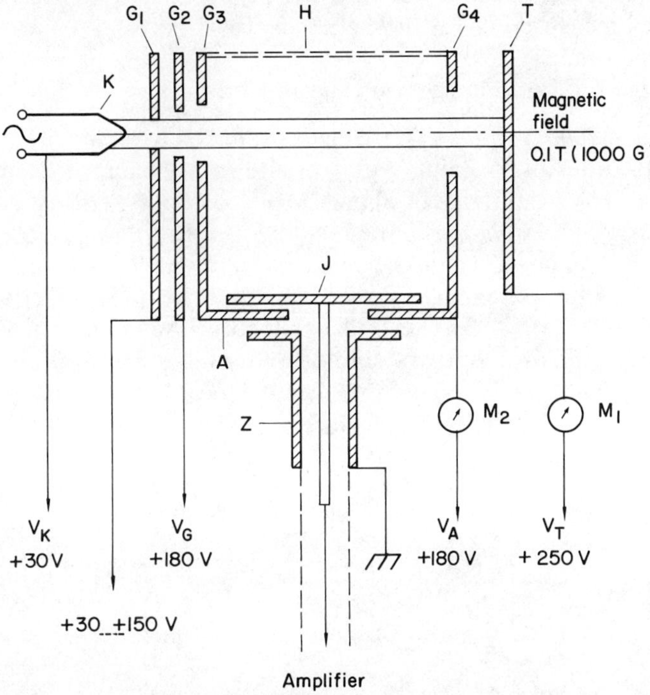

Fig. 4.58 The Klopfer gauge head. Reprinted with permission from *Trans. 8th Nat. Vac Symp.* 1961. A. Klopfer p. 439. Copyright 1962 Pergamon Press.

There are no published results regarding the x-ray limit of this gauge head. The N_2 sensitivity, when operated with the potentials indicated in Fig. 4.58, is 0.11 Pa^{-1} (15 $Torr^{-1}$). The normalized ion current (I_c/I_e) varies linearly with pressure between 1 and 10^{-9} Pa (10^{-2}–10^{-11} Torr).

(c) *The virtual cathode gauge.* Lloyd (1966) suggested the principle of a gauge in which electrons emitted from a thermionic cathode are confined to a well-defined beam shape by a magnetic field. The field also prevents electrons from moving in transverse directions with respect to the field lines. Electrons thus become concentrated in a so-called virtual cathode. Positive ions produced in the gas phase are trapped in the potential well of the electron beam where they neutralize the electron space charge. The displacement current when the virtual cathode suddenly dissipates is proportional to gas pressure.

Such gauges can measure pressure in the range 10^{-2}–10^{-6} Pa (10^{-4}–10^{-8} Torr) and are not affected by x-rays even in the measurement of very low pressures, because the ion current is not directly measured. Lloyd (1966) mentioned that linearity is preserved down to 5×10^{-7} Pa (4×10^{-9} Torr). At lower pressures, measurements are not reproducible. The lowest pressure Lloyd measured is 7×10^{-8} Pa (5×10^{-10} Torr).

4.4.9 Cold-Cathode Ionization Gauge Heads

(a) *The discharge tube.* This gauge is referred to as the pressure indicator. The order of magnitude of pressure in a vacuum system can be *appreciated* over the range of about 2666.6 to 10^{-1} Pa (20–10^{-3} Torr) by setting up a *glow discharge* in the residual gas. For that purpose, a glass tube with metallic electrodes at each end is connected to the vacuum system, and a dc potential is applied to the electrodes in series with a resistor (Fig. 4.59). A few secondary electrons set free from the electrodes by cosmic rays and stray radioactivity initiate ionization. Positive

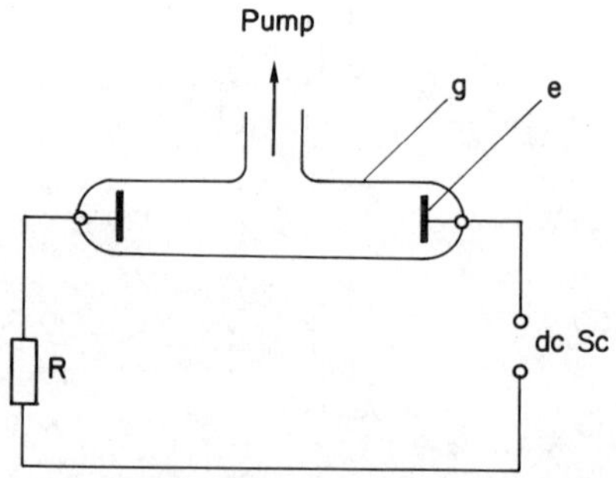

Fig. 4.59 Discharge tube. e, electrodes; g, glass tube; R, resistor.

ions possessing enough energy to cause secondary emission strike the electrodes with subsequent further ionization of the gas. A self-sustaining glow discharge is thus created. Stability of the discharge is achieved by using a large series resistor, which also prevents the flow of excessive currents. At pressures in the range 2666.4–400 Pa (20–3 Torr) a broad spark (streamer) discharge fills the space between the electrodes. As pressure is reduced, the glow discharge first widens towards the tube walls (at about 133.3 Pa, 1 Torr), then becomes sectioned by dark spaces in definite regions in the vicinity of the electrodes. At pressures of about 10^{-1} Pa (10^{-3} Torr) the walls of the glass tube begin to fluoresce. At pressures below 10^{-1} Pa (10^{-3} Torr) the fluorescence of the glass tube fades out giving rise to a state referred to as *black discharge*. In practice, a Tesla coil is used instead of a dc or ac potential for exciting the discharge.

Because the color of the discharge is dependent on the nature of the gas, a spectroscope can supply qualitative information with respect to the residual gas.

(b) *The Penning gauge (PIG).* The idea of determining pressure by measuring the current of a glow discharge in which the path length of electrons is increased due to a magnetic field was first implemented by Penning (1936, 1937a,b). He made use of crossed electric and magnetic fields to produce what is referred to as the *Penning discharge*. In his gauge a pair of cathode plates is arranged on either side of a ring-shaped anode (Fig. 4.60a). A 2-kV dc source in series with a resistor and an ammeter is connected to the electrodes, and a magnetic field is arranged parallel to the electrode axis.

The movement of electrons in the gas discharge between the electrodes is enhanced both by their travelling in helical paths along the magnetic flux lines and by their confinement within the anode ring caused by the magnetic field. Owing to this arrangement, electrons make many trips to and from the cathode plates before being collected at the anode. This therefore produces increased ionizing collisions. The positive ions are influenced little by the magnetic field, and they are collected at the cathodes. The total discharge current is used as a measure of pressure.

The gauge known as a Philips gauge could measure pressures in the range 10^{-1}–10^{-3} Pa (10^{-3}–10^{-5} Torr). At 10^{-3} Pa (10^{-5} Torr) the gauge performance was unstable and erratic.

Further studies of the gas discharge mechanism by Penning and Nienhuis (1949) led to the design of a gauge head where an anode cylinder replaced the anode ring of the older pattern (Fig. 4.60b). The Penning–Nienhuis gauge head, having a magnetron-like geometry of electrodes, was more stable in operation and could measure pressures in the range 6×10^{-1}–6×10^{-5} Pa (5×10^{-3}–5×10^{-7} Torr).

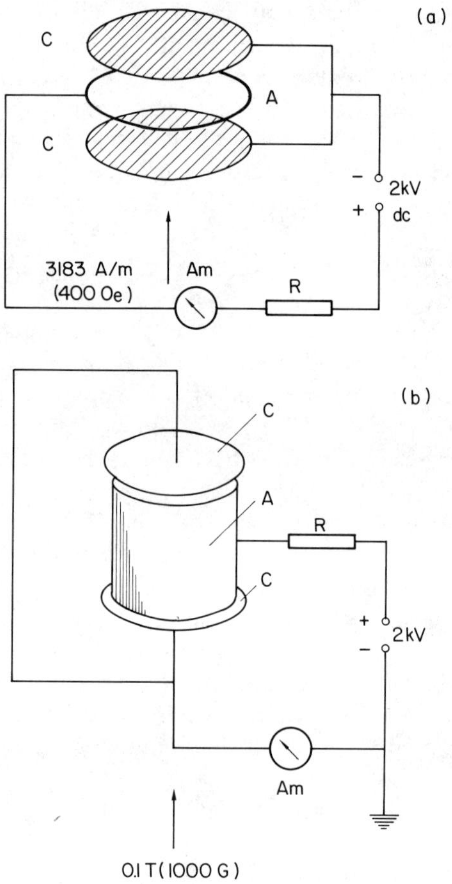

Fig. 4.60 The electrode arrangement and electrical control of the (a) Penning gauge and (b) Penning–Nienhuis gauge. A, anode; Am, ammeter; C, cathode; R, resistor.

A further modification of the Penning gauge head, permitting the measurement of pressure over 13–10^{-5} Pa (10^{-1}–10^{-7} Torr), was described by Evans and Burmaster (1950).

The advantages of cold-cathode gauges over thermionic gauges are

(i) simple and rugged construction, insensitive to air exposure because of the lack of a hot cathode;
(ii) absence of x-rays as the electron current is pressure independent;
(iii) no additional vapor pressure to the residual gas such as produced by the hot cathode in hot-cathode gauge heads;

(iv) very large sensitivity due to the long paths travelled by electrons; and

(v) reduced EID effects therefore causing little error in measurements.

Drawbacks of the Penning gauge are

(i) nonlinearity of the collected current as a function of pressure;

(ii) temporary low-frequency instabilities and high-frequency oscillations due to the trapped electron space charge in the plasma;

(iii) difficulties in starting the electrical discharge in gas at low pressure;

(iv) high pumping speed (one order of magnitude larger than that of hot-cathode ionization gauge heads) even for inert gases such as helium;

(v) field emission particularly at the cathode's edges caused by the high electric field at the cathode (*the resultant field emission current leaving the cathode cannot be distinguished in the measuring circuit from the positive ion current arriving at the cathode;* thus, there is a lower pressure limit of the measuring range analogous to the x-ray limit in thermionic gauge heads); and

(vi) extinction of the discharge at low pressures.

Starting the discharge at low pressures can be facilitated by providing electrons to initiate ionization. McIlwraith (1947) used a tungsten filament located close to the electrode structure. When flashed for a short time the filament provided the electrons necessary for "striking" the discharge. Young and Hession (1963) designed a conventional gauge with Penning electrode structure slightly differing from that of the parent gauge head. A *trigger filament* located just behind an off-axis hole in one of the cathode plates provides electrons when energized. The filament is also flashed for a short period of time. The construction is known as a *trigger gauge*.

Young (1966) calibrated the trigger gauge against a mass spectrometer and a BAG head in the pressure range 1×10^{-5}–1.3×10^{-10} Pa (1×10^{-7}–1×10^{-12} Torr) and found that the collected ion current varies with pressure at power $n = 1.2$ [see Eq. (2.17)]. At 10^{-9} Pa (10^{-11} Torr) sensitivity was found to be 7.5×10^{-3} A/Pa (1 A/Torr). The value of the sensitivity increased to 4.7×10^{-2} A/Pa (6.3 A/Torr) at 10^{-5} Pa (10^{-7} Torr).

Trigger gauges have also been calibrated by Bryant and Gosselin (1966) and Lange *et al.* (1966).

Another approach for triggering the gas discharge at low pressures is to incorporate a point or a sharp edge on one of the electrodes (Evans and Burmaster, 1950) or to use a radioactive substance emitting beta radiation (Hayashi, 1966). In the first case, with a sufficient voltage the field emission from the point or sharp edge provides electrons but leads to the pressure limitation discussed before. When radionuclides are used, the health hazard has to be considered.

Detailed analyses and discussions are available with respect to:

(i) the Penning discharge mode (Valle, 1950, 1952; Dumas, 1955; Jepsen, 1961; Knauer, 1962);
(ii) space-charge effects (Vasi'leva and Reichrudel, 1962);
(iii) radial field distribution (Dow, 1963; Knauer and Lutz, 1963);
(iv) performance with various electrode arrangements (Conn and Daglish, 1953, 1954b).

Gauge heads of the Penning type are largely used in applications where accuracy is not essential.

(c) *The inverted magnetron gauge (IMG).* Investigations aimed at overcoming the drawbacks of the Penning magnetron diode gauge resulted in the construction of the inverted magnetron gauge (Beck and Brisbane, 1952; Haefer, 1953, 1954, 1955). (The major difference between the magnetron and inverted magnetron is that the position of the anode and cathode are interchanged.) Thus, in Beck and Brisbane's gauge the anode is a long, fine wire stretched along the axis of a cylinder with multiple cathodes. The gauge could measure pressures from 10^{-2} to 10^{-6} Pa (10^{-4} to 10^{-8} Torr) and had a guad ring to the cathode in order to eliminate leakage currents assumed to be affecting determinations in the low pressure range.

Investigations of Townsend's discharge in crossed electric and magnetic fields (Redhead, 1958) led to the development of the inverted magnetron gauge of Hobson and Redhead (1958), illustrated in Fig. 4.61. The design overcomes the major drawbacks of the Penning gauge discussed before, namely the failure to start the discharge at low pressures and the low pressure limitation produced by the field emission current from the cathode. This is realized by arranging the electrodes such as to produce efficient electron trapping in the discharge region and adding an auxiliary cathode to shield the ion collector from the influence of high electric fields.

The gauge head consists of three electrodes, a rod anode, a closed-end cylinder (the ion collector) with center holes in the end plates, and a box-shaped auxiliary cathode enclosing the ion collector. Two shielding cylinders connected to the auxiliary cathode project inside the ion collector, thus preventing field emission from the collector end plates.

The main features of the gauge head are the following.

(i) The current to the auxiliary cathode, having as a component a part of the field emission current, does not pass through the ion amplifier.

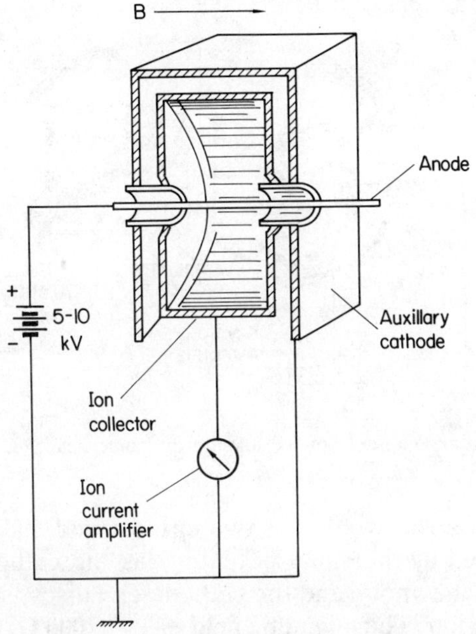

Fig. 4.61 Cutaway diagram of inverted magnetron gauge head and associated circuit. (After Hobson and Redhead, 1958.)

(ii) Electrons required to start the discharge in the residual gas at low pressure are supplied by the field emission of the auxiliary cathode.

The anode is operated at 6 kV while the cathodes (ion collector, auxiliary cathode) are grounded. The magnetic flux density is 0.2 T (2000 G).

The ion-current–pressure characteristics follow the predictions of Eq. (2.18) with $n \simeq 1.10$–1.15 within the range 10^{-2}–10^{-8} Pa (10^{-4}–10^{-10} Torr) (Hobson and Redhead, 1958). Feakes *et al.* (1965) found $n \simeq 1.25$ for N_2 and He in the pressure range 10^{-6}–3×10^{-11} Pa (10^{-8}–3×10^{-13} Torr).

The inverted magnetron gauge can measure pressures from 10^{-1} to 10^{-10} Pa (10^{-3}–10^{-12} Torr) and strikes the discharge even at 10^{-8} Pa (10^{-10} Torr). The sensitivity is 7.5×10^{-3} A/Pa (1 A/Torr).

(d) *The magnetron gauge (MG)*. A geometric variant of the Penning gauge suitable for pressure measurement from about 10^{-2} to 10^{-10} Pa (10^{-4}–10^{-12} Torr), with no limit on extending the measuring range both upward and downward, was developed by Redhead (1959). The gauge head has the magnetron configuration of electrodes as illustrated in Fig. 4.62. The anode is a cylinder with many perforations to permit the gas flow into the electrode structure. The cathode, which is in the form of a

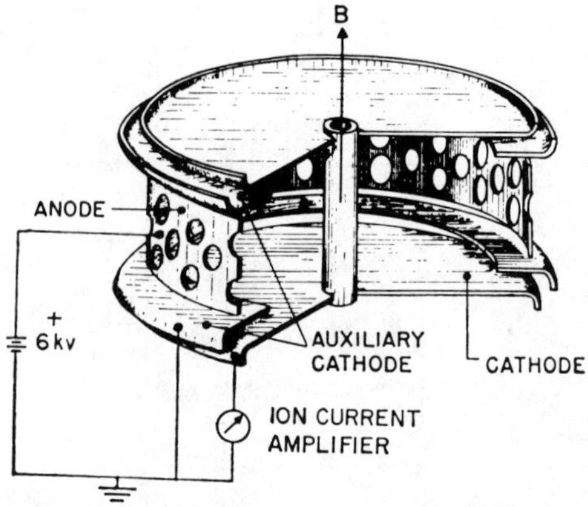

Fig. 4.62 Electrode arrangement of the magnetron gauge head. (After Redhead, 1959.)

spool, consists of a rod welded to two circular end disks. Two auxiliary cathodes, operated at the same potentials as the cathode (ground) and inserted between the anode and the cathode-end discs, shield the cathode against field emission. The magnetic field (0.1 T, 1000 G) and anode potential (4.5–6 kV) are selected so that electrons are trapped in the magnetron-like cavity. Electrons travel long distances between cathode and anode, producing enhanced ionization and setting up a negative space charge. Similar to the operation of the inverted magnetron, the field emission currents do not pass through the ion current amplifier, thus leaving unaffected the ion-collected current.

The advantages of this gauge head over the inverted magnetron gauge head are that the ion-current–pressure characteristic predicted by Eq. (2.18) is linear ($n = 1$) within the range 10^{-2}–5×10^{-8} Pa (10^{-4}–5×10^{-10} Torr) (Redhead, 1959; Hobson, 1959; Feakes and Torney, 1963; Torney and Feakes, 1963; Bryant *et al.*, 1963) and that the sensitivity for N_2 within the linear range is very high (7.5×10^{-2} A/Pa, 10 A/Torr).

Below 10^{-8} Pa (10^{-10} Torr) the ion-current–pressure characteristic is nonlinear, with n depending on the operating conditions (magnetic field density, anode voltage, type of gas).

A magnetron gauge head with no shield electrodes was reported by Bryant and Gosselin (1966). The gauge is nonlinear at all pressures.

The drawback of the magnetron gauge, common to all crossed-field cold-cathode gauges, is the very high pumping speed (about $10^{-3} m^3 s^{-1}$), more than one order of magnitude greater than that of a BAG.

(e) *The radioactive gauge.* Interest in using radioactive gauges dates back to 1946 as a result of the increasing demand for large-scale vacuum processing techniques. Bulk dehydration, molecular distillation, and vacuum metallurgy are areas where vacuum gauges are needed to permit continuous measurement of water vapors as well as nonpermanent or even corrosive gases. None of the gauges available at the time (McLeod, Pirani, thermocouple, and ionization gauges) were suitable. In addition, a range of pressures spanning from atmospheric pressure to 1 Pa (10^{-2} Torr) had to be measured, if possible, by the same gauge. These requirements thus led to the development of a new type of gauge, the radioactive gauge (also referred to as a radiological gauge).

Within the framework of these requirements, Mellen (1946) and then Downing and Mellen (1946) reported details of the development of a new ionization gauge. The gauge, whose trade name was Alphatron, made use of radium as an ionizing agent. The ions, produced as a result of the collisions between alpha particles emitted by the source and gas molecules inside the gauge head, were collected at an electrode by means of an electric field. The resulting weak current was amplified and provided an analog response to the gas density inside the gauge head. The gauge spanned the pressure range between 266 and 1 Pa (20–10^{-2} Torr).

Since then much work has been put into improving this type of gauge by

(i) diminishing the health hazard arising from the radioactive source,
(ii) extending the range of pressures within which the response of the gauge head remains linear,
(iii) obtaining an electronic circuit having a constant amplification factor over the whole measuring range, a reduced zero drift, and a high sensitivity, and
(iv) reducing the dimensions of both the gauge head and the associated electronic circuit.

The first radioactive gauges (Mellen, 1946, 1948; Downing and Mellen, 1946; Sibley and Roehring, 1953; Sicinski *et al.*, 1954; Lawrence, 1954; Beynon and Nicholson, 1956; Vacca, 1957) made use of alphas particles emitted in the decay of ^{226}Ra as the ionizing agent. The first decay product, the nobel gas radon ^{222}Rn, had to be prevented from escaping from the source, both for forther utilizing its series of decay products and for health hazard reasons. Alpha particles were emitted by isotopes produced in the natural decay series of radium, e.g., ^{218}Po (historical name radium A) and ^{214}Pb (historical name radium B). Beta and gamma particles are also emitted in the decay, but the specific ionization produced is negligible compared with that of alpha particles. The health hazard due to ^{222}Rn and gamma particles was claimed to be overcome by both sealing the radioactive source and shielding the ionization chamber with a thick lead wall. Evidence shows that even when the source is sealed, gaseous radon diffuses through the gauge head walls, thus jeopardizing personnel (Moreau, 1962). On the other hand, while the lead shield against gamma rays proved effective, it made the gauge head unwieldy and difficult to handle.

Attempts were made to avoid ^{226}Ra sources because of its first decay product, radon. Thus, a mixture of ^{210}Pb (historical name radium D) and ^{210}Bi (historical name radium E) in radioactive equilibrium with ^{210}Po was chosen as a source (Moreau, 1962). The main ionizing agents are alpha particles, although beta particles are also emitted. Gamma particles released in the decay process were entirely retained by a 4-cm thick stainless steel wall shielding the gauge head. Suggestions to use the pure alpha particle decay products of ^{144}Nd, ^{147}Sm, and ^{190}Pt, could not be implemented because of the unavailability of these isotopes at that time (Moreau, 1962). Since 1959 neither ^{256}Ra nor ^{210}Pb–^{210}Bi mixtures are utilized as a source because of either the radon release or the delicate stability of the isotope mixture.

Further efforts were aimed at finding isotopes emitting pure alpha particles, or alpha particles accompanied by low-energy gamma particles. Thus, ^{210}Po, ^{239}Pu, ^{240}Pu, and ^{241}Am were employed as sources by many workers. By utilizing these isotopes, the lead shielding could be replaced by moderately thick ionization chamber walls.

When the difficulties connected with the use of alpha-emitting isotopes were overcome, the development of radioactive gauges was given a push forward. First, a radioactive isotope of hydrogen, ^{3}H, was chosen as a pure beta particle source. The lower energy of the beta particles made it possible to reduce both the dimensions and the weight of the ionization chamber considerably. Tritium (^{3}H) forms a chemically stable bond as titanium or zirconium tritide and is not released from the source in a detectable amount, as long as the temperature remains below 200°C (Rème, 1963). The x-ray *bremsstrahlung* resulting from beta particle impingement on the metallic carrier of the source are completely adsorbed on the metallic wall of the ionization chamber. Owing to these features, gauges using ^{3}H as a source are considered safe.

Other isotopes like ^{147}Pm, which emits pure beta particles, reportedly have also been used as sources (Kirsanov, 1964).

Like other types of ionization gauges, radioactive gauges contain two parts: a gauge head to transduce molecular densities in terms of electric currents and an associated electronic circuit to amplify or use these currents for the display of analog or digital information.

Generally the sensing head is composed of an ionization chamber containing the electrodes and the source. The whole assembly is enclosed in a housing to fit the vacuum system. In some cases, the housing and the ionization chamber are one and the same; in other cases, additional repelling or reflecting electrodes are added to the ionization chamber.

The ionization chamber is generally a cylinder with the electrodes running parallel to the axis. The ionization chamber dimensions depend upon the distance within which *ionizing* particles, emitted by the source, are stopped by the gas molecules inside the chamber and lose their energy and the distance the *ionized* particles travel between their buildup point and the collector electrode. The first condition depends on the nature of the source chosen (Mellen, 1946; Downing and Mellen, 1946; Khavkin, 1961; Moreau, 1962). In the second condition (which in turn depends upon the magnitude of the collecting potential) the distance that *ionized* particles travel before being collected must be shorter than the distance within which these particles could recombine. Obviously, this condition determines the dimensions of the ionization chamber for high densities. (For high densities, the distance travelled by *ionized* particles of several tens of electron volts before recombination is on the order of magnitude of millimeters.) For low densities, the dimensions of the ionization chamber must be sufficient so that *ionizing* particles can still collide with gas molecules and ionize them.

Overcoming the health hazard, due to the source employed, and accommodating the gauge response over a large range of densities are further considerations in utilizing a radioactive gauge head. The choice of an isotope having a lifetime which insures the gauge stability, in amounts which yield enough ionizing particles for the gauge to function but not enough to be hazardous, is fundamental. A careful balance must be made when choosing the amount of the radioisotope and the electric field necessary for ion collection and the distance between the electrodes. This is in order to obviate the ignition of a self-sustaining Pashen-type discharge and to provide linearity of response within the range required.

For a correctly designed gauge head connected to an amplifier with a linear response within the working range, loss of linearity in the ionization chamber output usually occurs at the higher and lower limits of the range. Loss of linearity at high densities is unanimously recognized as the result of recombination phenomena. The limit of usefulness of the gauge by loss of linearity at low pressures has been explained in different ways. For gauge heads where sources yielding alpha particles are employed, the loss of linearity is due to a "dark" current, induced by direct collection of *ionizing* particles at the collector (Mellen, 1948; Sibley and Roehring, 1953; Sicinski *et al.*, 1954; Lawrence, 1954; Akishin, 1960; Vitkus, 1960). In the case of sensing heads employing beta particles as the ionizing agent, the "dark" current is explained by secondary electron emission from the collector, owing to the x-ray *bremsstrahlung* (Roehring and Vanderschmidt, 1960; Blanc and Dagnac, 1964; Vanderschmidt and Simons, 1967; Rigby and Wright, 1968) or by leakage currents (Howard *et al.*, 1968).

Currents resulting from the ionization process are exceedingly small, ranging between 10^{-13} and 10^{-8} A, and delicate techniques are required for their measurement. In the main, one or two methods are employed, i.e., an analog or a pulse technique. In one reported case the collected current was measured by using an electron amplifier (Akishin, 1960).

The ionization chamber is usually a metallic cylinder where the walls constitute one of the electrodes. Well insulated from the walls, the second electrode is situated along the geometrical axis of the cylinder (Fig. 4.63a).

The need to measure densities over many orders of magnitude with the same gauge head led to different designs. Some workers designed two ionization chambers, the volume of each appropriate for a given range of densities. A small chamber, with a volume of a few cubic millimeters, was

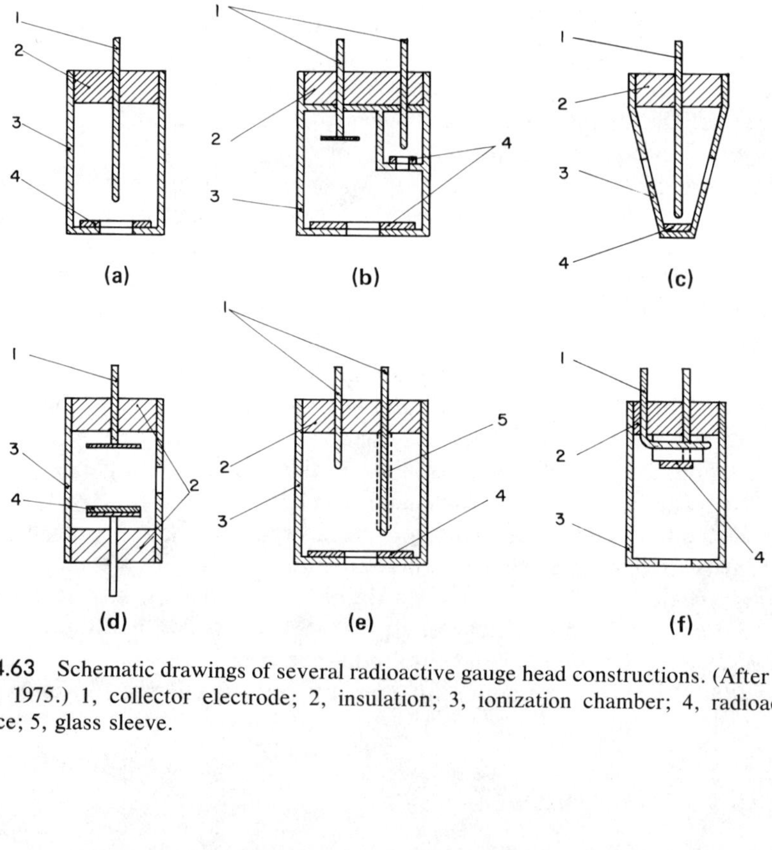

Fig. 4.63 Schematic drawings of several radioactive gauge head constructions. (After Berman, 1975.) 1, collector electrode; 2, insulation; 3, ionization chamber; 4, radioactive source; 5, glass sleeve.

intended for the measurement of high densities, whereas a large chamber, with a volume of tens of cubic millimeters was used for measurements of low densities (Fig. 4.63b). The small chamber was contained in the large one which fit directly into the vacuum system (Sibley and Roehring, 1953; Spencer and Bogges, 1959; Roehring and Vanderschmidt, 1960; Moreau, 1962; Andrieux and Iapteff, 1968). It must be emphasized that it is not feasible to use very small-volume ionization chambers to obtain a linear output response in the range of high densities up to atmospheric (Leck, 1964).

Other workers designed a unique ionization chamber housing two electrodes (Fig. 4.63e), one for collecting currents in the high density range and the other in low density ranges. The shape of these electrodes, as well as their position relative to the source, complied with the requirements for ion collection at high and low densities (Blanc and Dagnac, 1964; Boziau, 1965; Rigby and Wright, 1968). Finally other workers (Cambou *et al.*, 1964; Boziau, 1965) found an original way of obtaining a single ionization chamber with a unique collector electrode for a large range of densities. They shaped the ionization chamber like a frustum and placed the source on the small base of the frustum (Fig. 4.63c). The collector electrode was placed close to the source. For high densities, the small volume of the frustum in the vicinity of the small base fulfilled the same function as a small-volume ionization chamber. For lower densities, the continuously increasing volume along the axis of the cone gave the same effect as a large-volume chamber.

Generally, the ionization chamber contains the collector electrode and sometimes a reflecting screen or repelling electrode. In one reported case, however, blinds were built inside the ionization chamber to prevent the escape of chips originating from a possible breakage of the source (Boziau, 1965). Collected currents are too small to be read directly on a microammeter and require adequate amplification to be displayed. Two techniques are available: current measurement by the analog or pulse technique.

According to the first technique, the collected current is passed through a very high electrical resistance. The voltage drop across the resistor terminals is applied to a sensitive electrometer tube, constituting the first stage of amplification circuit. The output from the electrometer is then fed into an amplifier. Various current ranges can be measured by inserting resistors of different values in either the first or the second amplification stage.

In the pulse measuring technique, an electronic circuit measures the time necessary to discharge a capacitance. The capacitance is the ionization chamber whose electrodes are set to a predetermined potential difference. As negatively and positively charged particles from the ionizing process accumulate on the electrodes, the initial potential difference is slowly neutralized. After reaching a given low voltage value, the grid of an electrometer tube to which the collector electrode is connected triggers a circuit. The circuit generates a signal pulse, and at the same time the voltage between the ionization chamber electrodes is brought back to its initial value. The discharge time depends, among other parameters, upon the gas density inside the ionization chamber. The number of signal pulses is a measure of density. Both techniques of current measurements are suited for telemetry. For special applications, information can be stored by means of systems using the pulse technique.

Regardless of the current measurement technique employed, the electrometer tube and the first amplification stage are located on top of the ionization chamber. The electrical connection between the collector electrode and the grid should be as short as possible. Range-changing servocircuits providing for both selection and insertion of resistors for each particular range of densities are also located in the first amplification stage. Special attention is paid to the electrical insulation of the first stage so as to avoid leakage currents. The length of the connecting cables between the first stage and the other stages of amplification is not an important parameter.

Particular problems regarding the electronic circuit associated with the measurement of collected currents have been carefully considered (Vacca, 1957). Means of linearizing the amplifier response (Vitkus, 1960), preventing temperature variations in the trigger circuit in the pulse technique (Howard *et al.*, 1968), and compensating for the instability of the source with time (El-Moslimany, 1960), have been successfully employed.

Except for a few papers, nothing has been reported on gauge sensitivity. Unfortunately, the data of different researchers are not consistent with each other because information on neither calibration methods nor accuracy and reproducibility of experiments were reported. Calculated sensitivities from the current versus pressure curves for the linear characteristic of the response show that they range from 2.66×10^{-7} A/Pa (2×10^{-9} A/Torr) (Sicinski *et al.*, 1954) to 2×10^{-10} A/Pa (1.5×10^{-12} A/Torr) (Cotin, 1967).

Radioactive gauges retain the advantage of conventional ionization gauges (e.g., physical ruggedness, no burnout, nearly instantaneous response, can be operated in any position, no damage upon exposure to atmospheric pressure, and linear relationship between output of the gauge head and gas density over a wide range). Besides these, they have two unique advantages, namely, no pumping action and no change of the gas composition during operation. Owing to their advantages over other types of gauges, radioactive gauges are successfully used both in the laboratory and in industry. In industry they are used to measure corrosive gases like halides, the proportions of noble gas mixtures (Gimenez and Labeyrie, 1951), and unstable organic vapors (Rigby and Wright, 1968). Miniaturization of the sensing head and associated electronics has reduced the weight so that these gauges can be used for high-altitude research work.

Disadvantages of these gauges are that the source constitutes a potential health hazard, it is necessary to calibrate very often the electronic circuit, some nonlinearity in the gauge head output affects the gauge response at very high and very low densities, and for very low densities, the output current is rather low.

The most important shortcoming is the potential health hazard. This is very important when sources emitting alpha particles are used and decreases to a minimum when using tritium. Gauge heads using tritium are less hazardous as long as the outgassing temperature does not rise above the temperature at which the gas is released from its chemical compound.

To date, such important topics as field configuration between electrodes and its influence on the collection process, methods of compensating for spurious currents, correlation of sensitivity with molecular properties of gas, and the phenomenon of hysteresis at relatively high densities have not been thoroughly investigated yet.

4.5 The Brownian Motion Gauge

Small particles suspended in a fluid medium exhibit random motion due to collisions with the surrounding molecules. This effect, known as *Brownian motion,* occurs in all fluids and at all pressures and can be used as a measure of the number density of molecules in a gas. A pressure gauge using this principle can be considered a primary pressure standard since its calibration can be predicted only from the knowledge of the particle dimension and density. An outstanding feature is that it can determine pressures down to the lowest degree of rarefaction attainable under laboratory conditions.

Kappler (1931) used the Brownian motion of a mirror suspended in gas for the determination of Avogadro's number. He indirectly deduced the molecular collision rate since experimental conditions did not permit recording each molecular impact.

Morimura *et al.* (1974) measured the random fashion in which a small mirror suspended on a fine quartz fiber was deflected in a rarefied gas environment. They showed that the damping moment acting on the mirror is proportional to gas pressure provided that the mean free path of gas molecules is larger than the mirror size. The damping ratio varies almost linearly with pressure in the range 10^{-1}–1 Pa (10^{-3}–10^{-2} Torr) and is almost independent of pressure from 1 to 1066.6 Pa (10^{-2}–8 Torr). Morimura *et al.* assumed that nonlinearity in this region results from molecular viscosity. The gauge needs a complex electronic system and does not permit determinations below 10^{-1} Pa (10^{-3} Torr), because of the mirror suspension and vibrations imparted to the instrument by the vacuum system.

Kendall (1970) suggested the use of Brownian motion as a possible basis for a vacuum gauge by utilizing very small particles suspended in gas. Such a gauge is feasible if problems such as the particle suspension, damping of oscillations, and illumination and counting of particle collisions could be adequately solved.

Particles of 1-μm radius are launched from a vibrating surface mechanically or electrically in a levitation region (Butler *et al.*, 1977). Suspension of particles in the earth's gravitational field can be achieved electromagnetically (for conducting particles), magnetically (for diamagnetic particles), or electrostatically (for charged particles). It is essential to obtain three-axis stabilization of levitated particles and damping of their oscillation after disturbances in order to keep them in the field of observation and to comply with stability requirements. Particles can be illuminated either by conventional methods or by laser, provided that heating and diffraction effects are maintained at a minimum. A medium-power microscope is sufficient for the observation of the movement of particles.

The major drawbacks to the method are *clustering of particles* mainly when the pressure which is to be determined is below 10^{-1} Pa (10^{-3} Torr), *mechanical vibrations,* and *radiation-induced desorption.*

The Brownian gauge has not yet left the research laboratory.

4.6 Range of Pressures Covered by Vacuum Gauges

Many factors considered in the detailed analysis of this chapter limit the pressure range spanned by gauges measuring low total pressure. To measure pressure over a large range, say 10 decades, several gauges overlapping in range must be selected. Figure 4.64 illustrates the approximate range of gauges discussed in this chapter. The solid and dashed lines indicates the ranges attained by usual and special constructions, respectively.

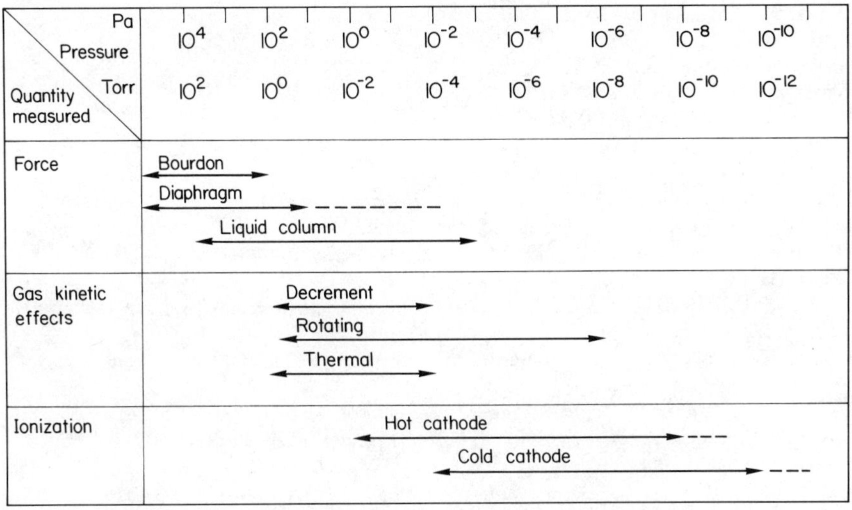

Fig. 4.64 Range of pressures covered by vacuum gauges.

In the selection of a gauge, parameters such as accuracy, short- or long-term stability, reliability, and nature of the rarefied environment must be considered in addition to the useful range.

References

Abey, D. E. (1969). *J. Appl. Phys.* **40,** 284.

Akishin, A. J. (1960). *Instrum. Exp. Tech.,* p. 260.

Alexeff, I. (1962). *Trans. 8th Natl. Vac. Symp. U.S.A. 1961,* p. 472. Pergamon Press, New York.

Alpert, D. (1953). *J. Appl. Phys.* **24,** 860.

Alpert, D. (1958). *In* "Handbuch der Physik," vol XII, p. 613. Thermodynamic der Gase. Springer-Verlag, Berlin.

Alpert, D. (1962). *Vide* **17,** 19.

Alpert, D., and Buritz, R. S. (1954). *J. Appl. Phys.* **25,** 202.

Altemose, V. O. (1961). *J. Appl. Phys.* **32,** 1309.

Amdur, I., and Guildner, L. A. (1957). *J. Am. Chem. Soc.* **79,** 311.

Anderson, J. R. (1958). *Rev. Sci. Instrum.* **29,** 1073.

Andrieux, P., and Iapteff, B. (1968). *Proc. 4th Int. Vac. Congr. Manchester 1968,* p. 675. Inst. of Phys. and Physical Soc. Conf. Series No. 6. Adlard and Son, Surrey, England.

Apker, L. (1948). *Ind. Eng. Chem.* **40,** 846.

Appelt, G. (1962). *Vak. Tech.* **11,** 174.

Armas, B. (1966). *These grade docteur de specialité.* Touluse University.
Armas, B., Blanc, D., and Dagnac, R. (1966). *Vide* **21,** 399.

Atkinson, H. H., (1963). *Trans. 10th Natl. Vac. Symp. U.S.A. 1963,* p. 213. Macmillan, New York.

Atkinson, H. H., and Banks, P. H. T. (1966). *J. Sci. Instrum.* **43,** 511.
Austin, W. E. (1969). *Vacuum* **19,** 319.
Austin, W. E., and Leck, J. H. (1972). *Vacuum* **22,** 331.
Avidenko, A. A., and Malev, M. D. (1977). *Vacuum* **27,** 583.
Baker, F. A. (1962a). *Vide* **17,** 256.
Baker, F. A. (1962b). *Trans. 8th Natl. Vac. Symp. U.S.A. 1961,* p. 395. Pergamon Press, New York.
Baker, R. A., and Giorgi, T. A. (1960). *Brit. J. Appl. Phys.* **11,** 433.
Ballard, H. N., Sellers, B., and Izquierdo, M. (1968). *3rd Natl. Conf. Aerospace Metrology,* New Orleans, Louisiana, May 6–9, p. 17. *Proceedings A 68-35067.*
Ballard, H. N., Sellers, B., and Izquierdo, M. (1969). *Amer. Astronautical Soc. National Meeting. New Mexico State University,* paper AAS 69-570. Las Cruces, New Mexico, Oct. 23–25.
Ballard, H. N., Bayers, N. J., Izquierdo, M., and Whitcare, J. (1970). *J. Geophys. Res.* **75,** 3501.
Barz, A., and Kocian, P. (1970). *J. Vac. Sci. Technol.* **7,** 200.
Bayard, R. T., and Alpert, D. (1950). *Rev. Sci. Instrum.* **21,** 571.
Beams, J. W., Young, J. L., and Moore, J. W. (1946). *J. Appl. Phys.* **17,** 886.
Beams, J. W., Spitzer, Jr., D. M., and Wade, Jr., J. P. (1962). *Rev. Sci. Instrum.* **33,** 151.
Becker, W. (1960). *Vak. Tec.* **9,** 48.
Becker, H. U., and Messer, G. (1980). *Proc. 8th Int. Vac. Congress. Cannes 1980,* p. 234. Suppl. Le Vide, Les Couches Minces, No 201.
Becker, J. A., Becker, E. J., and Brandes, R. G. (1961). *J. Appl. Phys.* **32,** 411.
Beck, A. H., and Brisbane, A. D. (1952). *Vacuum* **2,** 137.
Beck, U., and Reich, G. (1972). *J. Vac. Sci. Technol.* **9,** 126.
Beitel, G. A., and Gosselin, C. M. (1970). *J. Vac. Sci. Technol.* **7,** 580.
Benson, J. M. (1957). *Trans. Natl. Symp. Vac. Technol. U.S.A. 1956,* p. 87. Pergamon Press, New York.
Benvenuti, C., and Hauer, M. (1980). *Proc. 8th Int. Vac. Cong. Cannes 1980,* p. 199. Suppl. Le Vide, Les Couches Minces, No. 201.
Berman, A. (1975) Density sensitive gauges using radionuclides for rarefied gas content measurement. Lecture given at the Vth Italian Vacuum Congress, Peryia, Italy.
Bernardet, H., and Gourdon, C. (1963). *Vide* **18,** 472.
Bernardet, M. H., and Shaw, M. (1968) *Vide* **23,** 80.
Bernardet, H., and Choumoff, P. S. (1970). *Vide* **25,** 167.
Beynon, J. H., and Nicholson, G. R. (1956). *J. Sci. Instrum.* **33,** 376.
Bills, D. G., and Carleton, N. P. (1958). *J. Appl. Phys.* **29,** 692.
Blanc, D., and Dagnac, R. (1963). *Vide* **18,** 322.
Blanc, D., and Dagnac, R. (1964). *Vacuum* **14,** 145.
Blechschmidt, D. (1974). *Proc. 6th Int. Vac. Congr. 1974 Kyoto,* p. 105. *Jpn. J. Appl. Phys. Suppl. 2 (Part 1).*
Blodgett, K. B., and Vanderslice, T. A. (1960). *J. Appl. Phys.* **31,** 1017.
Blodgett, K. B., and Vanderslice, T. A. (1962). *Trans. 8th Natl. Vac. Symp. U.S.A. 1961,* p. 400. Pergamon Press, New York.
Bloomer, R. N. (1960). *Vacuum* **10,** 260.
Bloomer, R. N., and Haine, M. E. (1953). *Vacuum* **3,** 128.
Boziau, C. (1965). *Vide* **20,** 100.
Bretschi, J. (1976). *J. Phys. E* **9,** 261.
Briggs, W. E. (1955). *Trans. Vac. Symp. U.S.A. 1954,* p. 3. Committee on Vac. Tech, New York.
BS 2951 (1958). Glossary of terms used in high vacuum technology.

British patent (1969). No. 1.006.687.
Brombacher, W. G. (1967). *NBS Technical Note 298,* p. 27.
Brombacher, W. G. (1970). *NBS Monograph 114.*
Brown, E., and Leck, J. H. (1955). *Brit. J. Appl. Phys.* **6,** 161.
Bruining, H. (1954). "Physics and Applications of Secondary Electron Emission," p. 132. Pergamon, London.
Bryant, P. J., and Gosselin, C. M. (1966). *J. Vac. Sci. Technol.* **3,** 350.
Bryant, P. J., Taylor, L. H., and Gutshall, P. L. (1963). *Trans. 10th Natl. Vac. Symp. A.V.S. U.S.A. 1963,* p. 21. Macmillan, New York.
Buckingham, J. D. (1965). *Brit. J. Appl. Phys.* **16,** 1821.
Buckley, O. E. (1916). *Proc. Natl. Acad. Sci. U.S.A.* **2,** 683.
Butler, R. S., Kendall, B. R. F., and Morsnagel, S. M. (1977). *Vacuum* **27,** 589.
Cambou, F., Cotin, F., and Rème, H. (1964). *Ann. Geophys.* **20,** 346.
Carleton, N. P., and Bills, D. G. (1955). *Phys. Rev.* **99,** 1662.
Carmichael, J. H., and Knoll, J. S. (1959). *Trans. 5th Natl. Symp. Vac. Tech. U.S.A. 1958,* p. 18. Pergamon Press, New York.
Carmichael, J. H., and Trendelenburg, E. A. (1958). *J. Appl. Phys.* **29,** 1570.
Carcichael, J. H., and Waters, P. M. (1962). *J. Appl. Phys.* **33,** 1470.
Carter, G. (1959a). *Vacuum* **9,** 190.
Carter, G. (1959b). *Nature* **183,** 1619.
Carter, G. (1963). *Vak. Tech.* **12,** 47.
Carter, G. (1972). *Vacuum* **22,** 225.
Carter, G., and Leck, J. H. (1959). *Brit. J. Appl. Phys.* **10,** 364.
Carter, G., and Leck, J. H. (1961a). *Proc. Roy. Soc. London, A* **261,** 303.
Carter, G., and Leck, J. H. (1961b). *Trans. 7th Natl. Symp. Vac. Technol. U.S.A. 1960,* p. 339. Pergamon Press, New York.
Carter, G., Colligon, J. S., and Leck, J. H. (1962a). *Proc. Phys. Soc.* **79,** 299.
Carter, G., James, L. H., and Leck, J. H. (1962b). *Vacuum* **12,** 213.
Cavaleru, A., Comsa, G., and Iosifescu, B. (1964). *Brit. J. Appl. Phys.* **15,** 161.
Choumoff, P. S., and Iapteff, B. (1972). *SFITV Vacuum Technol. Meeting,* p. 95. *Vide* **27** Suppl.
Christian, R. G., and Leck, J. H. (1966). *Vacuum* **16,** 299.
Clay, F. P., and Melfi, L. T. (1966). *J. Vac. Sci. Technol.* **3,** 167.
Cleaver, J. S. (1967). *J. Sci. Instrum.* **44,** 969.
Cleaver, J. S. (1968). *Proc. 4th Int. Vac. Cong. Manchester 1968,* p. 651. Inst. Phys. and Phys. Soc. Conf. Series No. 6. Adlard and Son, Surrey, England.
Cleaver, J. S., and Zakrewski, W. H. (1968). *Vacuum* **18,** 73.
Close, K. J., and Yarwood, J. (1972). *Vacuum* **22,** 45.
Cobic, B., and Carter, G. (1962). *Vide* **17,** 321.
Cobic, B., Carter, G., and Leck, J. H. (1961a). *Brit. J. Appl. Phys.* **31,** 1017.
Cobic, B., Carter, G., and Leck, J. H. (1961b). *Vacuum* **11,** 247.
Comsa, G. (1966a). *J. Appl. Phys.* **37,** 554.
Comsa, G. (1966b). *Rev. Roum. Phys. (Part I)* **11,** 901.
Cosma, G. (1967). *Rev. Roum. Phys. (Part II)* **12,** 63.
Comsa, G., and Musa, G. (1957a). *J. Sci. Instrum.* **34,** 291.
Comsa, G., and Musa, G. (1957b). *Stud. Cercet. Fiz.* **8,** 119.
Comsa, G., Fremerey, J. K., and Lindenau, B. (1977). *Proc. 7th Int. Vac. Cong. Vienna 1977,* p. 157. Berger and Söhne, Horn, Austria.
Comsa, G., Fremerey, J. K., Lindenau, B., Messer, G., and Röhl, P. (1980). *J. Vac. Sci. Technol.* **17,** 642.
Conn, G. K. T., and Daglish, H. N. (1953). *Vacuum* **3,** 24.

Conn, G. K. T., and Daglish, H. N. (1954a). *J. Sci. Instrum.* **31,** 412.
Conn, G. K. T., and Daglish, H. N. (1954b). *Vacuum* **4,** 136.
Cotin, F. (1965). *Thèse grade docteur de specialite*. Touluse University.
Cotin, F. (1967). "Centre National d'Etudes Spatiales, Colloque International," Paris, vol. I, p. 90.
"CRC Handbook of Chemistry and Physics, 54th ed. (1973–1974). pp. B-250–B-253. Chem. Rubber Co. Press, Cleveland.
Dagnac, R. (1964). *Thèse grade docteur de specialite No. 273*. Touluse University.
Dalke, W. E. and Schultze, H. J. (1960). *20th Annu. Conf. Phys. Electron.* MIT, Cambridge, Massachusetts.
Davis, W. D. (1968). *Proc. 4th Int. Vacuum Cong. 1968 Manchester,* p. 605. Inst. Phys. and Phys. Soc. Conf. Series No. 6. Adlard and Son, Surrey, England.
Denison, D. R., Winters, H. F., and Donaldson, E. E. (1963). *Trans. 10th Natl. Vac. Symp. A.V.S. U.S.A. 1963,* p. 218. Macmillan, New York.
Dimeff, J., Lane, J. W., and Coon, G. W. (1962). *Rev. Sci. Instrum.* **33,** 804.
Dobrott, J. R., and Oman, R. M. (1970). *J. Vac. Sci. Technol.* **7,** 214.
Dow, R. G., (1963). *J. Appl. Phys.* **34,** 2395.
Downing, J. R., and Mellen, G. (1946). *Rev. Sci. Instrum.* **17,** 218.
Dressler, R. F. (1965). *Trans. Basic Eng.* **87,** 1072.
Dumas, G. (1955). *Gen. Electr. Rev.* **64,** 331.
Du Mond, J. W. M., and Pickels, W. M. (1935). *Rev. Sci. Instrum.* **6,** 362.
Dushman, S. (1915). *Phys. Rev.* **5,** 212.
Dushman, S. (1949). "Scientific Foundations of Vacuum Technique, second edition" p. 679. John Wiley and Sons, New York.
Dushman, S. (1962). "Scientific Foundations of Vacuum Technique." John Wiley and Sons, New York.
Dushman, S., and Found, C. G. (1921). *Phys. Rev.* **17,** 7.
Edelman, C. (1966). *Trans. 3rd Int. Vac. Congr. Stuttgart 1965,* p. 233. Pergamon Press, New York.
Edelman, C., and Engelman, P. (1974). *Exp. Tech. Phys.* **22,** 585.
Edwards, Jr., D. (1978). *J. Vac. Sci. Technol.* **15,** 128.
Edwards, Jr., D., and Lami, C. (1980). *J. Vac. Sci. Technol.* **17,** 355.
Eggelton, A. E., and Tompkins, F. C. (1952). *Trans. Faraday Soc.* **48,** 738.
El-Moslimany, M. A. (1960). *Theoretical and Experimental Investigation of Radioactive Ionization Gauges*. Univ. Michigan Sci. Rept. No. HS-1.
English, J., Fletcher, B., and Steckelmacher, W. (1965). *J. Sci. Instrum.* **42,** 77.
Evans, E. C., and Burmaster, K. E. (1950). *Proc. IRE* **38,** 651.
Evrard, R., and Boutry, R. (1969). *J. Vac. Sci. Technol.* **6,** 279.
Falkenberg, H. R., and Schumacher, B. W. (1962). *Can. J. Phys.* **40,** 1521.
Feakes, F., and Torney, F. L. (1963). *Trans. 10th Natl. Vac. Symp. U.S.A. 1963,* p. 257. MacMillan, New York.
Feakes, F., Torney, F. L., and Brock, J. (1965). NASA Rep. CR 167.
Feidt, M. L., and Paulmier, D. F. (1972). *Vacuum* **22,** 181.
Feidt, M. L., and Paulmier, D. F. (1978). *Vacuum* **28,** 333.
Feidt, M. L., and Petit, B. (1981). *J. Vac. Sci. Technol.* **18,** 987.
Fitch, R. K., and Rushton, G. J. (1980). *Vacuum* **20,** 535.
Fitch, R. K., Morris, T. J., and Thatcher, W. J. (1969). *Vacuum* **19,** 227.
Fletcher, B. (1970). *Vacuum* **20,** 381.
Fremerey, J. K. (1972). *J. Vac. Sci. Technol.* **9,** 108.
Fremerey, J. K. (1973). *Rev. Sci. Instrum.* **44,** 1396.

Fremerey, J. K., and Boden, K. (1978). *J. Phys. E* **11,** 106.
Fremerey, J. K., Comsa, G. H., and Comsa, G. (1974). *Electron. Fis. Apl.* **17,** 193.
Fryburg, G. C., and Simons, J. H. (1949). *Rev. Sci. Instrum.* **20,** 541.
Fujii, Y., Inoue, H., Mima, H., and Kanematsu, F. (1983). *J. Vac. Sci. Technol. A* **1,** 90.
Futch, Jr., A. H. (1961). *Rev. Sci. Instrum.* **32,** 1236.
Gabor, D. (1962). British Patent No. 887251.
Gallagher, H. E. (1969). *J. Appl. Phys.* **40,** 44.
Gammon, R. B. (1976). Vacuum **17,** 379.
Garbe, S., Klopfer, A., and Schmidt, W. (1960). *Vacuum* **10,** 81.
Gimenez, C., and Labeyrie, J. (1951). *J. Phys. Radium* **12,** 64A.
Gomer, R. (1953). *Rev. Sci. Instrum.* **24,** 993.
Gopalaraman, C. P., Armstrong, R. A., and Redhead, P. A. (1970). *J. Vac. Sci. Technol.* **7,** 195.
Görski, W., Litynski, K., and Smiech, A. (1977). *Proc. 7th Int. Vac. Cong. Vienna 1977,* p. 165. Berger and Sohne, Horn, Austria.
Grant, W. A., and Carter, G. (1965). *Vacuum* **15,** 477.
Grant, W. A., and Carter, G. (1966a). *Vacuum* **16,** 485.
Grant, W. A., and Carter, G. (1966b). *Proc. 7th Int. Conf. Phenom. Ioniz. Gases,* p. 200. Gradevinska Knyiga Beograd.
Grant, W. A., and Carter, G. (1967a). *Brit. J. Appl. Phys.* **18,** 527.
Grant, W. A., and Carter, G. (1967b). *Phys. Chem. Glasses* **8,** 35.
Groszkowski, J. (1965a). *Bull. Acad. Pol. Sci. Ser. Sci. Tech.* **13,** 177.
Groszkowski, J. (1965b). *Bull. Acad. Pol. Sci. Ser. Sci. Tech.* **13,** 397.
Groszkowski, J. (1965c). *Bull. Acad. Pol. Sci. Ser. Sci. Tech.* **13,** 261.
Groszkowski, J. (1966a). *Bull. Acad. Pol. Ser. Sci. Tech.* **14,** 169.
Groszkowski, J. (1966b). *Trans. 3rd Int. Vac. Congr. Stuttgart 1965,* p. 241. Pergamon Press, New York.
Groszkowski, J. (1968). *Proc. 4th Int. Vac. Cong. Manchester 1968,* p. 631. Inst. of Phys. and Physical Soc. Conf. Series No. 6, Allard and Son, Surrey, England.
Groszkowski, J. (1969a). *Vide* **24,** 226.
Groszkowski, J. (1969b). *Vide* **22,** 116.
Groszkowski, J. (1970). *Bull. Acad. Pol. Sci. Ser. Sci. Tech.* **18,** 101.
Groszkowski, J., Pytkowski, S., and Trzoch, W. (1977). *Proc. 7th Int. Vac. Congr. Vienna 1977,* p. 121. Berger and Sohne, Horn, Austria.
Grün, A. E. (1958). *Can. J. Phys.* **36,** 858.
Haefer, R. (1953). *Acta Phys. Austriaca* **7,** 251.
Haefer, R. (1954). *Acta Phys. Austriaca* **8,** 213.
Haefer, R. (1955). *Acta Phys. Austriaca* **9,** 200.
Hanley, T. E. (1948). *J. Appl. Phys.* **19,** 583.
Harbour, P. J., and Lord, R. G. (1965). *J. Sci. Instrum.* **42,** 105.
Hartman, T. E. (1963). *Rev. Sci. Instrum.* **34,** 1190.
Hayashi, C. (1966). *J. Vac. Sci. Technol.* **3,** 286.
Hayward, D. O., and Taylor, N. (1966). *J. Sci. Instrum.* **43,** 762.
Hayward, W. H., Jepsen, R. L., and Redhead, P. A. (1963), *Trans. 10th Natl. Vac. Symp. A.V.S. U.S.A. 1963,* p. 228. Macmillan, New York.
Heinje, L., and Vink, A. T. (1969). *Phillips. Tech. Rev.* **30,** 166.
Helbing, R., Jaeckel, R., and Pauly, H. (1962). *Trans. 8th Natl. Vac. Symp. U.S.A. 1961,* p. 525. Pergamon Press, New York.
Helmer, J. C., and Hayward, W. H. (1966). *Rev. Sci. Instrum.* **37,** 1652.
Herb, R. G., Pauly, T., and Fischer, K. J. (1963). *Bull. Am. Phys. Soc.* **8,** 336.

Herbert, B. K. (1976). *Vacuum* **26,** 363.
Hickmott, T. W. (1960). *J. Appl. Phys.* **31,** 128.
Hickmott, T. W. (1965). *J. Vac. Sci. Technol.* **2,** 257.
Hidber, H. R., and Süss, G. (1976). *Rev. Sci. Instrum.* **47,** 912.
Hirata, M., Ono, M., Hojo, H., and Nakayama, K. (1982). *J. Vac. Sci. Technol.* **20,** 1159.
Hobson, J. P. (1959). *Can. J. Phys.* **37,** 300.
Hobson, J. P. (1961). *Vacuum* **11,** 16.
Hobson, J. P. (1964). *J. Vac. Sci. Technol.* **1,** 1.
Hobson, J. P. (1970). *J. Vac. Sci. Technol.* **7,** 557.
Hobson, J. P., and Redhead, P. A. (1958). *Can. J. Phys.* **36,** 271.
Honig, R. E., and Kramer, D. A. (1969). *RCA Rev.* **30,** 275.
Hoverman, H. R. (1963). *J. Appl. Phys.* **34,** 3505.
Houston, J. M. (1956). *Bull. Am. Phys. Soc. II* **1,** 301.
Howard, R. L., Nelson, S. W., and Winkler, J. R. (1968). *Rev. Sci. Instrum.* **39,** 1013.
Hua, Z. Y., Lang, X. L., Qiu, S. X., and Zhuge, J. L. (1980). *J. Vac. Sci. Technol.* **17,** 351.
Hunt, A. L. (1968). *J. Vac. Sci. Technol.* **5,** 61.
Hurblut, F. C. (1959). *J. Appl. Phys.* **30,** 273.
Ishikawa, K. (1965). *Jpn. J. Appl. Phys.* **4,** 461.
Israel, M. G. (1962). *Vide* **17,** 148.
Jaeckel, R., and Teloy, E. (1962). *Trans. 8th Natl. Vac. Symp U.S.A. 1961,* p. 406. Pergamon Press, New York.
Jahrreiss, H., and Eckertová, L. (1967). *Z. Phys.* **199,** 98.
James, L. H., and Carter, G. (1962a). *Trans. 9th Natl. Vac. Symp. U.S.A. 1962,* p. 502. Macmillan, New York.
James, L. H., and Carter, G. (1962b). *Brit. J. Appl. Phys.* **13,** 2.
James, L. H., and Carter, G. (1962c). *J. Electron. Contr.* **12,** 63.
James, L. H., and Carter, G. (1963). *Brit. J. Appl. Phys.* **14,** 147.
Jepsen, R. L. (1961). *J. Appl. Phys.* **32,** 2619.
Jones, H., and Langmuir, I. (1927). *Gen. Electr. Rev.* **30,** 310.
Jones, G., and Roberts, G. T. (1978). *Vacuum* **28,** 13.
Kappler, E. (1931). *Ann. Phys.* **11,** 233.
Kanaji, T., and Urano, T. (1980). *Proc. 8th Int. Vac. Congr. Cannes 1980,* p. 230. Suppl. Rev. Le Vide, les Couches Minces No. 201.
Kanaji, T. Nakamura, S., Urano, T., and Nagata, S. (1977). *Proc. 7th Int. Vac. Congr. Vienna 1977,* p. 133. Berger and Sohne, Horn, Austria.
Kendall, B. R. F. (1970). *Vacuum* **20,** 540.
Khavkin, L. P. (1961). *Instrum. Exp. Tech.* **6,** 790.
Kirsanov, V. S. (1964). *Instrum. Exp. Tech.,* p. 1189.
Klopfer, A. (1962). *Trans. 8th Natl. Vac. Symp. U.S.A. 1961,* p. 439. Pergamon Press, New York.
Klumb, H., and Schwarz, H. (1944). *Z. Phys.* **122,** 418.
Knauer, W. (1962). *J. Appl. Phys.* **33,** 2093.
Knauer, W., and Lutz, M. A. (1963). *Appl. Phys. Lett.* **2,** 109.
Knudsen, M. (1910a). *Ann. Phys.* **31,** 205.
Knudsen, M. (1910b). *Ann. Phys.* **32,** 890.
Knudsen, M. (1910c). *Ann. Phys.* **33,** 1435.
Konjevic, R., Farell, G., and Grant, W. A. (1969). *J. Vac. Sci. Technol.* **6,** 413.
Kornelsen, E. V. (1964). *Can. J. Phys.* **42,** 364.
Kornelsen, E. V. (1966). *Trans. 3rd Int. Vac. Congr. Stuttgart 1965,* p. 65. Pergamon Press, New York.

Kreisel, W. (1976). *Vacuum* **26,** 339.
Lafferty, J. M. (1951). *J. Appl. Phys.* **22,** 299.
Lafferty, J. M. (1961a). *Trans. 7th Natl. Symp. Vac. Technol. U.S.A. 1960,* p. 97. Pergamon Press, New York.
Lafferty, J. M. (1961b). *J. Appl. Phys.* **32,** 424.
Lafferty, J. M. (1962a). *Trans. 8th Natl. Vac. Symp. U.S.A. 1961,* p. 460. Pergamon Press, New York.
Lafferty, J. M. (1962b). *Trans. 9th Natl. Vac. Symp. A.V.S. U.S.A. 1962* p. 438. Macmillan, New York.
Lafferty, J. M. (1963). *Rev. Sci. Instrum.* **34,** 467.
Lafferty, J. M. (1968). *Proc. 4th Int. Vac. Congr. Manchester 1968,* p. 647. Inst. of Phys. and Phys. Soc. Conf. Series No. 6. Adlard and Son, Surrey, England.
Lafferty, J. M. (1972). *J. Vac. Sci. Technol.* **9,** 101.
Lander, J. J. (1950). *Rev. Sci. Instrum.* **21,** 672.
Lange, W. J., and Eriksen, D. P. (1966). *J. Vac. Sci. Technol.* **3,** 303.
Lange, W. J., and Singleton, J. H. (1966). *J. Vac. Sci. Technol.* **3,** 319.
Lange, W. J., Singleton, J. H., and Eriksen, D. P. (1966). *J. Vac. Sci. Technol.* **3,** 338.
Langmuir, I. (1913). *Phys. Rev.* **1,** 337.
Lawrence, R. B. (1954). *Chem. Eng. Progr.* **50,** 155.
Lawson, R. W. (1967). *Brit. J. Appl. Phys.* **18,** 1763.
Leck, J. H. (1964). "Pressure Measurement in Vacuum Systems," p. 134. Chapman and Hall, London, England.
Leck, J. H., and Carter, G. (1960). *Proc. 1st Int. Cong. Vac. Tech. Belgium 1958,* p. 463. Pergamon Press, New York.
Lee, D. (1963). *Rev. Sci. Instrum.* **34,** 816.
Levi, R., and Espersen, G. A. (1950). *Phys. Rev.* **78,** 231.
Lloyd, O. (1966). *Brit. J. Appl. Phys.* **17,** 357.
Lockenwitz, A. E. (1938). *Rev. Sci. Instrum.* **9,** 417.
Loeb, L. L. (1958). *J. Appl. Phys.* **29,** 1396.
Martin Jr., G. D. (1962). *Trans. 8th Natl. Vac. Symp. U.S.A. 1961,* p. 476. Pergamon Press, New York.
Martin, T. H. (1963). Sandia Laboratory, SCTM 232-62 (13).
Marton, L. L. (1971). *J. Vac. Sci. Technol.* **8,** 293.
Marton, L. L. (1974). *Acad. R. Belg. Cl. Sci. 5 é serie-Tome LX,* pp. 676–685.
McCulloh, K. E., and Tilford, C. R. (1981). *J. Vac. Sci. Technol.* **3,** 994.
McIlwraith, C. G. (1947). *Rev. Sci. Instrum.* **18,** 683.
Melfy, Jr., L. (1969). *J. Vac. Sci. Technol.* **6,** 322.
Mellen, G. L. (1946). *Electronics* April, p. 142.
Mellen, G. L. (1948). *Ind. Eng. Chem.* **40,** 787.
Metson, G. H. (1951). *Brit. J. Appl. Phys.* **2,** 46.
Meyer, E. A., and Herb, R. G. (1967). *J. Vac. Sci. Technol.* **4,** 63.
Mizuno, H., and Horikoshi, G. (1977). *Proc. 7th Int. Vac. Congr. Vienna 1977,* p. 129. Berger and Sohne, Horn, Austria.
Moreau, J. (1962). *Vide* **17,** 222.
Morimura, M., Nakagawa, K., and Nezu, Y. (1974). *Proc. 6th Int. Vac. Congr. Kyoto 1974,* p. 135. *Jpn. J. Appl. Phys. Suppl. 2 (Part 1).*
Mouraud, W. G., Pauly, T., and Herb, R. G. (1964). *Rev. Sci. Instrum.* **35,** 661.
Mueller, K. G. (1962). *Vak. Tech.* **11,** 77.
Neher, H. V. (1962). *Rev. Sci. Instrum.* **33,** 803.
Nichiporovici, G. A., and Khanina, I. F. (1969). *Instrum. Exp. Tech.* **Jan.–Feb.,** 143.

Normand, C. E. (1962). *Trans. 8th Natl. Vac. Symp. U.S.A. 1961,* p. 534. Pergamon Press, New York.

Norton, F. J. (1962). *Trans. 8th Natl. Vac. Symp. U.S.A. 1961,* p. 8. Pergamon Press, New York.

Nottingham, W. B. (1947). *Rept. 7th Conf. Phys. Electron.*, pp. 14–15. Massachusetts. Inst. of Technol., Cambridge, Massachusetts.

Nottingham, W. B. (1955). *Trans. Vac. Symp. U.S.A. 1954,* p. 76. Committee on Vac. Technol. Inc., New York.

Nottingham, W. B. (1962). *Trans. 8th Natl. Vac. Symp. U.S.A. 1961,* p. 494. Pergamon Press, New York.

Novak, W. (1972). *J. Phys. E.* **5,** 73.

Ohsako, N. (1982). *J. Vac. Sci. Technol.* **20,** 1153.

Pacey, D. J. (1959/1960). *Vacuum* **9,** 261.

Pelz, D., and Newton, G. (1967). *J. Vac. Sci. Technol.* **4,** 239.

Pencko, E. A. (1961). *Instrum. Exp. Tech.* **Sept.,** 174.

Penning, F. M. (1936). *Physica* **3,** 873.

Penning, F. M. (1937a). *Physica* **4,** 71.

Penning, F. M. (1937b). *Philips Tech. Rev.* **2,** 201.

Penning, F. M., and Nienhuis, K. (1949). *Philips Tech. Rev.* **11,** 116.

Pierre, J. (1961). *Vide* **16,** 18.

Pittaway, L. G. (1970a). *Vide* **25,** 146.

Pittaway, L. G. (1970b). *J. Phys. D.* **3,** 1113.

Pittaway, L. G. (1973). *J. Vac. Sci. Technol.* **10,** 507.

Pittaway, L. G. (1974). *Philips Res. Rep.* **29,** 261.

Poulter, K. F. (1970). *Vacuum* **20,** 385.

Poulter, K. F. (1972). *J. Phys. E.* **5,** 267.

Poulter, K. F., and Sutton, C. M. (1981). *Vacuum* **31,** 147.

Poulter, K. F., Rodgers, M. J., and Ashcroft, K. W. (1980). *J. Vac. Sci. Technol.* **17,** 638.

Radiation Protection Procedures (1973). Safety Series No. 38, p. 38. Int. Atomic Energy Agency, Vienna.

Redhead, P. A. (1958). *Can. J. Phys.* **36,** 255.

Redhead, P. A. (1959). *Can. J. Phys.* **37,** 1260.

Redhead, P. A. (1960). *Rev. Sci. Instrum.* **31,** 343.

Redhead, P. A. (1961). *Trans. 7th Natl. Symp. Vac. Technol. U.S.A. 1960,* p. 108. Pergamon Press, New York.

Redhead, P. A. (1962). *Vacuum* **12,** 267.

Redhead, P. A. (1963). *Vacuum* **13,** 253.

Redhead, P. A. (1965). *Can. J. Phys.* **43,** 1001.

Redhead, P. A. (1966). *J. Vac. Sci. Technol.* **3,** 173.

Redhead, P. A. (1967). *J. Vac. Sci. Technol.* **4,** 57.

Redhead, P. A. (1969). *J. Vac. Sci. Technol.* **6,** 848.

Redhead, P. A. (1970). *J. Vac. Sci. Technol.* **7,** 182.

Redhead, P. A. (1975). Very low pressures and ultra low pressures (below 10^{-6} Torr). *In* "Experimental Thermodynamics" (Le Neindre and B. Vodar, eds.). Butterworths, London.

Redhead, P. A., and Hobson, J. P. (1965). *Brit. J. Appl. Phys.* **16,** 1555.

Redhead, P. A., Hobson, J. P., and Kornelsen, E. V. (1962). Ultrahigh vacuum. *In* "Advances in Electronics and Electronic Physics" (L. Marton and C. Marton, eds.). Academic Press, New York.

Redhead, P. A., Hobson, J. P., and Kornelsen, E. V. (1968). "The Physical Basis of Ultrahigh Vacuum," p. 131. Chapman and Hall Ltd., London.
Reich, G. (1966). *Z. Instrumentkunde* **74,** 254.
Rème, H. (1963). *These grade docteur de specialité No 224,* pp. 24, 27. Touluse University.
Rhodin, T. N., and Rowner, L. H. (1961). *Trans. 7th Natl. Symp. Vac. Technol U.S.A. 1960,* p. 228. Pergamon Press, New York.
Riddoch, A., and Leck, J. H. (1958). *Proc. Phys. Soc.* **72,** 467.
Rigby, L., and Wright, C. R. (1968). *Vacuum* **18,** 274.
Robinson, N. W. (1962). *Vide* **17,** 570.
Robinson, N. W., and Berz, F. (1959). *Vacuum* **9,** 48.
Roehring, J. R., and Vanderschmidt, G. F. (1960). *Trans 6th Natl. Symp. Vac. Technol. U.S.A. 1959,* p. 82. Pergamon Press, New York.
Schissel, P. O., and Trulson, O. C. (1965). *J. Chem. Phys.* **43,** 737.
Schuemann, W. C. (1962). *Trans. 9th Natl. Vac. Symp. A.V.S. U.S.A. 1962,* p. 428. Macmillan, New York.
Schuemann, W. C. (1963). *Rev. Sci. Instrum.* **34,** 700.
Schutze, H. J., and Stork, F. (1962a). *Vak. Tech.* **11,** 133.
Schutze, H. J., and Stork, F. (1962b). *Trans. 9th Natl. Vac. Symp. A.V.S. U.S.A. 1962,* p. 431. Macmillan, New York.
Schultz, G. J. (1957). *J. Appl. Phys.* **28,** 1149.
Schultz, G. J., and Phelps, A. V. (1957). *Rev. Sci. Instrum.* **28,** 1051.
Schumacher, B. W. (1960). *Nucleonics* **18,** 106.
Schumacher, B. W. (1968). *In* "Electron and Ion Beam Science and Technology," p. 31, 3rd conf. (R. A. Bakish, ed.). The Electrochemical Src. Inc., New York.
Schumacher, B. W., Aruja, E., and Falkenberg, H. R. (1967). *Trans. 3rd Int. Vac. Congr. Stuttgart 1965,* p. 543. Pergamon Press, New York.
Schwarz, H. (1940). *Z. Phys.* **117,** 23.
Schwarz, H. (1944). *Z. Phys.* **122,** 437.
Schwarz, H. (1957). *Trans. Natl. Symp. Vac. Technol. U.S.A. 1956,* p. 37. Pergamon Press, New York.
Schwarz, H. (1960). *Rev. Sci. Instrum.* **31,** 433.
Schwarz, H. (1961). *Vacuum* **11,** 151.
Schwarz, H. (1962). *Trans. 8th Natl. Vac. Symp. U.S.A. 1961,* p. 467. Pergamon Press, New York.
Sellers, B., and Morel, P. (1969). Panametrics, Inc. Rep. No. FAA-RD-69-38.
Sellers, B., Ballard, H. N., and Izquierdo, M. (1969b). *Int. J. Appl. Radiat. Isotop.* **20,** 341.
Shapiro, E. (1952). *J. Am. Chem. Soc.* **74,** 5233.
Shioyama, T., Takiguchi, T., and Ogawa, S. (1978). *J. Vac. Sci. Technol.* **15,** 761.
Sibley, C. B., and Roehring, J. R. (1953). *Electronics* March, 176.
Sicinski, H. S., Spencer, N. W., and Dow, R. G. (1954). *Appl. Phys.* **25,** 161.
Singleton, J. H. (1966). *J. Chem. Phys.* **45,** 2819.
Singleton, J. H. (1967). *J. Vac. Sci. Technol.* **4,** 103.
Smeaton, P. G., Carter, G., and Leck, J. H. (1962). *Trans 9th Natl. Vac. Symp. A.V.S. U.S.A. 1962,* p. 491. Macmillan, New York.
Spencer, H. W., and Bogges, R. L. (1959). *J. Am. Rocket Soc.* **29,** 68.
Steckelmacher, W. (1962). *Trans. 9th Vac. Symp. A.V.S. U.S.A. 1962,* p. 421. Macmillan, New York.

Steckelmacher, W. (1969). *Vuoto* **2,** 189.
Steckelmacher, W. (1973). *Vacuum* **23,** 307.
Steckelmacher, W., and Fletcher, B. (1972a). *J. Phys. E.* **5,** 405.
Steckelmacher, W., and Fletcher, B. (1972b). *J. Vac. Sci. Technol.* **9,** 128.
Suchannek, R. G., and Sheridan, J. R. (1974). *Rev. Sci. Instrum.* **45,** 19.
Sutton, C. M., and Poulter, K. F. (1982). *Vacuum* **32,** 247.
Szwemin, P. J. (1972). *J. Vac. Sci. Technol.* **9,** 122.
Szwemin, P. J., and Pytkowski, St. (1977). *Proc. 7th Int. Vac. Congr. Vienna 1977,* p. 141. Berger and Sohne, Horn, Austria.
Tate, J. T., and Smith, P. T. (1932). *Phys. Rev.* **39,** 270.
Thiele, O. W., and Beyers, N. J. (1967). *J. Atm. Sci.* **24,** 551.
Thomas, L. B., and Brown, R. E. (1950). *J. Chem. Phys.* **18,** 1367.
Thomas, L. B., and Shofield, E. B. (1955). *J. Chem. Phys.* **23,** 861.
Tilford, Ch.R. (1983). *J. Vac. Sci. Technol. A* **1,** 152.
Tilford, Ch.R., and McCulloh, K. E. (1982). *J. Vac. Sci. Technol.* **20,** 1140.
Torney, F. L., and Feakes, F. (1963). *Rev. Sci. Instrum.* **34,** 1041.
Turner, D. J. (1964). *Vacuum* **14,** 477.
Tuzi, Y., Okada, S., and Kobayashi, M. (1968). *Jpn. J. Appl. Phys.* **7,** 1415.
U.S. Patent (1969). No. 3.449.659.
Urbanek, K. (1967). *J. Vac. Sci. Technol.* **4,** 328.
Vacca, R. H. (1957). *Trans. Natl. Symp. Vac. Technol. U.S.A. 1956,* p. 93. Pergamon Press, New York.
Valle, G. (1950). *Nuovo Cim.* **7,** 174.
Valle, G. (1952). *Nuovo Cim.* **9,** 145.
Vandershmidt, G. F. (1959). *Electronics* June, 60.
Vandershmidt, G. F., and Simons, Jr, J. C. (1967). *Trans. 3rd Int. Vac. Congr. Stuttgart 1965,* p. 305. Pergamon Press, New York.
Van Oostrom, A. (1962). *Trans. 8th Natl. Vac. Symp. U.S.A. 1961,* p. 443. Pergamon Press, New York.
Van Oostrom, A. (1967). *J. Sci. Instrum.* **44,** 927.
Varicák, M. (1962). *Trans. 8th Natl. Vac. Symp. U.S.A. 1961,* p. 483. Pergamon Press, New York.
Varicák, M., and Safic, B. (1959). *Rev. Sci. Instrum.* **30,** 891.
Varnerin, L. J., and Carmichael, J. H. (1955a). *Phys. Rev.* **99,** 1662.
Varnerin, L. J., and Carmichael, J. H. (1955b). *J. Appl. Phys.* **26,** 782.
Varnerin, L. J., and Carmichael, J. H. (1957). *J. Appl. Phys.* **28,** 913.
Vasil'eva, M. N., and Reikhrudel, E. M. (1962). *Sov. Phys. Tech. Phis.* **7,** 528.
Venable, D., and Kaplan, D. E. (1955). *J. Appl. Phys.* **26,** 639.
Venema, A. and Bandringa, M. (1958). *Philips Tech. Rev.* **20,** 145.
Visser, J. (1967). *Vacuum* **17,** 73.
Vitkus, P. L. (1960). *Trans. 6th Natl. Symp. Vac. Technol. U.S.A. 1959,* p. 89. Pergamon Press, New York.
Von Gerhard Strozer. (1958). *Z. Angew. Phys.* **10,** 207.
Von Gerhard Strozer. (1959). *Z. Angew. Phys.* **11,** 223.
Von Ubish, H. (1952). *Analit. Chem.* **24,** 931.
Von Ubish, H. (1957). *Vak. Tech.* **6,** 175.
Wagener, S. (1951). "The Oxide-Coated Cathode," p. 271. Chapman and Hall, London.
Walker, R., and Pacey, D. J. (1968). *Proc. 4th Int. Vac. Congr. Manchester 1968,* p. 638. Inst. of Phys. and Phys. Soc. Conf. Series No. 6. Adlard and Son, Surrey, England.

Webb, D. R. A. (1974). *J. Phys. E.* **7,** 453.
Weinman, J. A. (1966). *Rev. Sci. Instrum.* **37,** 636.
Weinreich, O. A. (1951). *Phys. Rev.* **82,** 573.
Werner, J. G., and Leck, J. H. (1969). *Vacuum* **19,** 317.
Weston, G. F. (1979). *Vacuum* **29,** 277.
Winters, H. F., Denison, D. R., Bills, D. G., and Donaldson, E. E. (1963). *J. Appl. Phys.* **34,** 1810.
Wu, Y., and Birendra, D. (1972). *J. Vac. Sci. Technol.* **9,** 1248.
Young, J. R. (1956). *J. Appl. Phys.* **27,** 926.
Young, J. R. (1959). *J. Appl. Phys.* **30,** 1671.
Young, J. R. (1966). *J. Vac. Sci. Technol.* **3,** 345.
Young, J. R. (1973). *J. Vac. Sci. Technol.* **10,** 212.
Young, J. R., and Hession, F. P. (1963). *Trans. 10th Natl. Vac. Symp. A.V.S. U.S.A. 1963,* p. 234. Macmillan, New York.

CHAPTER 5

Methods for Vacuum Gauge Calibration

5.1 General Considerations

Vacuum gauges display readings related to force or other physical quantities such as viscosity, thermal conductivity, or density of gas. So that these indications can be related to pressure, gauges require calibration. In the pressure range where primary or secondary standards are available (see Chapter 3), calibration consists of subjecting the gauge head to a pressure P_G accurately measured with such standards and relating the indication of the gauge P_g to the pressure P_G by

$$P_g/P_G = f(P_G) \tag{5.1}$$

Calibration performed in this way is referred to as *absolute* since it is based on measurements with absolute pressure standards. Upon combining absolute calibration with linearity measurements (see Section 7.2) calibration can be extended to the lowest pressures measured with vacuum gauges (see Fig. 4.64). An alternative to the extrapolation of absolute calibration in the UHV range by linearity measurements consists of producing fixed pressure points on the low pressure scale by using partial pressure standards. Such standards are based either on the thermodynamic equilibria of a gas–solid solution in a mixed phase region or on the vapor pressure of certain substances.

Two methods, the *static* and the *dynamic* methods, have evolved for the absolute calibration of vacuum gauges. Each one provides two measurement approaches according to whether P_g and P_G are related directly or by calculations based on the kinetic gas theory.

Calibration according to the *static method* is performed either directly or indirectly by means of *static expanders*.

In the direct method, both the gauge being calibrated and the pressure standard are attached to the calibration chamber and subjected to the required pressure points. The technique, suitable for calibration in the medium vacuum range with accuracies of $\pm 1\%$ or better, is also used for the determination of the absolute sensitivity of ionization gauge heads (see Section 7.3).

In the indirect method, known as the *expansion method,* calibration consists of expanding in the calibration chamber an accurately known amount of gas contained in a small volume. After expansion the calibration chamber is sealed off. If the temperature is maintained constant during the calibration process and the volume ratio of the vessels involved in gas expansion as well as the original pressure are known, the final pressure in the calibration chamber can be related to the original pressure by the equations of the kinetic theory of gas. Low pressures in the calibration chamber are obtained by repeating the process of gas expansion according to a technique referred to as *multiple expansion*.

Static expanders are slow to operate and sensitive to factors such as sinks of gas (especially in the UHV range), small volume changes in the vessels or equipment of the calibration system, or chemical composition of the calibration gas. The operational range of static expander systems is set by the gas load resulting from outgassing in the calibration chamber. Best results are obtained in the medium and high vacuum range.

In the *dynamic method* pressure is generated in the calibration chamber by balancing the rate of flow of gas into the chamber against the rate of pumping off that gas. The direct approach consists of reading the gauge being calibrated and the pressure standard, both attached to the calibration chamber, after reaching steady-state conditions of gas flow. Calibration in the high pressure range can be accomplished with an uncertainty of $\pm 10\%$ at the lower end of the range. Outgassing from the walls of the calibration chamber sets the calibration limit at 10^{-4} Pa (10^{-6} Torr).

The indirect approach consists of using *dynamic expanders* or *orifice flow* techniques, which permit generation of *steady calculable pressures* in the calibration chamber. This is done by passing a known flow of gas through an orifice of known conductance. Usually orifices of about 1-cm diameter are used, which set the upper pressure limit of the calibration system to about 10^{-2} Pa (10^{-4} Torr). Lower calibration pressures can be obtained by gas flow through multiple orifices.

Dynamic expanders have the advantages of generating quickly the required pressure point and of maintaining it for long periods of time. The use of such expanders with active gases is allowed, and sinks and sources of gas within the calibration system are less harmful than in the case of static expanders. Calibration in UHV is possible provided that the conflict between obtaining a sufficient pressure drop across the orifice and the need to generate base pressure points is resolved. If the tested gauge head has a significant pumping speed, the dynamic expander is to be preferred to other calibration approaches because the flow rates of the calibration gas is higher than the outgassing flow rates from the walls of the calibration chamber and higher than the pumping speed of the gauge heads undergoing calibration.

The adjustable conductance approach and the reference-transfer method are variants of the dynamic method of gas expansion through fixed orifices.

The pressure ranges of various calibration methods are listed in Table 5.1. Attempts have been made to compare calibration methods and systems used in various countries, but the lack of suitable transfer standards has limited these activities. It is hoped that the spinning-rotor viscosity gauge (see Section 3.6.3) will promote this work.

Vacuum gauges must be subjected to an overall calibration including both the gauge head and the control unit. Three approaches can be taken for this purpose:

(a) *The gauge head is calibrated, but not its control unit.* It is tacitly assumed that the accuracy of the control unit is much greater than that of the gauge head. Such an outlook is not correct, since the accuracies of the control unit and gauge head are the same order of magnitude (about ±10%).

(b) *The set gauge head-control unit is used throughout the calibration of only the gauge head.* The following drawbacks result: the gauge head and the control unit *must always* be used together because the accuracy of the control unit has not been evaluated. The reliability of the set is questionable after time has elapsed between two successive calibrations of the gauge head.

(c) *The gauge head and the control unit are independently calibrated.* This approach is more flexible in that any calibrated gauge head may be used with any calibrated control unit as long as they are designed to be used with each other. The control unit can be independently calibrated without requiring a vacuum system. The overall calibration accuracy is the combined accuracy of the set. A damaged gauge head can be replaced by a new calibrated one without requiring calibration of the control unit, since a damaged control unit can similarly be replaced without requiring recalibration of the gauge head.

Gauge heads are calibrated with an inert gas (e.g., Ar) or, usually, with nitrogen, their indications in the latter case being referred to as *nitrogen equivalent*. Pressure indications for other gases are then calculated from the absolute sensitivity of the gauge head for the inert gas or nitrogen and the knowledge of the relative sensitivity for the gas of interest (see Section 7.3).

Routine work in vacuum metrology is accomplished with secondary (transfer) standards. Thus, the conventional triode, the capacitance manometer, and the spinning-rotor viscosity gauge are used for the range of pressures $0.3–10^{-5}$ Pa ($2 \times 10^{-3}–10^{-7}$ Torr); $10^{5}–10^{-2}$ Pa ($750–10^{-4}$ Torr); and $1.0–10^{-4}$ Pa ($10^{-2}–10^{-6}$ Torr), respectively (see Sections 3.6–3.8).

Calibration to accuracies of 1% or less is restricted to a few national institutions which also set up pressure standards and use for that purpose rather expensive and sophisticated equipment (see Section 3.5). Discussions in this chapter are based on the requirement for calibration accuracies of ±10% for vacuum users who periodically check large numbers of gauges.

Table 5.1

PRESSURE RANGE OF VARIOUS CALIBRATION METHODS

Code[a] Pressure Range [Pa(Torr)]	Figure	Section	Reference
Static			
A $10\text{–}10^{-2}(10^{-1}\text{–}10^{-4})$	5.1	5.4.1	Utterbach and Griffith (1966)
B $10^{3}\text{–}10^{-4}(10\text{–}10^{-6})$	5.2	5.4.2	Elliot *et al.* (1967)
Dynamic			
A $10^{2}\text{–}10^{-4}(1\text{–}10^{-6})$	5.3	5.5.1	AVS 62(1969), AVS 64(1969)
BC $10^{-1}\text{–}10^{-4}(10^{-3}\text{–}10^{-6})$	5.6	5.5.2	Bannenberg and Tip (1968)
$10^{-1}\text{–}10^{-7}(10^{-3}\text{–}10^{-9})$	5.7		Normand (1962)
$10^{-2}\text{–}10^{-6}(10^{-4}\text{–}10^{-8})$	5.8		Fletcher and Watts (1967)
$10^{3}\text{–}10^{-6}(10\text{–}10^{-8})$	5.10		Close *et al.* (1977)
BCD $<10^{-8}(<10^{-10})$	5.14	5.5.2	Davis (1968)
$10^{-3}\text{–}10^{-10}(10^{-5}\text{–}10^{-12})$	5.15		Bernardet *et al.* (1970a)
$10^{-3}\text{–}10^{-7}(10^{-5}\text{–}10^{-9})$	5.16		Messer (1977a)
BE $10^{-2}\text{–}10^{-7}(10^{-4}\text{–}10^{-9})$	5.19	5.5.2	Roehring and Simons (1962, 1963)
BEU $10^{-5}\text{–}10^{-12}(10^{-7}\text{–}10^{-14})$	5.20	5.5.2	Feakes and Torney (1963)
$<10^{-9}(<10^{-11})$	5.21		Poulter (1974)
BD $10^{-2}\text{–}10^{-4}(10^{-4}\text{–}10^{-6})$	5.22	5.5.3	Henry (1956)
BU $\sim10^{-10}(\sim10^{-12})$	5.24	5.6	Berquist (1968)
$10^{-4}\text{–}10^{-13}(10^{-6}\text{–}10^{-15})$	5.25		Fowler and Brock (1970)
$\sim10^{-10}(\sim10^{-12})$	5.26		Grosse and Messer (1981)
Reference transfer			
BC $10^{-2}\text{–}10^{-6}(10^{-4}\text{–}10^{-8})$	5.11 5.12	5.5.2	Morrison (1966a, 1969)
Dissociation equilibrium pressure			
F $1\text{–}10^{-18}(10^{-1}\text{–}10^{-20})$	—	5.9.2	Taborek and Goodstein (1979)
G $\sim10^{-6}(\sim10^{-8})$	—	5.9.3	Titcomb and Wallace (1972)

[a] The readings of the gauge undergoing calibration are related to the calibration pressure P by

A: measurement of pressure P with a pressure standard (*direct technique*),
B: calculation of P based on kinetic theory of gas (*indirect technique*).

Other symbols used in the table are

C: single stage of flow divider using fixed conductance,
D: single stage of flow divider using variable conductance,
E: multiple stage of flow divider using fixed conductance,
F: equilibrium pressure technique,
G: gas-metal solution equilibrium technique.
U: UHV range

5.2 The Main Components of a Calibration System

5.2.1 The Calibration Chamber

It must be possible to achieve in the calibration chamber a uniform and stable distribution of calibration gas at the ports of the chamber where the gauge heads undergoing calibration are attached. Satisfactory working conditions in this regard are obtained if sources or sinks of gas mainly due to the outgassing load from the walls, the pumping of gauge heads, beaming effects due to the leak of the calibration gas in the calibration chamber, etc., are kept to a minimum (see Sections 2.3.1–2.3.3). It is also necessary that the volume of the calibration chamber should be large enough so that the gas flow from the internal leak in its volume should not cause a measurable pressure rise.

All these conditions can be met by properly designing the shape of the calibration chamber, the location of the gauge heads undergoing calibration, and the location of the orifice or pipe for the calibration gas leak, as well as by suitably selecting the materials for the calibration chamber and the fixtures inside the chamber.

The *shape* of the calibration chamber should be designed so that the ratio of the surface of the walls to the volume is minimal. The *spherical* or *right cylindrical* geometry, with height equal to four times the diameter of the cylinder, satisfies this requirement. However, other shapes where the ratio mentioned is kept at a minimum can be utilized. Thus the calibration chamber used at the German Vacuum Gauge Calibration Service (DKD) has a eliptical cross section (Reich, 1980).

In a spherical geometry, molecular beams are minimized, as shown by Holland and Priestland (1967), by means of an ionization gauge head rotated on a circular path in a spherical chamber (see Section 8.2.2). For reasonable accuracies at lower pressures, large spheres (60–100-cm diameter) are used. Such large sizes and the rather high purchase prices make the spherical shape a choice for vacuum metrology only.

For right cylindrical geometry, the volume of the cylinder must be at least 20 times the total volume of the tested gauge heads and any associated plumbing connected to it (ISO/DIS 3568).

The *construction material* of the calibration chamber is stainless steel with a low outgassing rate (e.g., stainless steel series 300, 321, 344, or AISI grade 304, 304 L).

The *ports* which connect the gauge heads and other auxiliary instruments (pressure standard, residual gas analyzer) with the calibration chamber must be sized so as to match at least the diameter of the tubing connecting the gauge head (tubulated) to the calibration chamber. In order to avoid gas sorption (ionization gauge heads) and desorption phenomena, the conductance of this tubing to the calibration gas should be about 100 times the volume of the gas flow corresponding to the combined effect of sorptions and desorptions (ISO/DIS 3568). According to another criterion, the ratio of the pressure drop across the tubing to the pressure in the calibration chamber should be less than 5% (Redhead, 1961). Tubings designed according to either of these criteria have pumping speeds of several liters per second.

The ports also must be located so as to minimize pressure differentials between the gauge heads tested and other measuring instruments connected to adjacent ports. This means that the rate of arrival of gas molecules at the aperture of the port should be the same as the rate of calibration gas entering the calibration chamber. Pressure differentials discussed are practically minimized by symmetrically distributing the ports around the periphery of the calibration chamber. Such an arrangement also helps in minimizing the mutual influence of gauge heads. Nash and Thompson (1983), however, distributed the ports around the periphery of a right cylindrical chamber used for calibration according to an odd number. Pairs of opposite gauge heads do not face each other, thus limiting gauge head interactions when conducting simultaneous calibration.

The uniformity of the pressure distribution at different ports can be checked by calibrating the same ionization gauge head at different ports and by examining the resulting scatter of sensitivities.

The interaction between gauge heads which occurs during their simultaneous calibration can be detected by cross-control checks. In steady operation, no appreciable interaction is to be noticed. If, however, such interactions are detected, precautions discussed in Section 7.5.6 must be taken.

The tubing connecting the tested gauge to the port should have a geometry permitting a gas molecule coming from the calibration chamber to collide at least once with the walls of the tubing before entering the active zone of the gauge head. This requirement is fulfilled if the length of the tubing is twice the tubing diameter. The leak-tight seal between the calibration chamber and the pumping system or between the ports and the instruments attached to them is achieved by elastomers or metallic gaskets, according to their compatibility with bake procedures.

The *leak of calibration gas* into the calibration chamber is accomplished through an orifice or a tubing. Both their design and their relative location with respect to the ports must be carefully considered in order to avoid inaccurate measurements. Thus, in the case of an orifice the distribution of molecular velocities at any point on the surface of the calibration chamber is deficient in molecules coming from the solid angle subtended by the orifice. Gas leaked into the calibration chamber from an incorrectly located tubing may not have a Maxwell–Boltzmann distribution of molecular velocities, or in the worst case, may be streamed directly into the calibration ports. Both situations lead to large errors in the calculation of pressure in the calibration chamber, calculations based on the assumption that gas is in a chaotic state. Inaccuracies of the type discussed are minimized by designing the orifice area to be small compared with the internal area of the calibration chamber or by locating the tubing such that incoming molecules are randomized by collisions before reaching the calibration or measuring ports.

When the flow rate of the gas leaked into the calibration chamber is very small compared to the flow rate of gas out of that chamber, an imbalance occurs which renders calibration work very difficult. Several systems using automatic leak valves controlled with feedback signals from a monitor gauge or from the gauge head undergoing calibration help in avoiding this drawback.

The calibration gas is usually supplied from the gas bottle via a *buffer* whose role is to damp pressure fluctuations of this gas in the calibration chamber.

5.2.2 The Pumping System

The pumping system must be able to evacuate the calibration chamber from atmospheric pressure down to the pressure where the calibration process is carried out and to maintain that pressure throughout the process. The *pump size* is determined by the pumping speed required for the calibration process and the ultimate pressure to be achieved in the calibration chamber. Thus, when calibration is performed according to the static method the base pressure in the calibration chamber, before admitting the calibration gas, must be about two orders of magnitude lower than the lowest limit of the calibration range. Calibration according to the dynamic method requires pumping the calibration chamber with *constant speed,* higher than the rate of flow of calibration gas through the calibration orifice. Other pertinent considerations, such as the requirement for an organic-free rarefied environment, compatibility with the calibration gas, and cost, contribute to the selection of the *pump type*.

The main pumps may be one or a combination of rotary and diffusion, turbomolecular, or cryogenic pumps. Diffusion pumps operated with high-quality fluid, backed by two-stage rotary pumps, are most used since they do not show variation in speed with the type of gas (except with hydrogen or helium).

Every pumping system generates impurities which can contaminate the calibration chamber. Thus, for ultrahigh vacuum conditions, oil diffusion pumps are a triple source of contaminants resulting from the backstreaming of

(a) permanent gases C_2H_4, CH_4, and H_2, generated as a result of the thermal degradation of the diffusion pump fluid,
(b) fluid from the diffusion pump, and
(c) fluid from the rotary pump.

Backstreaming from sources (a) and (b) contaminate only the calibration chamber whereas that from source (c) contaminates the roughing line. [For a detailed discussion of backstreaming produced by diffusion pumps, see Power and Crawley (1954).] Mercury diffusion pump systems contaminate both the calibration chamber and the roughing line with mercury vapor and fluid, respectively, from the rotary pump. Turbomolecular pumping systems usually contaminate the calibration chamber and the foreline with backstreaming hydrogen or with hydrogen which has not been pumped. Cryogenic pumping systems, properly operated, are not a source of contaminants.

Contamination can be reduced to an acceptable level by proper design of the pumping system, the use of high-quality, low-vapor-pressure anti-creep fluids for the diffusion pump, baffles, and traps. Baffles and traps, together with the ducts of the pumping system, may in turn become backstreaming sources (Hablanian, 1969). For an optimum design, the section of the pumping system from the calibration chamber inlet to the pumps must be considered as a whole, and the separate elements must be examined for compatibility.

5.3 General Procedures for Calibration

A calibration system is reliable when it maintains the level of measurement accuracies over a long operating period (about 6 months). To maintain such a level it is necessary to keep the system free of contamination and to adhere to the operational procedures established as a result of a long calibration routine. The following suggestions are intended only as a guide. Standards giving full information with regard to calibration procedures will be quoted in each subsection. Specific operational procedures, if considered necessary, will also be discussed in the following subsections.

Contaminants may adversely influence the performance of the pressure standard and of other measuring instruments, e.g., residual gas analyser (RGA), as well as the behavior of the calibration chamber as a source of gas. Contaminants may be brought into the measuring volume by the gauge heads undergoing calibration, the sealing gaskets, the gas used for either calibrating or venting the system, or the pumping system. The gauge heads and the gaskets must be cleaned and assembled under clean conditions. The calibration gas must contain a maximum 0.1% impurities and be fed into the measuring volume in such a way that its purity is maintained. It is advisable to monitor the purity of the calibration gas in the calibration chamber. Both the calibration gas and the gas used for venting the system must be dry. The use of mainly nitrogen for venting the volume of the calibration chamber is recommended because each pump-down cycle will be shorter than if performed with air. During venting, the pressure standard must be valved off from the calibration chamber. Baffles and traps must be operated so as not to be turned into sources of gas.

Temperature effects on the calibration process must be held at a minimum by controlling the temperature of the gas supply, the calibration chamber, the gauge heads, and other auxiliary instruments connected to the calibration chamber. A record of the ambient air temperature is necessary for further corrections.

Temperature effects arise basically from the temperature difference between the gauge heads attached to the ports and (a) the calibration chamber and (b) the gauge heads undergoing calibration.

(a) The response of gauge heads which measure quantities other than force (e.g., density) is highly related to both density and temperature of the gas. Thus, in the high vacuum range a *nude* ionization gauge head responds in proportion to the number of gas molecules in its active zone. In the case of a *tubulated* ionization gauge head also operated in the high vacuum range, gas molecules entering the tubulation quickly come to thermal equilibrium with the temperature of the active zone. This results in a difference between the temperatures of the gauge head and of the calibration chamber. The pressure and density of gas in the gauge head would therefore be higher and lower, respectively, than those in the calibration chamber. In addition to this effect, variations in the calibration chamber temperature (while the temperature of the leaked calibration gas is maintained constant) give rise to variations in both the gas density and the pressure, notwithstanding the fact that the molecular incidence of the calibration gas on internal surfaces is constant. For such conditions, nude ionization gauges show temperature sensitivity, whereas tubulated ones do not.

(b) The temperature of gauge heads, mainly of the hot ionization type, is affected both by changes in the emission current setting and by the "thinning" of the cathode as a result of its evaporation. In the first case, a change in the emission current of, e.g., 10 : 1 will theoretically modify the ion current in the same proportion. The resulting change in the temperature of the gauge head tubulation will result, however, in a calibration error. The reduction of the cathode surface produced by evaporation, as evidenced by a colored layer on the glass, is followed by a change in its emissivity. To maintain a constant emission current, a change in the operating temperature of the cathode is necessary, affecting in turn the temperature of the gauge head.

The temperature uniformity of the gauge heads undergoing calibration (an operation usually performed in batches) must be carefully controlled. For that purpose, the cathodes of the gauge heads attached at the calibration chamber must be held at the same operational temperature as the gauge head being measured. Hot-cathode ionization gauge heads must be calibrated at the emission current at which they are going to be used.

Before starting calibration, tests must confirm that the residual pressure in the calibration chamber is below the value at which calibration can be started. If several gauge heads are simultaneously calibrated, they must be allowed to stabilize with respect to ambient temperature.

During calibration the gauge head must be held in the position prescribed by the manufacturer. This is very important for thermal gauge heads since their readings might be greatly influenced by the position assumed during calibration.

Calibration must start with the lowest pressure the gauge head measures and be carried out by increasing the pressure stepwise. The sequence of steps must permit the generation of at least three pressure points for each decade of the control-unit range. The sequence of calibration points within a decade should be such as to give approximately 30, 60, and 100% full linear deflection (FSD) at the output (analog) of the instrument operated by the control unit. In the case of an instrument with a logarithmic scale, one point per range is sufficient. The overall accuracy of the control unit is improved by using a digital instead of an analog meter.

Changes with time in the accuracy level of the calibration system can be monitored by recording the results of every calibration.

5.4 The Static Method

5.4.1 Direct Comparison with a Pressure Standard

The *principle* of direct comparison can be stated as follows: The readings of the gauge head undergoing calibration are compared with those of a pressure standard, both subjected to a known *stable* pressure in a sealed-off vessel. The pressure range of this approach is 10–10^{-2} Pa (10^{-1}–10^{-4} Torr).

The calibration system illustrated in Fig. 5.1 consists of a calibration chamber, a pumping system, and means to supply the calibration gas to the calibration chamber. All the flange seals of the calibration are elastomers, thus making bake-out procedures impossible. The calibration chamber is chemically cleaned in order to reduce the outgassing yield from the surface of its walls.

The pressure standard used in the early days was the McLeod gauge. Its use required a cold trap to protect the calibration chamber against mercury vapor and a large measuring volume to ensure that no significant pressure changes follow when mercury rises and falls in the gauge (see Section 3.3.4). Thus, a 340-L calibration chamber was used by Utterbach and Griffith (1966) to intercalibrate a capacitance manometer and a McLeod gauge. Due to many drawbacks associated with this primary standard its use was superseded by the capacitance manometer, assumed to be linear for all gases and vapor (Leck, 1964, p. 168). This, however, is not true when the manometer is operated with its sensors maintained at

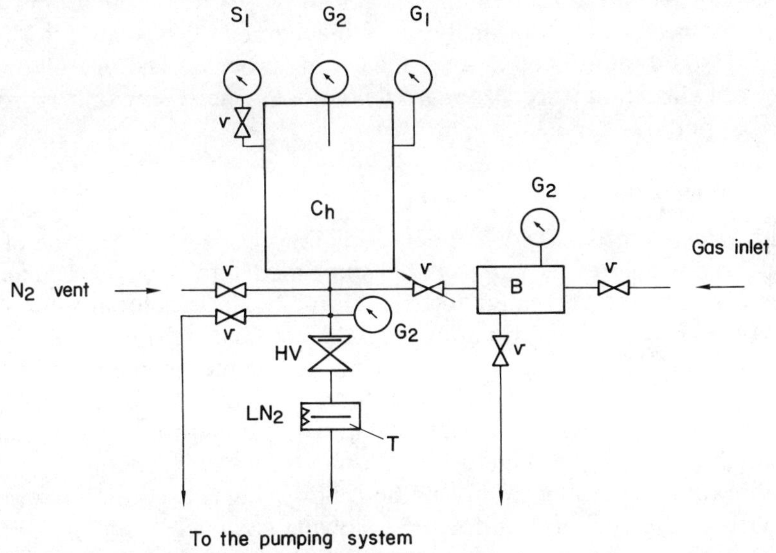

Fig. 5.1 Calibration system for direct comparison with a pressure standard according to the static method. B, buffer volume; Ch, calibration chamber; G_1, gauge head undergoing calibration; G_2, monitor gauge head; S_1, pressure standard; T, cooled trap; HV, high vacuum valve; v, valve. The valve v in series with S_1 is shut off when the pressure standard is not suitable for measuring condensable gas.

constant temperature (about 45°C), because of thermal transpiration phenomenon (Poulter *et al.*, 1983). Therefore, corrections accounting for thermal transpiration must be made when performing a calibration with this pressure standard. For calibration at the upper end of the pressure range (10–1 Pa, 10^{-1}–10^{-2} Torr), a U-tube oil manometer can be used as an alternative.

The operational procedures for calibration usually comply with the following schedule: The calibration chamber is pumped down until reaching a base pressure of at least two orders of magnitude lower than the lowest calibration point required. Since bake-out procedures cannot be used, prolonged pumping periods are necessary to reduce the gas load from the walls of the calibration chamber. After reaching the base pressure required, the calibration chamber is sealed off from the pumping system and checked for a low rate of rise of the residual gas. The calibration gas is then introduced via a leak in the calibration chamber until the scheduled pressure point is reached. Simultaneous readings of the gauge undergoing calibration and of the pressure standard are taken for *stable* pressure conditions at an ambient temperature of 23 ± 3°C. Usually at least three readings are taken for each pressure point checked, and the average deviation is noted. The weighted average deviation represents to a first approximation the readings uncertainty. Further evaluation is obtained by performing the analysis of inaccuracies associated with the pressure standard utilized and the measuring process. The overall uncertainty of calibration (three standard deviations) when using this approach is 1% or better.

5.4.2 Static Expanders

The *principle* of statics expanders is as follows: A small volume of gas accurately measured is *isothermally* expanded into a larger evacuated volume containing the gauge to be calibrated. The resulting pressure in this volume is calculated by using Boyle's law. The method, first used by Knudsen (1910) for the calibration of his radiometer gauge (see Section 4.3.5) and described in detail in further publications (Knudsen, 1914, 1934, p. 34), is discussed in Section 3.4. The pressure range covered by this method is 10^3–10^{-4} Pa (10–10^{-6} Torr).

Calibration according to this method, however, is used on a restricted scale for routine calibration for the following reasons.

(a) The gauge heads used in industrial processes in the medium and high vacuum ranges can release considerable amounts of gas during calibration, even if they have been thoroughly cleaned. This occurs because the elastomer sealings used are incompatible with bake-out procedures.

(b) The volume ratio changes resulting from progressive deformation, especially of moving elements, need careful and periodic checks which cannot be complied with in industry.

(c) The method is suitable only for inert gases or nitrogen.

The quantities which must be known precisely in the design of a calibration system are the range of pressures to be generated in the calibration chamber and the pressure at which the calibration gas is admitted into the high-pressure side of the system. These values are then used as a guide in selecting the main characteristics of the system, such as the gas expansion technique (single or multiple), the volume ratios, the material for the transfer samplers and the calibration chamber, and the pressure standard for the measurement of high pressure of the calibration gas.

Since data referring to the main characteristics of the calibration system reported in the literature (see Table 3.4) are slightly different from worker to worker, it appears difficult to obtain design rules such as those given in standards dealing with direct comparison methods. However, as an aid in the design of a calibration system the following suggestions can be used as guide lines.

(a) The choice of whether to use a single- or multiple-stage expansion technique is made on the basis of the following factors:

(i) The pressure increment in the calibration chamber after performing a step (see Section 3.4.3) is a function of the volume ratios and the high pressure of the calibration gas. Thus, for a single-stage expansion, the volume ratio recommended is 10^7, whereas for multiple expansion in series volumes, the volume ratio is about 500.

(ii) Multiple or cascaded steps contribute to increase the quantity of the residual gas, especially in the low pressure range. The sum of uncertainties involved in the performance of a step becomes cumulative in reaching the required pressure point. A large number of steps to generate a pressure point becomes time consuming.

(b) It is recommended that the transfer samplers for small volumes be cylindrically shaped in the bulk of a metal in order to minimize the volume changes resulting from temperature drifts.

(c) The valves which connect the different sections of the system should be such that the volume change introduced by their operation is minimum.

(d) Cold traps used with primary pressure standards employing mercury must be of the symmetrical U-type in order not to contribute to thermal transpiration effects. Secondary pressure standards, such as the capacitance manometer or the spinning-rotor viscosity gauge, do not need to be used with a cold trap.

(e) The geometrical size and shape of the inner volume of the calibration chamber contributes to randomizing the molecular velocities of the calibration gas. The larger this volume, the smaller the effect of the calibrated gauge heads (gas sink) on the rarefied environment. A spherical chamber would be desirable since it offers the minimum inner surface for a given volume. However, the high cost of spherical calibration chambers limits their use. Instead, a right cylindrical chamber, designed as discussed in Section 5.2.1, may be used.

(f) A residual gas analyzer (RGA), tunable to the calibration gas, must be connected to the calibration chamber. The analyzer may also be used for leak detection.

Figure 5.2 schematically illustrates a three-stage series expansion calibration system. The large right cylindrical vessels TR and Ch (about 6 L each), the small samplers Tr (about 0.025 L each), the tubing connecting the large vessels, the samplers, and the isolation valves are stainless steel, permitting bake-out procedures. Before assembling the system the vessels and interconnecting tubing are chemically cleaned.

The measurement of relative temperatures of the vessels and temperature variations at various parts of the equipment is made with sensitive thermocouple procedures or with thermometers in good thermal contact with the parts. The thermometers used for the ambient temperature mea-

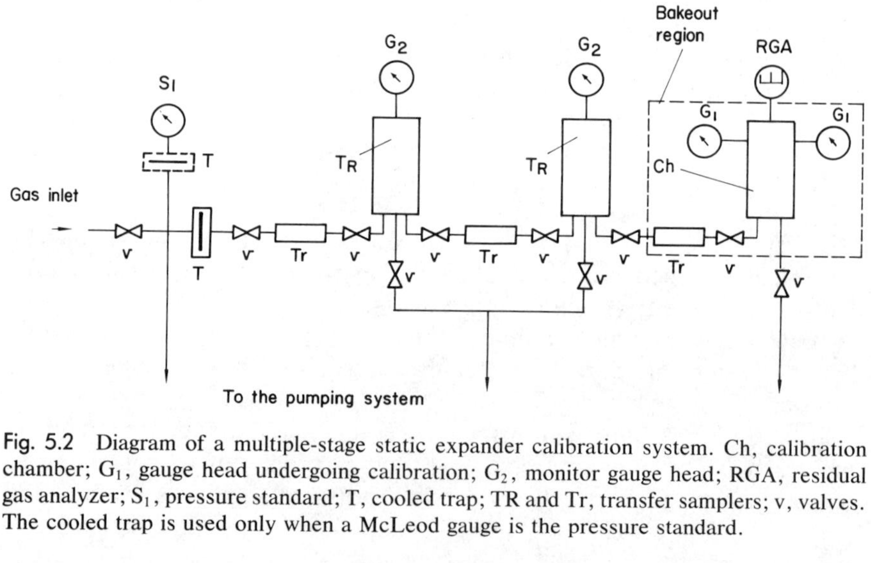

Fig. 5.2 Diagram of a multiple-stage static expander calibration system. Ch, calibration chamber; G_1, gauge head undergoing calibration; G_2, monitor gauge head; RGA, residual gas analyzer; S_1, pressure standard; T, cooled trap; TR and Tr, transfer samplers; v, valves. The cooled trap is used only when a McLeod gauge is the pressure standard.

surement must be shielded from heat radiation. Thermometers must be accurate to ±1°C.

Usually, oil diffusion-pump stacks are used to evacuate the calibration assembly.

After venting the system to dry nitrogen or argon, the following operational procedures are usually performed.

(a) The calibration chamber is evacuated and baked, and the gauge heads undergoing calibration as well as auxiliary instruments are outgassed. During evacuation all valves connecting the transfer samplers with the calibration chamber (single-stage expansion) or with the pumping system (multiple-stage expansion) are left open.

(b) The calibration chamber (or the individual sampler) is isolated from the pumping system and allowed to reach equilibrium with room temperature.

(c) The cold traps are emptied, moderately heated, and refilled with coolant. Then the calibration system is baked and outgassed again as in (a), permitted to cool down, and separated from the pumps. Some workers recommend flushing the calibration system with the calibration gas, the resulting effect being the adsorption of a monolayer of gas which prevents further adsorption.

(d) The transfer samplers and the calibration chamber are isolated from each other, and the background pressure and the residual gas composition are determined in the calibration chamber.

(e) The calibration gas is admitted into the first sampler, and its pressure is measured after temperature equilibrium with the sampler is attained.

(f) The gas is expanded according to the expansion schedule. After each expansion the calibration gas must reach temperature equilibrium conditions with the container. Then the temperature of the container is recorded.

(g) The pressure and gas composition are then determined in the calibration chamber. A record of both the high and the low pressure of the calibration gas as well as of the composition of the calibration gas in the calibration chamber is helpful in the calculation and interpretation of results.

5.5 The Dynamic Method

5.5.1 Direct Comparison with a Pressure Standard

The *principle* of direct comparison is as follows: The gauge head undergoing calibration and an appropriate pressure standard are attached at the calibration chamber and subjected to a *steady pressure* generated by balancing the rate of flow of calibration gas into the chamber against the rate of removal of gas. This approach, which permits calibration in the range 10^2–10^{-4} Pa (1–10^{-6} Torr), is used for the calibration of both thermal and ionization gauge heads.

The calibration system designed to meet this principle is illustrated in Fig. 5.3. The calibration chamber Ch has the shape of a right cylinder with a rather large volume (usually about 100 L). The advantages resulting from the use of such a large volume are that

(a) the flow rate of gas yielded by outgassing from the gauge heads and associated elastomers seals (in the case of thermal gauge heads) is reduced compared with the flow rate of the calibration gas, and
(b) changes in the measuring volume resulting from the variable number of gauge heads calibrated in a run are minimized.

The ports to the calibration chamber are arranged around a ring on the periphery of the chamber with a right angle bend in each connecting tube.

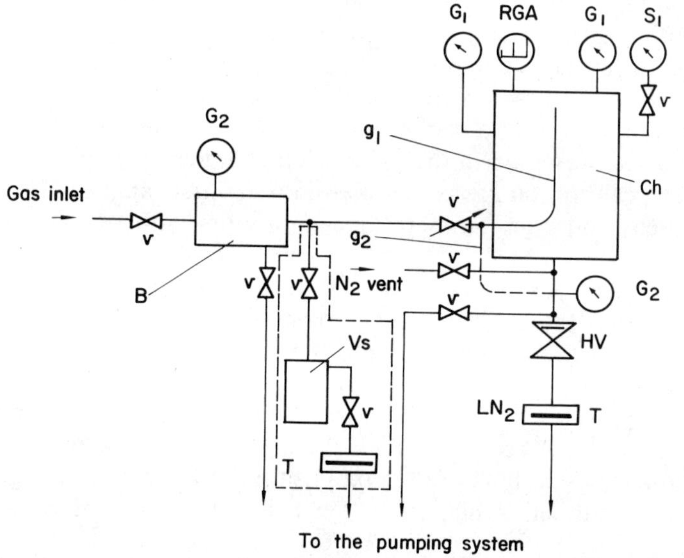

Fig. 5.3 Calibration system for direct comparison with a pressure standard according to the dynamic method. B, buffer volume; Ch, calibration chamber; G_1, gauge head undergoing calibration; G_2, monitor gauge head; g_1 and g_2 tubings for the leak of the calibration gas; HV, high vacuum valve; RGA, residual gas analyzer; S_1, pressure standard; T, cooled trap; Vs, vapor source; v, valve. The valve v in series with S_1 is shut off when the pressure standard is not suitable for measurement of condensable gas. The section drawn within the broken line is used only for calibration with condensable vapor.

This insures that gas within the gauge head has a random distribution of molecular velocities and is in thermal equilibrium with the calibration chamber walls. Thus, each gauge head "sees" identical samples of the calibration gas.

The calibration gas is supplied to the calibration chamber via a buffer volume (B) (3–15-L volume) at a pressure 1000 times larger than the pressure required for calibration.

Two approaches to leaking the calibration gas into the calibration chamber are recommended according to whether measurements are performed for the calibration of thermal conductivity or ionization gauge heads. According to one approach (AVS 6.2, 1969) regarding the calibration of thermal gauge heads, the calibration gas is admitted into the measuring volume through a tubing (g_1 in Fig. 5.3) pointing at the center of the calibration chamber top wall. For the calibration of ion gauge heads, it is recommended that the calibration gas be admitted from below the calibration chamber as illustrated in Fig. 5.3 by the broken bent line (g_2) (AVS 6.4, 1969).

Ruthberg (1969) used this last approach, with the aim of minimizing pressure gradients in the calibration chamber. Since the calibration gas inlet was asymmetrically located with respect to the geometrical axis of the calibration chamber, uniform pressure distribution at the calibration ports was insured by incorporating an internal baffle screen such as illustrated in Fig. 5.4. The length and location of the baffle are selected so as to prevent calibration gas molecules from entering the ports directly. However, according to the ISO/DIS 3568 recommendation for calibration of ionization gauge heads by direct comparison with a reference gauge head, the calibration gas is admitted through a length of tubing protruding into the calibration chamber.

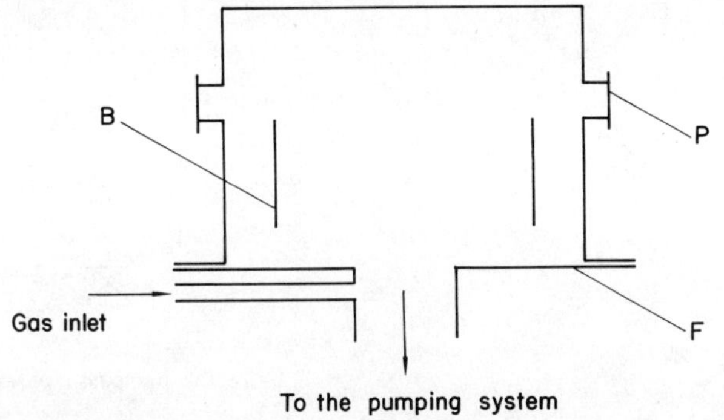

Fig. 5.4 Schematic diagram of a cylindrical calibration chamber. B, baffle; F, flange; P, port.

A manually operated needle valve permits, in conjunction with the suitable operation of the throttle and quarter-swing valves at the gas outlet from the calibration chamber, the achievement of the required pressure point. At the lower end of the calibration range (10^{-3} Pa, 10^{-5} Torr), the pressures generated in the calibration chamber by operating both inlet and outlet gas valves become unstable and sometimes uncontrollable over the time required for reaching equilibrium conditions. This is due to factors such as incompatibility in correctly setting *both* the needle valve and the throttle (quarter-swing) valves for the reduced amount of calibration gas required, decreasing the pressure of the calibration gas in the buffer volume, and fluctuations of the pumping speed due to the pumping system. Such a situation was prevented by Ruthberg (1969) who used an electrically controlled leak valve driven in a servo loop by the signals from a gauge head, as illustrated in Fig. 5.5. The electrical output of the control unit, which receives signals from the gauge head (G) measuring the gas pressure in the calibration chamber, is compared with a reference voltage in a unit (Ra). The difference signal amplified in the same unit (Ra) is then fed into a solenoid-driven leak valve, which opens or closes so as to reduce to zero the difference signal. The gauge head (G) is required to be stable and to have a continuous response throughout the range of pressures monitored. Throttling the pumping system to appropriate speeds in conjunction with the automatic operation of the leak valve for calibration gas improves pressure stability.

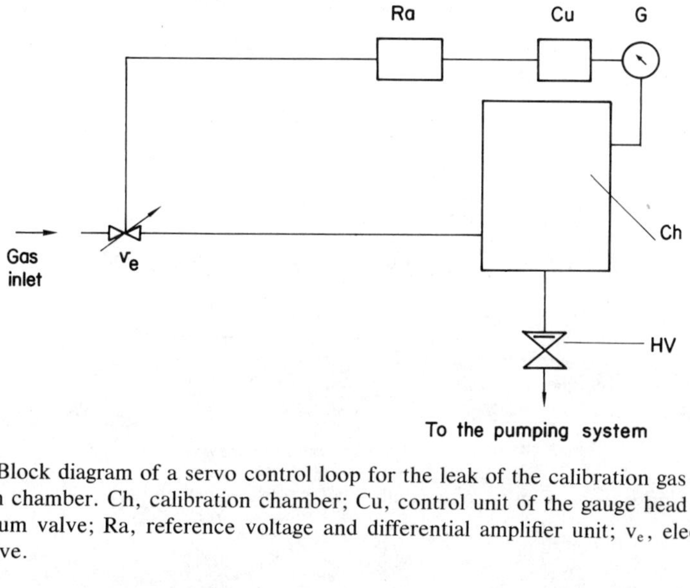

Fig. 5.5 Block diagram of a servo control loop for the leak of the calibration gas into the calibration chamber. Ch, calibration chamber; Cu, control unit of the gauge head G; HV, high vacuum valve; Ra, reference voltage and differential amplifier unit; v_e, electrically driven valve.

The temperatures of the different parts of the calibration system (mainly the calibration chamber) and of the calibration gas must be accurately controlled. Thermocouple junctions inserted in aluminum or copper disks, or thermometers accurate to better than 1°C, are set in good thermal contact where temperature must be known. The temperature of ambient air is measured in the vicinity of the calibration chamber. The thermal sensor for the measurement of ambient temperature must be shielded from heat radiation.

The pressure standard is selected according to the pressure range spanned by the gauge head undergoing calibration. The capacitance manometer, the spinning-rotor viscosity manometer, and the conventional triode gauge head are recommended. When using a McLeod gauge care should be taken to avoid the mercury vapor flow from the instrument to the calibration chamber. The use of a bent U trap cooled in a mixture of dry CO_2 and alcohol or acetone is better than the use of a LN_2-cooled trap (see Section 3.3.5). For calibration with a condensable vapor, gauges which have a relatively large surface exposed to the rarefied environment (e.g., diaphragm, capsule) must be avoided. The reason for this is that adsorption and desorption at and from the surface in contact with the calibration gas may introduce relatively large measurement errors. A simple U-tube manometer, calibrated *in situ* against a secondary pressure standard, can be used as a transfer standard as long as the vapors do not dissolve or chemically combine with the manometric fluid.

The use of a residual gas analyzer (RGA) is necessary when calibration is performed below 10^{-2} Pa (10^{-4} Torr) for information with regard to the residual gas in the calibration chamber.

The pumping system selected must be capable of bringing the base pressure in the calibration chamber down to 10^{-6} Pa (10^{-8} Torr) and the pumping speed downstream from the calibration chamber, before reaching the throttle valve, to at least 100 L/s.

The operational procedures for the calibration of (a) thermal and (b) ionization gauge heads (to be discussed below) are similar in certain respects. However, it was decided that separate descriptions would be less complicated although they would involve some repetition. More details can be found in standards (ISO/DIS 3568; ISO/DIS 3570) as well as in tentative standards AVS 6.2 (1969) and AVS 6.4 (1969).

(a) The procedure for the *calibration of thermal gauges* is as follows.

(i) The calibration system is vented with nitrogen and the gauge heads (cleaned according to the manufacturer's instructions), and the other instruments (pressure standard, RGA) are mounted in the calibration chamber. The gauge heads are then connected to the corresponding control units, and traps are cooled.

(ii) The calibration system is pumped down while watching the readings of the master gauge G_2 in the pumping line to confirm that no gross leakage exists (Fig. 5.3).

(iii) After reaching the base pressure required for starting calibration (10^{-2} Pa, 10^{-4} Torr) the calibration chamber is isolated from the pumping system. The zero reading of the thermal gauge head over 30 s indicates that the total outgassing is below the permitted level. This is the pressure rise test. It must be noted that bake-out procedures cannot be performed, since in nearly all cases the gauge heads undergoing calibration are joined to the calibration chamber with elastomer O rings.

(iv) All gauge heads which are going to be simultaneously calibrated are maintained in continuous operation. Sufficient time must be allowed for the gauge heads to reach temperature equilibrium with the ambient air.

(v) The buffer volume, evacuated in a previous operation (ii), is filled with calibration gas at the required pressure. Sufficient time must be allowed for the gas temperature to stabilize with respect to the ambient air.

(vi) With the calibration chamber evacuated at the ultimate pressure, calibration gas is leaked into the measuring volume until reaching the lowest pressure point at which calibration is required. At least 30 s must be allowed for the readings of the control unit to stabilize at a steady value within $\pm 1\%$ of the full scale electrical range (AVS 6.2, 1969).

(vii) The readings of both the gauge head undergoing calibration and the pressure standard are taken and recorded. The temperature of the ambient air (under stationary conditions) at that time must be 23 ± 3°C.

For calibration with condensable vapor, a special unit (broken-line box in Fig. 5.3) permitting the introduction of the required vapor into the calibration chamber is added to the system.

(b) The procedure for the calibration of ionization gauges is as follows.

(i) With the gauge heads to be calibrated and the calibration system assembled and evacuated as specified for thermal gauges [(i)–(ii)], the calibration chamber is baked out.

(ii) When the required base pressure (10^{-6} Pa, 10^{-8} Torr) is reached, baking is stopped, and the calibration chamber is permitted to attain ambient temperature.

(iii) Degassing of the gauge heads (see Section 4.4.7) undergoing calibration is started when the calibration chamber temperature is about 150°C.

(iv) The isolating resistance of the electrodes, checked after the calibration chamber temperature has reached ambient temperature, must correspond to the manufacturer's specifications.

(v) The calibration gas from the buffer [see step (v) under thermal gauges] is leaked into the calibration chamber so as to establish the lowest measuring pressure required, and time is allowed for stabilization.

(vi) The emission current and other potentials are set according to the values specified by the manufacturer. Readings are taken after a period of about 5 min (during which the measuring meter stabilizes to indicate the collected current) along with readings of the pressure standard and ambient temperature. Each determination must be repeated at least twice for each of the three calibration points within a decade.

Detailed procedures for setting the potentials when several gauge heads are simultaneously calibrated are specified in ISO/DIS 3568.

The overall uncertainty in the calibration of thermal and ionization gauge heads is the sum of the uncertainties in the pressure points determination and uncertainties accumulated in the various steps of the procedure. The calibration accuracy is evaluated by carefully taking into account all the factors involved in the pressure points determination.

5.5.2 Dynamic Expanders with Fixed Conductance

The *principle* of dynamic expanders with fixed conductance is as follows: Under adequate conditions a *steady, calculable* pressure drop ΔP can be generated in a leak-tight vessel by balancing a *known* gas throughput Q into the vessel against the rate of flow of gas out of the vessel across an orifice of *known* conductance C. The equations relating ΔP, Q, C, and the pressures of gas P_1 and P_2 on the two sides of the orifice (inside and outside the vessel), respectively, are

$$Q = \Delta PC = (P_1 - P_2)C \tag{5.2}$$

$$Q = P_1 S_e = P_2 S \tag{5.3}$$

where S_e and S are the volume rates of the gas flow upstream and downstream from the orifice, respectively. The conductance has the dimensions of a volume flow and may be considered as the volume flow per unit pressure difference.

Equation (5.2) holds true when

(a) the gas flow develops under free molecular conditions and is steady all over the vessel;

(b) the distribution of molecular velocities of gas inside the vessel and the direction of molecules entering the orifice is not controlled by the gas flow from other sections of the vessel;

(c) during the generation of the pressure drop across the orifice, the temperature of the vessel, of the gas flowing into the vessel, and the speed with which gas is evacuated through the orifice are constant;

(d) the gas throughput is maintained constant, i.e., it remains unaltered by sinks or sources of gas from the vessel.

Lower pressures P_1 generated in the vessel are obtained by reducing Q or increasing S_e.

Q can be reduced by one of the following techniques:

(a) admitting into the vessel small, controllable volume flow rates of gas; practically, pressures lower than 10^{-4} Pa (10^{-6} Torr) cannot be accurately generated due to the large uncertainties which affect the measurement of reduced gas throughputs;

(b) admitting a gas throughput in a stack of vessels separated by orifices such that the pressure generated in each one corresponds to a fixed ratio of the orifice conductances; by carefully selecting such a ratio it is possible to generate pressures as low as 10^{-6} Pa (10^{-8} Torr) in the vessel located close to the pumping system; the limiting parameter to the achievement of lower pressures is the speed S of the pumping system which is selected such that $S \geq 100S_e$;

(c) dividing the gas throughput into successive stages to produce smaller but known flow rates into the vessel in which the generation of pressure P_1 is required; upon connecting several stages, pressures as low as 10^{-6} Pa (10^{-8} Torr) can be achieved; the technique is limited at the lower end of the pressure range by the errors which accumulate proportional to the number of stages.

S_e can be increased by augmenting the size of the pumping system so as to provide a large speed S downstream from the orifice. Usually pumping speeds S of about 1000 L/s are required to maintain satisfactorily low pressures downstream from the orifice (about 10-mm diameter) close to the pumping system.

Typically, ΔP is generated by the gas flow through an orifice or tube with circular cross section having a fixed conductance. Orifices with adjustable conductances are also used.

Systems based on the techniques (a) and (b) are referred to as *single stage of flow (pressure) divider* systems. Those relying on the technique described under (c) are referred to as *multiple stage of flow divider systems*.

A variant to the single stage of flow divider is the *reference transfer method,* which permits generation of pressures in the range 10^{-7} Pa (10^{-9} Torr). Lower pressures in the ultrahigh vacuum, down to 10^{-11} Pa (10^{-3}

Torr), can also be obtained by generating the rarefied environment in a system using the single stage of flow divider technique.

The calibration of gauge heads by the dynamic expander technique is standardized (ISO/DIS 3570; DIN 28416, 1973; ASTM 297/70).

We now discuss *single stage of flow divider systems*. A typical calibration system consists of a calibration chamber (Ch) and a large-volume chamber (Vc) joined by an *orifice* or a *tube* of conductance *C* connected to a pumping system, such as is illustrated in Fig. 5.6.

The calibration chamber is usually a right cylinder (diameter about 250 mm, height about 500 mm) made of stainless steel, electropolished inside. Spherical chambers with diameters of 500 to 600 mm or more are used for metrological purposes (Hojo *et al.*, 1977; Poulter, 1978).

The gauge heads to be calibrated as well as other instruments considered necessary (e.g., McLeod gauge) are attached to the ports by bakable seals. The calibration gas, whose throughput or quantity (*PV*) (McCulloch, 1983) are accurately measured, is introduced into the calibration chamber through a pipe via a control valve (v). Beaming effects due to gas leak are avoided either by baffling the gas outlet or by pointing the pipe at the upper wall center of the calibration chamber.

Circular orifices (Barre *et al.*, 1956; Normand, 1962; Bennewitz and Dohmann, 1965; Das, 1966; Bannenberg and Tip, 1968) as well as a long tube with circular cross section (Florescu, 1962; Hayward and Jepsen,

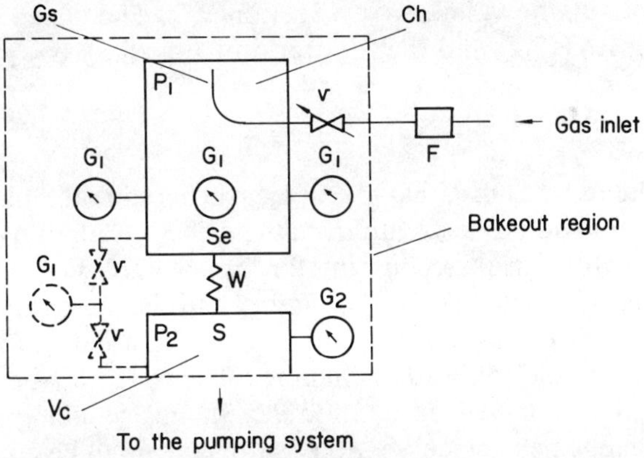

Fig. 5.6 Single-stage flow divider using the measurement of gas throughput Q. Ch, calibration chamber; F, gas flowmeter; G_1, gauge head undergoing calibration; G_2, monitor gauge head; Gs, tubing for gas inlet; v, valve; Vc, large volume chamber; W, constriction (orifice or tube) with conductance C. S_e and S are pumping speeds up- and downstream from W; P_1 and P_2 pressures in Ch and Vc.

1962) are used for joining the calibration and large volume chambers. The role of the large-volume chamber is to insure a proper distribution of molecular velocities on the low-pressure side of the orifice (tube).

The calibration system must be pumped at a sufficiently high speed for the partial pressure of gas evolved from the system to form a negligible part of the total gas pressure. To reduce the effect of the background pressure in the calibration chamber the rate of removing gas by the pumping system must exceed the combined desorption and leakage rates. Pumping systems of the oil diffusion type with pumping speed of about 1500 L/s are usually used for the evacuation of the calibration system, but diffusion pumps (Poulter, 1978) and ion pumps (Hoffman, 1963) are also utilized.

Calibration can be performed over the pressure range 10^{-1}–10^{-4} Pa (10^{-3}–10^{-6} Torr) with an estimated accuracy of better than 2% (Bannenberg and Tip, 1968). Upon making corrections in the calculation of the orifice conductance for the near molecular flow region, Poulter (1978) designed a calibration system permitting the generation of steady pressures ranging from molecular flow pressures to 10 Pa (10^{-1} Torr), with accuracies of ±4% in the molecular region and ±6% at 10 Pa (10^{-1} Torr).

Chemically active gases such as SF_3, BH_3, and HCl have also been used instead of N_2 as calibration gas (Bennewitz and Dohmann, 1965).

The *pressure* P_1 generated in the calibration chamber under steady-state conditions of gas flow is calculated by replacing in the first term of Eq. (5.3) both of the measured quantities Q and S_e. The throughput is measured by using flowmeters (see Section 5.7). The speed S_e expressed by combining Eq. (5.2) and the first term of Eq. (5.3) as

$$S_e = (1 - P_2/P_1)C \tag{5.4}$$

can be calculated by computing C and measuring the ratio of pressures P_1 and P_2 upstream and downstream from the orifice. Equation (5.4) must be supplemented with terms accounting for the background pressure P_0 due to residual gases in the calibration chamber and the pumping due to gauge heads (ionization) during calibration. The background pressure is measured after baking the calibration chamber. The gas removed by the gauge head can be determined (see Section 7.5.3), but reducing the pumping speed of the gauge heads (see Section 7.5.6) appears to be a more convenient approach.

The *conductance* C_o of an orifice in a thin plate to the molecular flow of gas, given by Knudsen (1910), is

$$C_o = A(R_0T/2\pi M)^{1/2} \tag{5.5}$$

where A is the cross-sectional area of the orifice and R_0, T, and M have their usual meaning in kinetic theory of gas (see List of Symbols).

Equation (5.5) strictly is valid only when the plate is infinitely thin and large compared with the diameter of the orifice. In practice the thickness and width of the plate is finite so that the conductance of the orifice must be calculated as

$$C = C_o K_{1-4} \tag{5.6}$$

The correction factors K_{1-4} to Eq. (5.5) account for the finite thickness of the plate, the mounting of that plate in a short tube (its width being thus reduced), the chamfer on the orifice edge (to increase its mechanical stability), and the case when the diameter of the orifice and the mean free path of gas molecules are of the same magnitude (Clausing, 1932, 1971; DeMarcus, 1956; Steckelmacher, 1966; Bureau et al., 1952; Iczkowski *et al.*, 1963; Berman, 1956; Willis, 1965; Peggs, 1973).

The calculation of the conductance of a tube requires considerable mathematical manipulation involving the use of Monte Carlo–type statistical methods and/or semi-empirical formulas.

The calculated values of C are usually compared with experimentally determined values (see Section 5.8). This is done in order to check the calculation validity and to determine changes occurring during extended periods of operating the calibration system, which result from distortions produced by bake-out procedures or clogging.

The maximum pressure difference which can be generated with an aperture with circular cross section in a thin plate is about 5×10^{-5} Pa (3.75×10^{-7} Torr). Still lower pressure differences can be obtained with materials such a fritted glass, unglassed porcelain, sintered steel, or carbide used to restrict the gas flow in the calibration system. Such materials formed into a plug, referred to as a porous plug, contain a large number of fine-bore capillaries whose dimensions are comparable to the mean free path λ of gas molecules, even at pressures of some tens of pascals. The conductance of porous plugs is not amenable to calculation from physical dimensions and therefore must be experimentally determined (see Section 5.8).

The *operational procedures* indicated by Normand (1962) are basic for calibration with single flow divider systems. The system (Fig. 5.6) is evacuated (valve controlling the calibration gas closed) and baked. The resulting residual pressure in the calibration and large volume chambers is $P_0 = 10^{-5}$ Pa (10^{-7} Torr). After being degassed (see Section 4.4.7) the gauge heads (G_1) (ionization) are operated at the values recommended by the manufacturer. The valve controlling the gas leak is adjusted to allow gas into the calibration chamber at a constant flow rate while the pumping system is operating. After a transient period, an equilibrium pressure P_1 is established in the calibration chamber (Ch). At equilibrium

$$\Delta P = P_1 - P_0 = Q/s \tag{5.7}$$

The corresponding differential gauge response Δg deduced from the gauge (G_1) indications g_1 and g_0 at P_1 and P_0, respectively, can be written

$$\Delta g = g_1 - g_0 \tag{5.8}$$

for each equilibrium pressure.

Provided that the ratio $f = \Delta P/\Delta g$, referred to as the *calibration factor,* is constant, the response of the gauge heads is linear, and calibration is reduced to a comparison of the gauge head response with pressure changes over the pressure range of pressure checked.

The pressure P_1 in the calibration chamber resulting from the gas flow under isothermal conditions can be expressed by writing Eq. (5.2) as

$$P_1 = Q/C(1 - P_2/P_1) \tag{5.9}$$

To calculate this pressure Normand (1962) computed C according to Bureau *et al.* (1952) and measured both Q and P_2/P_1. The throughput Q was measured by timing the displacement of an oil manometer column produced by the calibration gas admitted into chamber (Ch) (see Section 5.7).

The measurement of P_2/P_1 was performed bearing in mind that only the ratio of these pressures must be known accurately, their absolute values being of no interest. Therefore, the indications of the two gauge heads G_1 and G_2 identically biased and mounted in chambers Ch and Vc were first checked for good agreement. This exists when identical readings are obtained by interchanging the location of the two gauge heads. With these conditions satisfied the readings of both gauge heads were taken for different pressure points generated in the calibration range 10^{-1}–10^{-4} Pa (10^{-3}–10^{-6} Torr). The average of more than 10 readings was then used to derive a single value of the ratio P_2/P_1 for the calculation of S_e [Eq. (5.4)]. Calibration with different gases is possible provided that the value of the conductance is accordingly calculated.

Errors in determining the ratio P_2/P_1 result in errors in the pressure P_1 determination. Thus, an error of 5% in the determination of this ratio in the calibration performed by Normand (1962) would lead to a 0.9% error in the calculation of P_1 with Eq. (5.9). It is therefore advisable to select a large value for P_1 compared with P_2 so that errors in determining their ratio should be negligible.

Instead of using two gauge heads for the measurement of the pressure ratio in the two chambers of the calibration system, a manifold between these chambers (broken line in Fig. 5.6) permits valving the same gauge head (G_1) in each of the chambers (Ch and Vc) (Kündig and Mennenga, 1980; McCulloch, 1983).

By changing the input rate of the calibration gas, the calibration factor f could be determined over the whole range of generated pressures. The plot of ΔP as a function of Δg made it possible to determine P_1 over that range. In the case where a large number of calibration runs were performed it was convenient to consider an "average factor" applied to the range of pressures considered, provided that $f = \text{Ct}$.

Before using the calibration system for routine work, Normand (1962) checked the generated pressures with those measured using a McLeod gauge mounted in the calibration chamber. He found agreement between the generated and measured pressures within about 1% in the range 10^{-1}–10^{-3} Pa (10^{-3}–10^{-5} Torr).

Hayward and Jepsen (1962) also used a calibration system based on the same technique of generating calibration pressures, differing, however, in that the calibration and the large volume chambers were joined by a tube with circular cross section instead of by an orifice. Their approach to calculating S_e differed from that used by Normand (1962) in that they deduced the pumping speed of the gas in the calibration chamber by considering the volume flow rates of the calibration gas crossing the tube (of conductance C) and entering the pump (of speed S), expressed as

$$S_e = \frac{C}{(1 + C/S)} \tag{5.10}$$

Provided that $S \geq 100C$, the ratio $C/S \leq 1\%$ and P_1 can be calculated directly from Eq. (5.3).

Uncertainties in the calculation of P_1 arise from

(a) nonuniform and nonisothermal distribution of the calibration gas in the calibration chamber (see Section 8.2);

(b) reduced rates of flow of the calibration gas as compared with the rates of flow of gas outgassed from the calibration chamber walls;

(c) errors in the measurement of Q due to temperature variations during determinations (see Section 5.7);

(d) nonuniformities in the cross section of the orifice or tube joining Ch and Vc as well as their partial clogging or distortion over time;

(e) errors in the measurement of the ratio P_2/P_1 due mainly to sensitivity changes of the gauge heads (ionization, see Section 4.4.7) used for the determination of that ratio.

To achieve calibration pressures lower than 10^{-4} Pa (10^{-6} Torr), the single stage of a flow divider system is connected, via a constriction to gas flow, to a chamber containing the calibration gas at high pressure. This arrangement permits generating pressures in the range 10^{-1}–10^{-7} Pa (10^{-3}–10^{-9} Torr). In systems based on this approach (illustrated in Fig. 5.7), the calibration gas is supplied to a chamber (Cp) at high pressure, wherefrom it is leaked into the calibration chamber (Ch) through a constriction of conductance C_1 in series with a valve (v) for the control of the gas inlet. The gas pressure in chamber Cp can be measured accurately with a pressure standard (S_1). The rest of the system is similar to that illustrated in Fig. 5.6.

At equilibrium and constant throughput Q, the gas flow from chamber Cp across the constrictions W_1 and W_2 to the pumping system is expressed by using the notations in Fig. 5.7 as

$$Q = (P_1' - P_1)C_1 = (P_1 - P_2)C_2 = P_1(1 - P_2/P_1)C \tag{5.11}$$

Here C_1 and C_2 are the conductances of the constrictions W_1 and W_2.

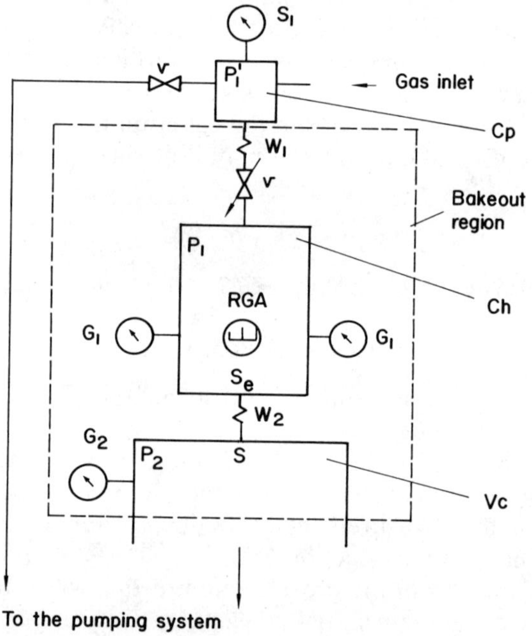

Fig. 5.7 Single-stage flow divider using the measurement of gas pressure P_0. Cp, high-pressure chamber; Ch, calibration chamber; G_1, gauge head undergoing calibration; G_2, monitor gauge head; RGA, residual gas analyzer; S_1, pressure standard; v, valve; Vc, large volume chamber; W_1, constriction (porous plug or capillary tube) with conductance C_1; W_2, constriction (orifice) with conductance C_2. S_e and S are pumping speeds up- and downstream from W_2; P_1', P_1, and P_2 are pressures in Cp, Ch, and Vc.

Upon selecting the conductance C_1 such that $P_1 \geq 10^3 P_1$, the expression of P_1 resulting from considering the first and the last terms of Eq. (5.11) is

$$P_1 = P_1'(C_1/C_2)|1/(1 - P_2/P_1)| \tag{5.12}$$

An analysis of Eq. (5.12) shows that for $P_1' = \mathrm{Ct}$, and if C_1 and C_2 do not change their values due to operation conditions (clogging, distortion), P_1 changes with P_2/P_1 due to variations of the pumping speed S of the system [see Eqs. (5.10) and (5.4)]. Assuming S reasonably constant over the calibration pressure range, it is necessary to ensure conditions such that $P_2 \geq P_1$ in order to minimize errors in the calculation of P_1 due to inaccuracies in the determination of the ratio P_2/P_1.

Upon taking into account the theoretical value C_o of the conductance of an orifice [Eq. (5.5)], the ratio of conductances in Eq. (5.12) becomes

$$C_1/C_2 = A_1/A_2 \tag{5.13}$$

It therefore results that P_1/P_1' [Eq. (5.12)] is independent of the molecular mass and absolute temperature of the calibration gas, provided that the gas temperature remains constant during its flow across C_1 and C_2. Theoretically that means that the method is "absolute" in that pressure P_1 is derived from the physical dimensions of the conductances C_1 and C_2 (see Section 3.1). The advantage arising from this fact is that once the ratio of the two conductances had been determined for any gas, other gases can be calibrated.

Equation (5.12) also shows that with P_1' having values in the medium vacuum range (where it can be measured accurately with a pressure standard) and $P_2/P_1 \cong 0.15$, it is necessary to have $C_1/C_2 \leq 10^{-6}$ in order to obtain $P_1 \leq 10^{-6}$ Pa (10^{-8} Torr). Using the McLeod gauge, errors due to mercury vapor flow to the calibration chamber and thermal transpiration are negligibly small (see Section 3.3.5). If the capacitance manometer is used for the measurement of P_1', its readings must be corrected for thermal transpiration.

The condition $C_1/C_2 \leq 10^{-6}$ is met by using a porous plus having a very small conductance.

Owens (1965) calibrated ionization gauge heads over the pressure range 10^{-7}–10^{-1} Pa (10^{-9}–10^{-3} Torr) in the calibration system illustrated in Fig. 5.7. Two porous plugs ($C_1 = 6.3 \times 10^{-4}$ and 1.26 L/s) were selected to cover the calibration range mentioned. The conductances of the porous plugs were experimentally determined *in situ* upon using the calibration system (see Section 5.8). The pumping system ($S = 1000$ L/s) evacuated the calibration chamber through an orifice of conductance $C_2 = 247.8$ L/s. The background pressure of the stainless steel calibration chamber, with an inner area of 1915 cm^2, was 5×10^{-8} Pa (3.75×10^{10} Torr) after an overnight pump down. The ratio P_2/P_1 was determined by resorting to Normand's technique, previously discussed. A 10% variation in determining this ratio resulted in a 2.82% change in the quantity $(1 - P_2/P_1)$. This was attributed to variations in the gauge head sensitivity. Calibrations were made in succession with nitrogen and other gases, the onlv requirements after shifting from a run with one gas to another being to flood the calibration chamber with nitrogen and to allow enough pumping down of that chamber to remove the residual gas. The same ultimate pressure was obtained for each gas used for calibration. Changes in leak rates of the calibration gas due to possible clogging of pores in the porous plug were periodically checked by comparing the plot $P_2 = f(t)$ with the plot obtained when the porous plug was used for the first time. The readings of a McLeod gauge mounted in the calibration chamber were the same as the calculated pressure P_1, from 10^{-1} to 10^{-3} Pa (10^{-3}–10^{-5} Torr), and differed by about 12% in the pressure range 5×10^{-3}–10^{-5} Pa (3.75×10^{-5}–10^{-7} Torr). The accuracies obtained were less than ±10% down to 10^{-6} Pa (10^{-8} Torr) and about 15% in the 10^{-7} Pa (10^{-9} Torr) range.

Fletcher and Watts (1967) used another approach in calculating the calibration pressure. They wrote P_1' by equating the first and second terms of Eq. (5.11) as

$$P_1 = P_1'C_1/(C_1 + C_2) + P_2C_2/(C_1 + C_2) \tag{5.14}$$

Provided that $P_1 \geq 100P_2$ and $C_1 \ll C_2$, Eq. (5.14) becomes

$$P_1 = P_1'C_1/C_2 \tag{5.15}$$

The error resulting in calculating P_1' with Eq. (5.15) instead of with (5.14) is thus 1% or less.

In the system they designed (Fig. 5.8) the calibration chamber (Ch) was a 61-cm diameter stainless steel sphere joined with another chamber (Vc) by means of two valves (v), one is series with an orifice of conductance C_2 having a diameter of 10.44 cm. The inlet of the calibration gas into chamber Ch is via two restrictions of conductance C_1 and C_1' each one contained in the sealing flange (bakeable), connected in turn to the reservoir (R). The calibration gas was introduced into the reservoir through a bakeable needle valve. The pressure standard for the measurement of the high pressure of the calibration gas was a McLeod gauge. Provisions were made to also use the expansion method as a measurement alternative.

The pumping system, an oil diffusion pump, backed by another diffusion rotary pump, produced a pumping speed of 2700 L/s. Cheveron baffles, one of which was cooled to about −30°C, protected chamber Vc against contamination from the pumping system. The calibration chamber

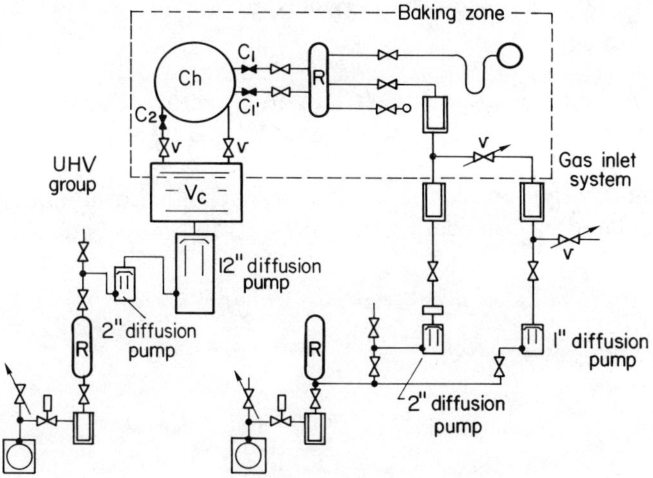

Fig. 5.8 Calibration system. Ch, calibration chamber; R, reservoir; Vc, large volume chamber. Reprinted with permission from *Vacuum* **17,** 447, B. Fletcher and F. Watts. Copyright 1967 Pergamon Press.

could be quickly evacuated through a valve bypassing the valve connected to C_2.

The background pressure in the calibration chamber after bake out was 5.33×10^{-8} Pa (4×10^{-10} Torr). A pressure rise of the order of 1.06×10^{-5} Pa/h (8×10^{-8} Torr/h) due to an outgassing rate of 4×10^{-11} Pa L/s cm^2 (3×10^{-13} Torr L/s cm^3) in the calibration chamber was usually obtained. That pressure has a negligible effect on the measurement accuracy.

The ratio C_1/C_2 was determined *in situ* by a method requiring the measurement of time and a parameter of the gauge head mounted in the calibration chamber. Thus, upon assuming that pressure is an unknown time-independent function of a measurable parameter of the gauge head $P = f(x)$, the pressure reduction in Ch evacuated through the orifice of conductance C_1 from a level $P_a = f(x_a)$ to a level $P_b = f(x_b)$ in a time t_1 is

$$C_1 = (V/t_1) \ln|f(x_a)/f(x_b)| \tag{5.16}$$

where V is the calibration chamber volume.

If the gauge head being calibrated has a linear response, the only requirement for C_1/C_2 determination is to know the ratio of two gauge readings separated by a time interval t.

Repeating the determination with the same values of x for the pump down of Ch through the orifice of conductance C_2 gives

$$C_2 = (V/t_2) \ln|f(x_a)/f(x_b)| \tag{5.17}$$

Finally, the conductance ratio becomes

$$C_1/C_2 = t_2/t_1 \tag{5.18}$$

Because of the large value of C_1/C_2 involving rather long measuring times it is recommended to measure C_1/C_1' and C_1'/C_2 and to then multiply these ratios to obtain C_1/C_2.

The time t_e required for the system to reach equilibrium pressure P_1 in Ch is calculated by considering the gas flow rates in and out of the calibration chamber

$$P_1'C_1 = P_1C_2 + V\,dP_1/dt_e \tag{5.19}$$

which by integration gives

$$\int_0^t dt_e = V \int_{P_0'}^{P_1} dP_1/(P_1C_1 - P_1C_2) \tag{5.20}$$

Solving Eq. (5.20) for $t_e = 0$ (when $P_1 = P_0'$) and $t_e = t$ (when $P_1 = P_1$) one obtains

$$t_e = (V/C_2)\ln|(P_1'C_1 - P_0'C_2)/(P_1'C_1 - P_1C_2)| \tag{5.21}$$

The time t_e taken for P_1 to reach 1/1000 of its equilibrium value calculated by Eq. (5.21) is 69 s (Fletcher and Watts, 1968).

To estimate the accuracy of the calibration process, Fletcher and Watts (1968) equalized the volume rate of flow of the calibration gas in and out of the calibration chamber and took into account both the desorption rate O_{Ch} from the walls of the calibration chamber and the pumping speed S_g of the gauge heads undergoing calibration. Thus,

$$C_1(P_1' - P_2) + O_{Ch} = C_2(P_1 - P_2) + S_gP_1 \tag{5.22}$$

which, solving for P_1 with $K = C_1/C_2$, gives

$$P_1 = (KC_2P_1' + C_2P_2 + O_{Ch})/C_2(1 + K) + S_g \tag{5.23}$$

The estimated accuracy of the system calculated according to Eq. (5.23) is shown in Fig. 5.9.

Close *et al.* (1977) designed a system in which the calibration gas is expanded by a measured volume ratio of about 10^2 in a high-pressure chamber. The gas pressure in this chamber is then reduced by a 10^5 ratio by using the conventional single stage of flow divider technique discussed. The system which permits calibration in the 10^3–10^{-6} Pa (10–10^{-8} Torr) range, consists of a stack of three cylindrical chambers, Cp, Ch, and Vc, each of about 3 L as illustrated in Fig. 5.10a.

The calibration gas is introduced in the expansion volume V of about 30

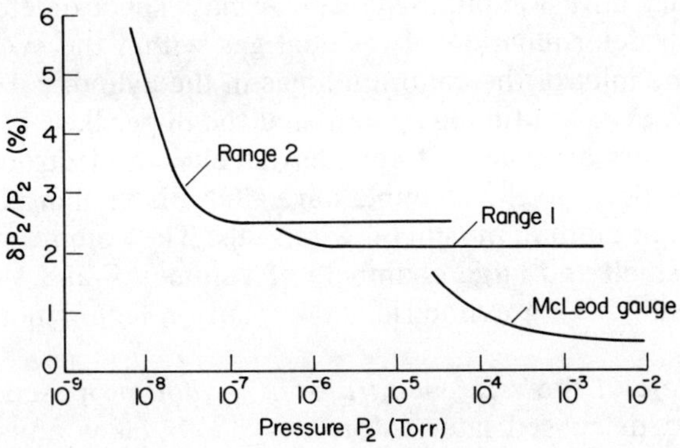

Fig. 5.9 Accuracy as a function of pressure P_1 generated in Fletcher and Watts' calibration system (after Fletcher and Watts, 1967).

mL (Fig. 5.10b) at pressures larger than 1 Pa (10^{-1} Torr) and expanded in the chamber Cp of the system. A McLeod gauge protected by a liquid nitrogen trap is used to measure the pressure in volume *V*. Chambers Cp and Ch as well as Ch and Vc are joined by a porous plug (W_1) and an orifice in a thin plate (W_2), respectively. The chamber Vc is evacuated

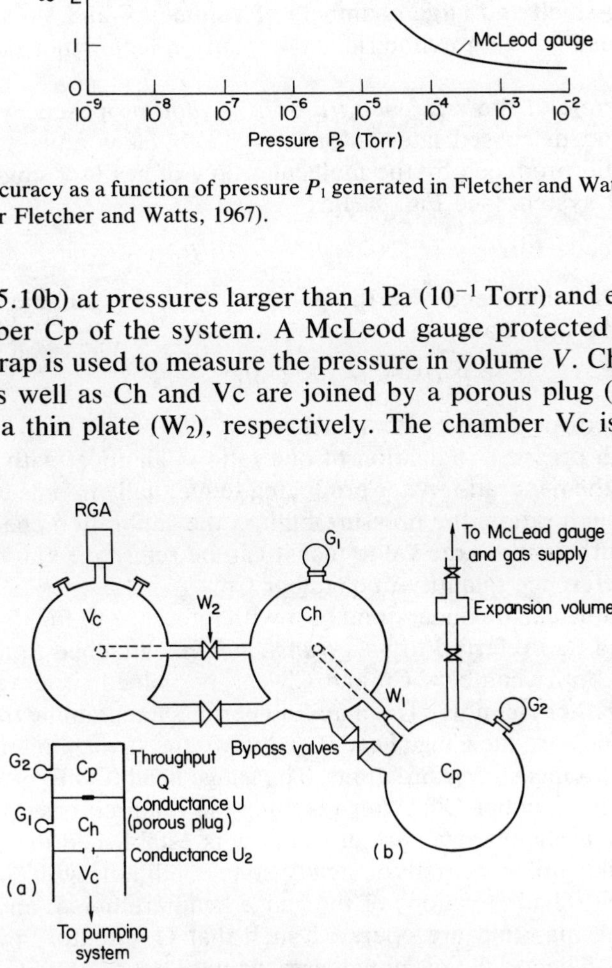

Fig. 5.10 Diagram of the calibration system designed by Close *et al.* (1977). Cp, high pressure chamber; Ch, calibration chamber; G_1, gauge head to be calibrated; G_2, monitor gauge head; RGA, residual gas analyzer; V_c, large volume chamber; W_1, constriction (porous plug) with conductance C_1; W_2, constriction (orifice) with conductance C_2. Reprinted with permission from *Vacuum* **27,** 512. K. J. Close *et al.* Copyright 1977 Pergamon Press.

with a mercury diffusion pump system. A mass spectrometer (RGA) was used for both determination of residual gas within the system and leak detection. The inlet of the calibration gas in the cylindrical chambers Ch and Vc is achieved via tubings pointing at the upper base of the cylinder. All the chambers are made of stainless steel with electropolished inner surfaces. The ports to the chambers are CONFLAT flanges with OFCH copper, gold, or aluminum–silicon wire seals. The volume ratio measurement of the small and large chambers of volumes V and V_1 was accomplished by using both gravimetric and expansion techniques (see Section 3.4.4).

The principle of the *reference transfer method* proposed by Morrison (1966a,b) and discussed later (Morrison, 1969) takes advantage of the pressure ratio produced by the molecular flow of gas in a single stage of flow divider system (see Fig. 5.2) expressed as

$$P_1'/P_1 = (C_1 + C_2)/C_1 - (P_2/P_1)(C_2/C_1) \tag{5.24}$$

Assuming that the speed S of the pumping system is sufficient to make $P_1 \ll P_2$, it results that

$$P_1'/P_1 = (C_1 + C_2)/C_1 = U \tag{5.25}$$

Calibration consists in repeatedly generating a pressure ratio P_1'/P_1 such that the high pressure indication of one ratio is identical with the lower pressure of the next ratio. After producing sequentially in time a sufficient number of such ratios, the pressure built in the calibration chamber (Ch) reaches a sufficiently large value that it can be referred to an accurately measured reference quantity in chamber Cp.

The principle can be understood best with reference to Fig. 5.11, where a gauge head (G_t) referred to as a *transfer gauge* attached to a manifold mounted in both chambers Cp and Ch can be valved so as to measure pressure in either chamber. The manifold permits maintaining the symmetry of transfer, i.e., the gauge head (G_t) "sees" the same plumbing configuration in both measuring situations. The gauge head (G_1) to be calibrated in mounted in chamber Ch. After reaching the required base pressure in the calibration chamber, a certain gas flow is established in Cp and the pressure builds up, measured for steady-state conditions with G_t valved in this chamber. The indications of the gauge head are noted, and then the valves of the manifold are operated such that G_t monitors pressure in chamber Ch. The gas flow is now increased until both G_t and G_1 indicate, under steady-state conditions, the same pressure as was formerly measured in Cp. Upon repeating the process, one can reach a point where the flow of gas through the system is large enough to be accurately measured with a flowmeter. The pressure P_n built in Ch after n transfer series is

$$P_n = P_r U^{n-r} \tag{5.26}$$

Here P_r is the reference pressure in Ch calculated from the gas flow in this chamber, and $n - r$ is the number of times the ratio U was applied between the generation of P_n and the measurement of P_r.

The relative error dP_n/P_n in the calibration pressure [Eq. (5.26)] is

$$dP_n/P_n = dP_r/P_r + (dU/U)(n - r) \tag{5.27}$$

where dP_r/P_r is the relative error in the measurement of the reference and dU/U is the relative error affecting the conductances and the transfer operation.

Readings of the gauge head G_t must remain consistent for the same measured pressure when being transferred from one chamber to another. Its accuracy in reading pressures is not essential for calibration accuracy. This statement is based on the assumption, difficult to quantify, that the manifold influences the measuring process with G_t in the same way when valved in either volume.

The absolute accuracy of calibration using the reference transfer method can be greatly improved by careful operations and by reducing the background pressure.

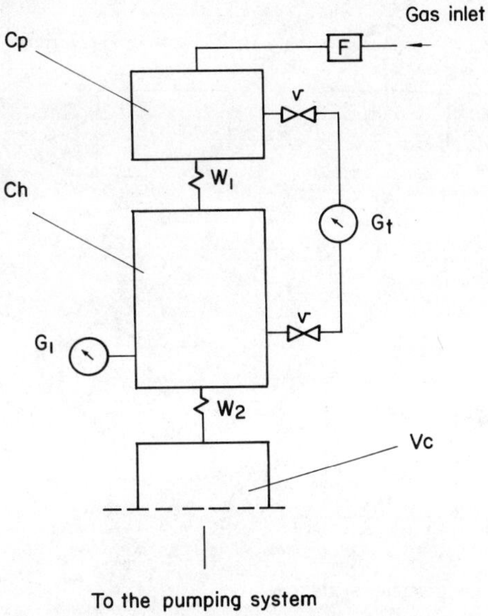

Fig. 5.11 Calibration system using the reference transfer method. Ch, calibration chamber; Cp, high-pressure chamber; F, flowmeter; G_1, gauge head under calibration; G_t, transfer gauge head; v, valve; Vc, large volume chamber; W_1 and W_2, constrictions (orifices) with conductance C_1 and C_2.

In order to minimize the errors in the transfer ratio U, Morrison (1966a) used the configuration illustrated in Fig. 5.12 where the gauge head (G_1) undergoing calibration is moved from the calibration chamber Ch (as in Fig. 5.11) to chamber Vc (in Fig. 5.12). The resulting advantage is that calibration can start with lower pressures and the gas load in chamber Ch produced by the outgassing of its walls is much reduced compared to the former system.

Blado and Lilienkamp (1972) described the application of this technique in the pressure range 10^{-2}–10^{-6} Pa (10^{-4}–10^{-8} Torr). They measured pressure in chamber Cp by means of a capacitance manometer instead of using as reference the flow rate of the calibration gas for the calculation of the pressure in that chamber.

Flow divider techniques providing a well-distributed Maxwellian gas have been used with *ultrahigh vacuum calibration systems* for the calibration of gauge heads down to 10^{-10} Pa (10^{-12} Torr). Thus, Davis (1963) used a system in which the flow of the calibration gas was divided by two orifices (W_1 and W_2) with small (C_1) and large (C_2) conductance, respectively (Fig. 5.13). The pressure ratio across C_2 was 7 : 1. The calibration

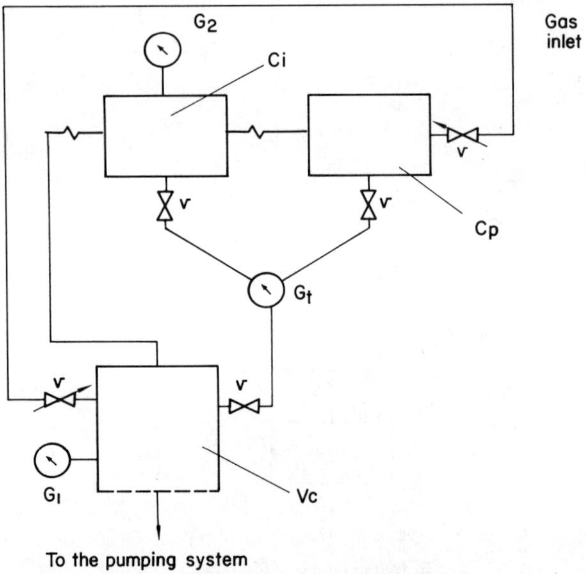

Fig. 5.12 Schematic of the arrangement used for calibration with the reference transfer method. Ci, intermediary chamber; Cp, high-pressure chamber; Vc, large volume chamber; G_1, gauge head under calibration; G_2, monitor gauge head; G_t, transfer gauge head; v, valve.

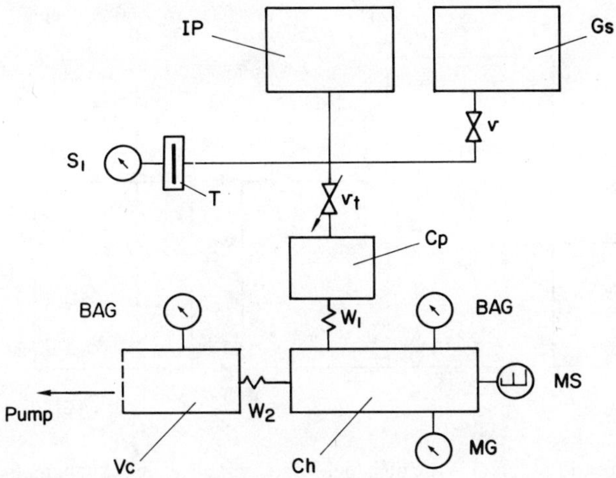

Fig. 5.13 Calibration system in the ultrahigh vacuum range pumped by a diffusion pump. BAG, Bayard–Alpert gauge head; Ch, calibration chamber; Cp, high-pressure chamber; Gs, gas supply chamber; IP, ion pump; MG, magnetron gauge head; MS, mass spectrometer; S_I, pressure standard (McLeod gauge); T, cooled trap; v, valve; v_t, throttling valve; Vc, large volume chamber; W_1 and W_2, constrictions (orifices) with conductance C_1 and C_2.

gas was introduced in Cp via a bakeable valve (v_t) which could be throttled so that it became the limiting conductance. Chamber Cp was pumped by an ion pump before introducing the calibration gas from chamber Gs. A small glass mercury pump and a liquid nitrogen trap were used to evacuate the calibration system. The pressure standard (S_I) was a McLeod gauge. The system, excepting the pressure standard and gas supply system, could be baked at 425°C, thus making it possible to obtain a residual pressure of about 10^{-10} Pa (10^{-12} Torr). The main residual gases were N_2 and small amounts of He and H_2. Calibrations in the 10^{-1}–10^{-5} Pa (10^{-3}–10^{-5} Torr) range were made by using the McLeod gauge to measure pressure on the high-pressure side of the small orifice (W_1) and by further calculating the pressure on the low-pressure side of that orifice by using the known value of C_1. Pressures lower than 10^{-5} Pa (10^{-7} Torr) were obtained by throttling v_t. The main difficulty was caused by the adsorption of calibration gas at the wall surfaces at a rate comparable with the rate of flow of gas through the system.

Davis (1968) designed a calibration system based on the single flow divider principle for calibrating in the ultrahigh vacuum region down to 10^{-8} Pa (10^{-10} Torr) (Fig. 5.14). In this system a liquid-nitrogen-baffled, liquid-helium-cooled cryopump (LHe) produced a baffled speed of 300 L/s, thus permitting the achievement of a pressure drop of 8 : 1 across the

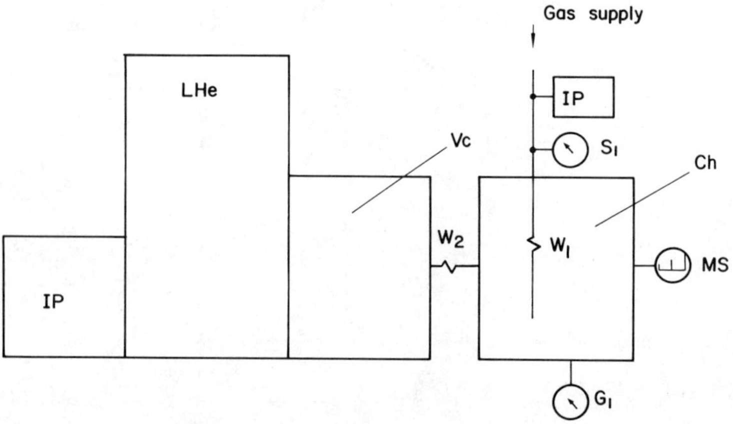

Fig. 5.14 Calibration system in the ultrahigh vacuum range using cryopumping. Ch, calibration chamber; G_1, gauge head undergoing calibration; IP, ion pump; LHe, liquid He pump; S_1, pressure standard (McLeod gauge); Vc, large volume chamber; W_1, constriction (small-diameter tube) with conductance C_1; W_2, constriction (orifice) with conductance C_2.

orifice close to the pump. The system was built of stainless steel (type 304 and 247) with stainless steel flanges and copper gaskets. The restrictions to the gas flow were an orifice (W_2) with circular cross section with a calculated conductance $C_2 = 373$ L/s (for N_2 at 25°C) and a stainless steel tube (W_1) 10-cm long, 0.111-cm diameter with a calculated conductance $C_1 = 1.7 \times 10^{-3}$ L/s. The calibration gas (N_2, O_2, or Kr) was introduced into this tube via a special throttling valve which became the limiting conductance to the gas flow. The valve permitted monitoring the flow rates of gas which could maintain the pressure, measured by a McLeod gauge, within reasonable limits of the instrument. Pumping during the bake-out and standby operation was performed with a triple-stage mercury diffusion pump baffled by a liquid N_2 trap. After baking, the background pressures were 0.2×10^{-9} Pa (1.5×10^{-12} Torr) (calibration chamber at ambient) and 13×10^{-12} Pa (9.75×10^{-14} Torr) (calibration chamber at 253 K). A mass spectrometer (MS) was used to check the composition of the residual gas. The calibration chamber walls were maintained at ambient temperature during determinations, but gas in chamber Vc was cooled at about 253 K in the proximity of the pump. The system was used for the calibration of a magnetron gauge head of the Lafferty type (see Section 4.4.8) and a BAG head.

The calibration system designed by Bernardet *et al.* (1970a) permits calibration in the ultrahigh vacuum range in a double-walled chamber (Ch) which can be operated at either ambient or low temperature (Fig. 5.15).

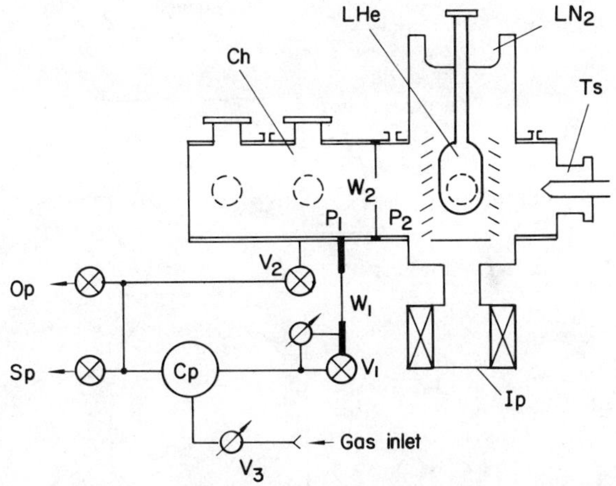

Fig. 5.15 Calibration system in the ultrahigh vacuum range (after Bernardet *et al.*, 1970a). Ch, calibration chamber; Cp, high-pressure chamber; LHe, liquid He pump; Ip, ion pump; Op, orbitron pump; Sp, sorption pump; Ts, titanium sorption pump; V_{1-3} valves; W_1, constriction (capillary tube) with conductance C_1; W_2, constriction (orifice) with conductance C_2. Pressures P_1 and P_2 in Ch and in the manifold containing the UHV pump assembly are indicated.

An UHV pumping assembly made of an ion pump, a Ti sublimation pump, and a LHe pump with an overall speed of about 2000 L/s is mounted in a short manifold separated from Ch by an orifice (W_2) with conductance C_2 100 L/s for N_2 at room temperature. The calibration gas supplied to a chamber Cp previously evacuated can be flown into Ch via a capillary ($C_1 = 7 \times 10^{-4}$ L/s) or a variable valve (C_x about 7×10^{-7} L/s) according to whether a BAG head or a McLeod gauge is used as a pressure standard for measuring pressure in Cp.

An assembly of sorption (S_p) and orbitron (O_p) pumps permits evacuating the whole calibration system. After several cycles of pumping down and flushing with dry nitrogen, pressure in Ch is reduced to 10^{-7} Pa (10^{-9} Torr). The calibration chamber and the UHV pumping assembly can be baked at 350°C. The lowest pressure in Ch with the UHV assembly on, after bake-out procedures, is 10^{-10} Pa (10^{-12} Torr). Calibration pressures in the range 10^{-3}–10^{-10} Pa (10^{-5}–10^{-12} Torr) can be obtained by maintaining Ch at ambient temperature.

Experience with cryopumps in calibration systems at room temperature (Grosse and Messer, 1970) has shown that under certain circumstances a Maxwellian gas distribution can be obtained in the calibration chamber. Messer (1977a) designed a calibration system (Fig. 5.16) pumped with an

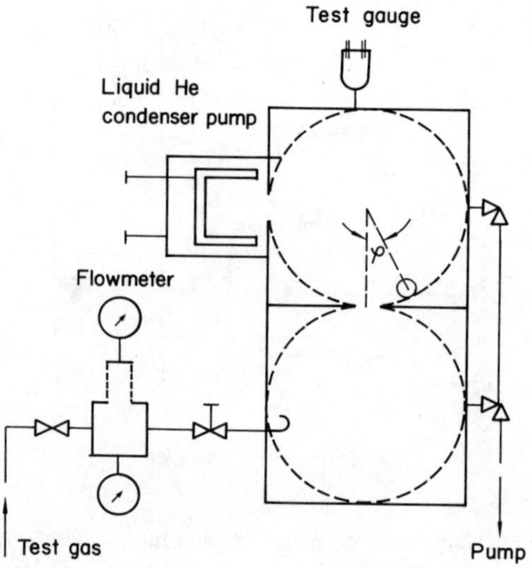

Fig. 5.16 Diagram of the calibration system used by Messer (1977a) for calibration in the UHV range showing the orientation of a rotatable pressure transducer.

unbaffled liquid-He-fed cryocondenser (T = 27 K) located behind a constriction to the gas flow with a large conductance (orifice with circular cross section). The calibration gas is introduced in a high-pressure chamber via a pipe bent so as to beam the gas against the wall of this chamber. Either the flow rate of gas (Messer, 1977a) or its pressure (Messer, 1977b) can be measured. The high-pressure and calibration chambers are joined by an orifice with circular cross-section with conductance $C \cong 8$ L/s (for nitrogen at room temperature). The gas distribution in the calibration chamber illustrated in Fig. 5.17, checked with a rotatable pressure transducer according to the method of Holland and Priestland (1967) (see also Chapter 8), shows that the molecular distribution of gas is nearly uniform. The effects due to the radiation of the calibration walls or heat conductance to cryocondenser surfaces as well as the influence of back-streaming effects from the cooled surface to the calibration gas were checked and found to be reduced. The residual pressure after baking the system is 4×10^{-9} Pa (3×10^{-11} Torr). The system can be used to generate pressures in the range 10^{-3}–10^{-7} Pa (10^{-5}–10^{-9} Torr) with an uncertainty of ± 6 and $\pm 1\%$ at the lower and higher end of the range of the measured pressure, respectively. The calibration system discussed cannot be used for calibrations with H_2 and He.

Gosselin and Beitel (1970) proposed the use of a calibration system combining the single flow divider and the reference transfer methods for

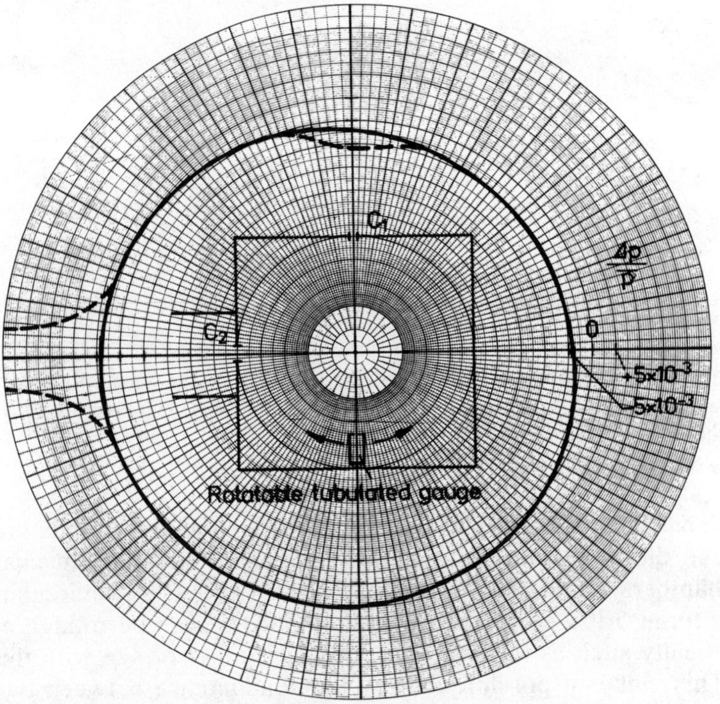

Fig. 5.17 Pressure readings of a rotatable pressure transducer in the calibration chamber of the system designed by Messer (1977b).

the generation of pressures in the 10^{-9} Pa (10^{-11} Torr) range. To obtain this performance the calibration chamber is cooled to about 77 K.

We now discuss *multiple stage of flow divider systems*. In this type of system, the flow rate of gas into a vessel is divided by driving it into two sinks with a distribution proportional to the volume rate of flow of each sink, as shown schematically in Fig. 5.18. To achieve sufficiently low pressures several vessels are connected in series (Fig. 5.19) such that the throughput Q of gas entering in one vessel is divided (Fig. 5.18), a small fraction rQ being passed on to an adjacent vessel and a major fraction $(1 - r)Q$ being pumped off by an auxiliary pumping system. It is thus possible, by arranging such vessels in series, to generate a sufficiently low pressure drop by means of an accurately measurable high pressure.

Roehring and Simons (1962, 1963) proposed the use of a multiple stage of flow divider in a calibration system such as that schematically illustrated in Fig. 5.19. Six cylindrical stainless steel chambers ($\phi = 25.4$ cm)

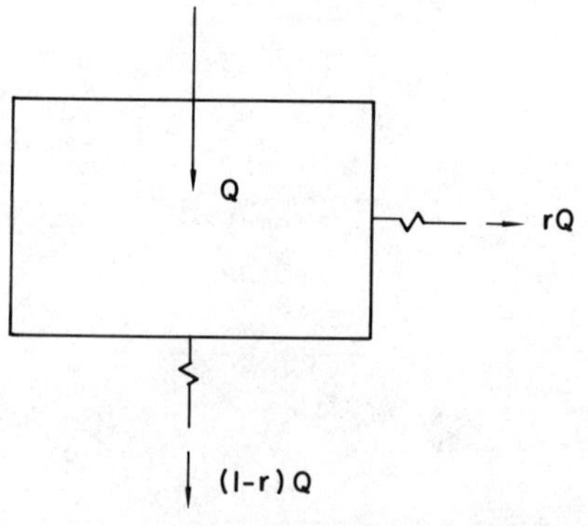

Fig. 5.18 Principle of multiple stage flow divider. Q, gas throughput entering the vessel; rQ, small fraction of Q being passed on to an adjacent vessel; $(1\text{-}r)Q$, major fraction of Q pumped out of the vessel.

are arranged in series and pumped both individually (by a pumping system) and differentially (through an orifice from the adjacent chamber). The chambers are joined by apertures in front of which a circular plate, having three orifices each of a certain diameter, can be rotated electromagnetically such as to align concentrically either orifice with the aperture. This makes it possible to vary the conductance between two adja-

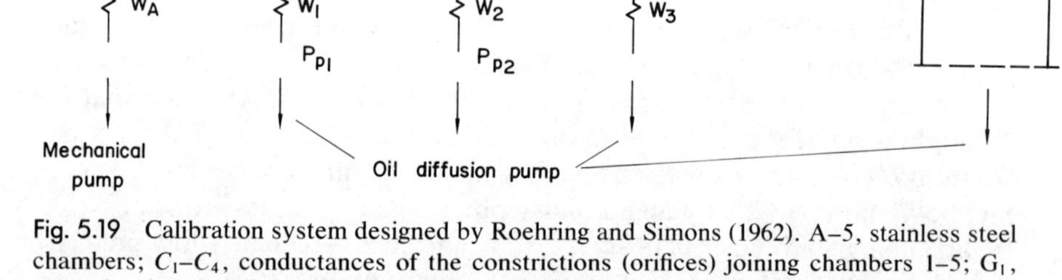

Fig. 5.19 Calibration system designed by Roehring and Simons (1962). A–5, stainless steel chambers; C_1–C_4, conductances of the constrictions (orifices) joining chambers 1–5; G_1, gauge head undergoing calibration; MS, radio-frequency mass spectrometer; P_1'–P_5, pressures of gas in chambers 1–5; P_{p1} and P_{p2}, pressures downstream from the constrictions W_1 and W_2 with conductances C_1' and C_2'; W_A and W_3, constrictions to the gas flow to the pump evacuating the chambers A and 3.

cent chambers. In the following description, the word *stage* refers to a chamber and its associated pump, the constriction to the gas flow between the chamber and the pump, and the constrictions by which the chamber is connected with the chambers immediately preceeding and following it. Stage A is pumped by a mechanical pump through a tubular restriction to such a degree as to reduce the high inlet pressure to a convenient level. Each of the chambers 1–3 is pumped by a three-stage oil diffusion pump (375 L/s) in series with a constriction (W_1, W_2, and W_3 with conductances C_1', C_2', and C_3' respectively), to minimize fluctuations in the speed of the pump. As it is known, the pumping speed of the pump is largely determined by the conductance of the restriction, usually an orifice. Stage 4 is the calibration chamber, pumped from stage 5 through an orifice with conductance C_4. Since outgassing from the walls of stage 5 can introduce large errors in calibration (if the volume rate of flow of the gas load is larger than that of the calibration gas in chamber 4), a large oil diffusion pump (1550 L/s) is used to pump off this final chamber.

A McLeod gauge and a rf mass spectrometer are attached to stage 1. A McLeod gauge mounted in stage 2 permits checking both the pressure in this stage and the value of the conductance connecting the first and second stages. The gauge head to be calibrated and another rf mass spectrometer are mounted in chamber 4. Both mass spectrometers can be used for leak detection and for checking the contamination of the calibration system. Joints are sealed with Viton A gaskets and cooled during the operation of the system. Chambers 3 and 4 and the adjoining restrictions can be enclosed in an oven permitting bake-out temperatures of 250°C.

With the notation of Fig. 5.19, if the throughput Q of the calibration gas in and out the first stage is

$$Q = (P_1' - P_2)C_1 = (P_2 - P_3)C_2 - (P_2 - P_{p1})C_2' \tag{5.28}$$

then

$$P_1'/P_2 = (C_1 + C_2 + C_2')/C_1 - (C_2/C_1)(P_3/P_2) - (C_1'/C_1)(P_{p1}/P_2) \tag{5.29}$$

The pressure ratios in the second and third terms of Eq. (5.29) can be reduced to negligible values by selecting very low values for the ratio of conductances C_2/C_1 and C_1'/C_1. Thus

$$P_1'/P_2 = (C_1 + C_2 + C_2')/C_1 \tag{5.30}$$

Progressing from stage 3 to 4 (neglecting the gas load produced by outgassing from the walls)

$$P_2/P_3 = (C_2 + C_3 + C_3')/C_2 \tag{5.31}$$

$$P_3/P_4 = (C_3 + C_4)/C_3 + (C_4/C_3)(P_5/P_4) \tag{5.32}$$

Again the last term of Eq. (5.32) can be reduced to negligible values by a proper choice of the conductance ratio C_4/C_3.

The pressure P_4 can be calculated by measuring P_1' and by computing the corresponding terms in Eqs. (5.30–5.32).

The *operational procedures* consist in baking the system until obtaining the ultimate pressure in the calibration chamber (2×10^{-8} Pa, 2×10^{-10} Torr) and outgassing the gauge heads to be calibrated. The selection of orifices in the rotatable plate in front of the aperture joining two adjacent chambers permits maintaining pressure ratios in these chambers such as to enable, as discussed, the calculation of pressure P_4.

The pressure range of the system is 10^{-2}–10^{-7} Pa (10^{-4}–10^{-9} Torr), the accuracy attainable by a skilled operator being $\pm 10\%$. Calibration can be performed with atmospheric gases, either singly or mixed.

The calibration system has the advantage of permitting pressure generation upon using higher initial throughputs of the calibration gas than would be required to generate the same pressure in a single flow divider system. However, the following drawbacks limit the use of Roehring and Simon's setup:

(a) Considerable complexity due to the rather large number of orifices, pumps, cold traps, and associated controls results in high investment costs and complex maintenance.

(b) The speed of gas across an orifice of conductance C is well defined *only* for the molecular flow of gas. Therefore the pressure of the calibration gas in the first stage must be low enough to ensure molecular flow conditions. This conflicts with the use of the McLeod gauge as a pressure standard since the instrument is rather inaccurate at the extreme lower limit of the pressure it can measure in the molecular range of flow (see Section 3.3.4).

Ultrahigh vacuum calibration systems based on the multiple stage of flow divider principle have been designed as an alternative to single stage of flow dividers. Thus, Feakes and Torney (1963) designed a two stage flow divider system for calibrating magnetron and suppressor gauge heads in the range 10^{-5} to about 10^{-12} Pa (10^{-7} to about 10^{-14} Torr). The system, schematically illustrated in Fig. 5.20, consists of a calibration chamber (Ch) connected to a high-pressure chamber (Cp) and a large-volume (Vc) chamber by a capillary tube (W_1) and an orifice (W_2) with conductances C_1 and C_2, respectively. Chamber Vc is evacuated by two oil diffusion pumps (D_2 and D_3) via an intermediate chamber (C_M). The pumping speed of the pumps (D_2 and D_3) at chambers Vc and C_M are S_2 and S_3. The flow of the calibration gas (highly purified He) introduced into chamber Cp through a variable leak is divided by being pumped both by a diffusion

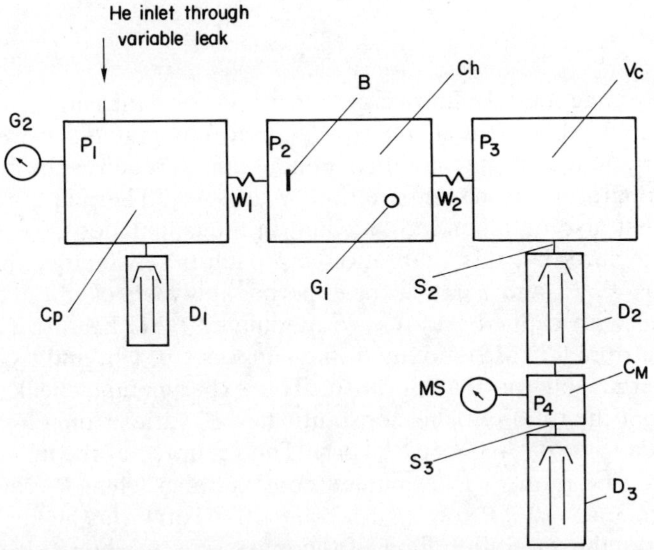

Fig. 5.20 Two-stage flow divider system for calibration in the ultrahigh vacuum range. B, baffle; Ch, calibration chamber; C_M, intermediate chamber; Cp, high-pressure chamber; D_1–D_3, oil diffusion pumps; G_1 and G_2, gauge head undergoing calibration and monitor, respectively; MS, mass spectrometer; Vc, large volume chamber; W_1 and W_2, constrictions to the gas flow with conductances C_1 and C_2. The pumping speeds in chambers Vc and C_M produced by the pumps D_2 and D_3, respectively, as well as the pressures P_1–P_3 in different chambers are indicated.

pump (D_1) and by chamber Ch through W_1. Beaming effects produced by the He inlet into chamber Ch are minimized by a baffle (B). An ionization gauge head (G_2) used to monitor pressures (10^{-2}–10^{-7} Pa, 10^{-4}–10^{-9} Torr) on the high side of the system and a mass spectrometer (MS) are mounted in chamber C_M. Pressure in Ch is reduced to values estimated below 10^{-13} Pa (10^{-15} Torr) by combining the pumping through the diffusion pumps D_2 and D_3 with cryopumping. Chamber Ch is operated at a temperature of approximately 10 K thus permitting negligible cryosorption of He.

If the flow rate of He in the various chambers of the calibration system is

$$Q = (P_1 - P_2)C = (P_2 - P_3)C_2 = P_3S_2 = P_4S_3 \qquad (5.33)$$

with the symbols as defined on Fig. 5.20 and if the system is designed so that

$$P_2 \ll P_1 \quad \text{and} \quad C_2 \ll S_2 \quad \text{or} \quad P_3 \ll P_2 \qquad (5.34)$$

then

$$P_2/P_1 \cong C_1/C_2 \tag{5.35}$$

If C_1/C_2 is accurately known and P_1 accurately measured, P_2 can be calculated. At this point it must be pointed out that *the system is not isothermal,* hence the calculation of conductances requires the knowledge of the temperature of the constriction W_1 and W_2. Thermal transpiration effects must also be taken into account in the calculation of P_2.

To make sure that this calibration approach is satisfactory, both pressure ratios P_1/P_2 and P_1/P_4 were experimentally checked. A tubulated BAG head with capped grid (G_2), a modulated BAG head (MG_2), and a mass spectrometer (MS) mounted in chambers Cp, Ch, and C_M, respectively, were used for that purpose. The experimental check of P_1/P_2 showed that the ratio remains constant when P_2 varies from 6.6×10^{-7} to 4×10^{-9} Pa (5×10^{-9}–3×10^{-11} Torr). The accuracy of the measurement was ±7%. The ratio P_1/P_4 remained constant also when P_1 varied from 1×10^{-7} to 3.6×10^{-4} Pa (2.7×10^{-6}–1×10^{-9} Torr), this higher pressure representing the saturation limit of the mass spectrometer (MS).

Poulter (1974) used the principle of multiple stage of flow divider in the arrangement shown in Fig. 5.21 for calibration in a gas with a Maxwell–

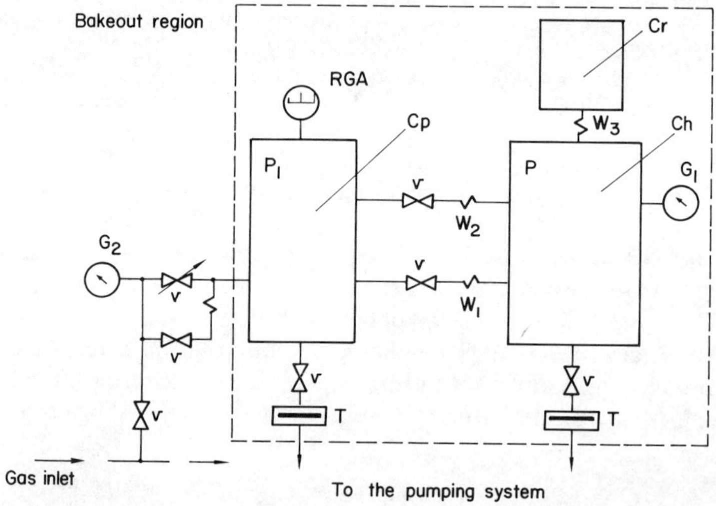

Fig. 5.21 Ultrahigh vacuum calibration system designed by Poulter (1974). Ch, calibration chamber; Cp, high-pressure chamber; Cr, cryopump; G_1, gauge head undergoing calibration; G_2, monitor gauge head; RGA, residual gas analyser; T, cooled trap; v, valves; W_1–W_3, constrictions with conductances C_1–C_3. P_1 and P are the pressures in Cp and Ch.

Boltzmann distribution of velocities. The calibration gas at high pressure flows into chamber Cp, wherefrom a large part is driven into the diffusion pump evacuating Cp, and a small fraction is permitted to enter chamber Ch through either of the constrictions (orifices) W_1 and W_2 of conductances C_1 and C_2, respectively. The calibration gas in Ch is pumped through a constriction W_3 (orifice with conductance C_3) in a reentrant cylinder fitted to a cryopump of the liquid-He-pool type.

Upon assuming that the gas flux returning through W_3 does not give rise to pressure gradients in Ch (i.e., that the rates of arrival of calibration gas at the calibrated gauge head ports and at the constriction W_3 are the same), the calibration pressure can be expressed as

$$P = C_1 P_1/(\Lambda + C_1) \tag{5.36}$$

Here Λ is a function of the thermodynamic quantities R_0, T, M and the condensing probability ξ of the calibration gas as well as of the transmission probability α of the orifice W_1 of area A.

Both chambers Cp and Ch are made of stainless steel specially treated according to the method used by Calder and Lewin (1967) for the reduction of hydrogen concentration in the bulk of the material. Each chamber is operated by an oil diffusion pump system in conjunction with both water and liquid nitrogen traps. During calibration, the calibration chamber is isolated from the pumping system. The base pressure (6×10^{-8}–8×10^{-8} Pa, 4.5×10^{-10}–6×10^{-10} Torr) obtained in the calibration chamber after bake-out and outgassing of the gauge heads being calibrated is reduced to 2×10^{-9}–3×10^{-9} Pa (1.5×10^{-11}–2.25×10^{-11} Torr) when the cryopumping system is cooled. The highest pressure which can be generated is 10^{-4} Pa (10^{-6} Torr). The uncertainty in the calculated pressure is ±7%.

5.5.3 Dynamic Expanders with Adjustable Conductance

The relationship

$$P_1 = Q/C(1 + C/S) \tag{5.37}$$

obtained by combining the first term of Eq. (5.3) with Eq. (5.10) shows that there are two ways to get different values for P_1: one is by changing Q while maintaining C and S constant, and the other is to vary C while maintaining Q and S constant. For the latter case, Eq. (5.37) represents a straight line, as illustrated in Fig. 5.22, giving $P_1 = f(1/C)$. This approach was used by Oatley (1954) to determine the pumping speed of a vacuum pump. When P_1 is measured with an ionization gauge head having I_c a

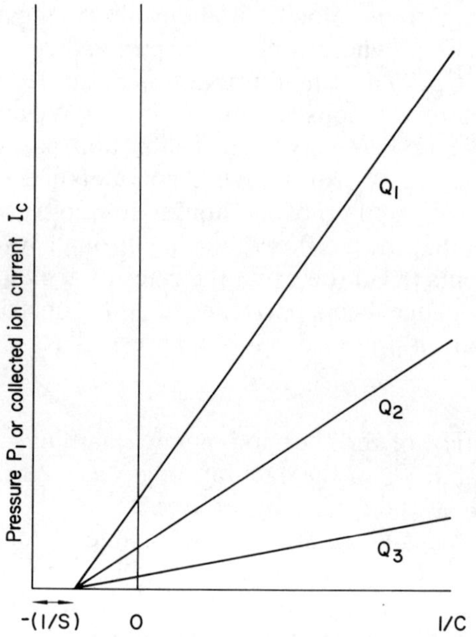

Fig. 5.22 Pressure variation versus reciprocal of conductance for different values of the throughput. $Q_1 > Q_2 > Q_3$.

linear function of P_1, the relation $I_c = f(1/C)$ also yields a straight line in an I_c versus $1/C$ plot. Plots of P_1 or I_c as functions of $1/C$ when the throughput is varied converge to one point on the $1/C$ axis corresponding to $-1/S$. Once S is determined that way and Q measured, P_1 can easily be calculated. Provided that an ionization gauge head with linear response is used for measurements, the determination of only one throughput value is necessary. Practically, only a high value of Q is measured, smaller values being related to it by a plot $I_c = f(1/C)$. This technique of throughput determination is rather accurate ($\pm 2\%$) but requires a large number of determinations [about 30, according to Choumoff *et al.* (1970)].

Equation (5.37) does not take into account the gas load produced in the calibration chamber by the outgassing from its walls. The main source of error is the conductance C, values of the ratio C/S less than 1 greatly affecting the accuracy of P_1 determinations. The pressure range spanned with calibration systems operated according to this method is 10^{-2}–10^{-4} Pa (10^{-4}–10^{-6} Torr) with an accuracy of $\pm 2\%$.

Calibration systems operated according to the adjustable conductance method are designed as single stage of flow dividers (see Fig. 5.6). The cross-sectional area of the orifice (W) joining chambers Ch and Vc can be varied according to one of the following techniques.

(a) A disc with several circular apertures of various diameters can be rotated flush with the orifice W (by means of a shaft and conical gears, magnetically, etc.). Each aperture in the disc can be concentrically aligned with the orifice. The apertures are so shaped that no matter which one is in position opposite W, the leak of gas between the disk and the orifice is negligible (Henry, 1956).

(b) The aperture in the wall separating Ch and Vc is shaped as an equilateral triangle (Fig. 5.23a). A wide plate moved flush, parallel with one of the triangle sides, permits variation of the cross-sectional area of the triangle (Choumoff and Aubry, 1965).

(c) Chambers Ch and Vc are separated by a large-diameter orifice. The movement of a frustum-shaped body concentrically with the orifice permits varying the cross-sectional area of the orifice (Fig. 5.23b) (Bernardet *et al.*, 1970a; Choumoff *et al.*, 1970). Calcatelli *et al.* (1974a) studied the variation of the conductance of a circular orifice whose cross section could be changed by the concentrical movement of a body with parabolic or conic profile (Fig. 5.23c).

(d) A multielement iris could also be used (Gourjault *et al.*, 1972).

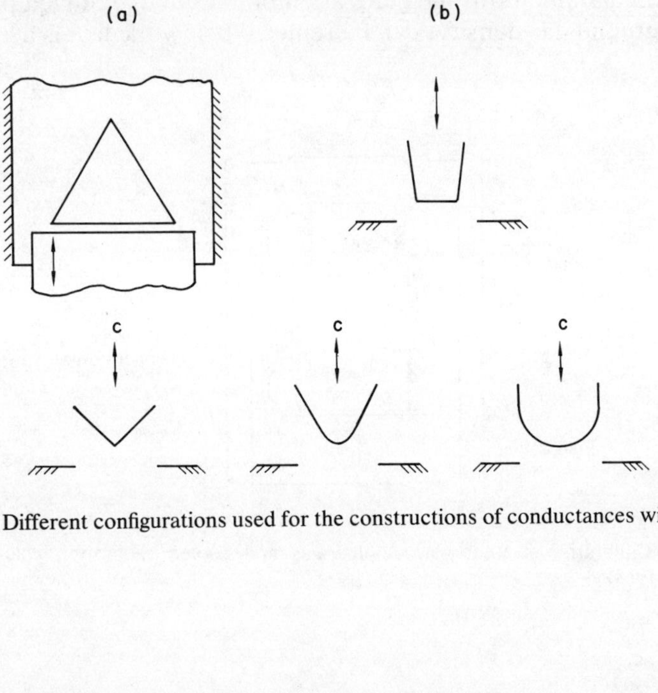

Fig. 5.23 Different configurations used for the constructions of conductances with variable geometry.

5.6 Other Methods for Calibration in Ultrahigh Vacuum

Special systems have been designed for gauge head or mass spectrometer calibration under simulated space conditions or in the lowest part of the ultrahigh vacuum range. Characteristics of such systems are cooled gas and cooled calibration chamber as well as cooled system and preferential direction of molecular flow.

Bergquist (1968) calibrated a nude quadrupole gas analyzer and a Helmer gauge head at a pressure of about 10^{-10} Pa (10^{-12} Torr) in the system illustrated in Fig. 5.24, where both the calibration gas and the calibration chamber are cooled. The calibration chamber (aluminum) and a sublimation pump, joined by a chevron baffle, are located in a volume pumped by two diffusion pumps of 18,000 L/s. A baffled box joined to the sublimation pump by an orifice of known conductance permits the inlet of the calibration gas from a guard vacuum into the calibration chamber. The calibration chamber, sublimation pump, and baffled box are cooled to 14 K with a Santeler (1960) array. A needle valve controls the calibration gas inlet into the guard vacuum. For calibration, a known gas flow is admitted into the guard vacuum, wherefrom, after cooling, most of it is evacuated. The small amount of gas which remains flows into the calibration chamber. The equilibrium pressure is then calculated.

Fowler and Brock (1970) investigated a calibration technique in which the limiting parameter to the generation of extremely reduced pressures, the background gas density, can be reduced below the levels achieved by

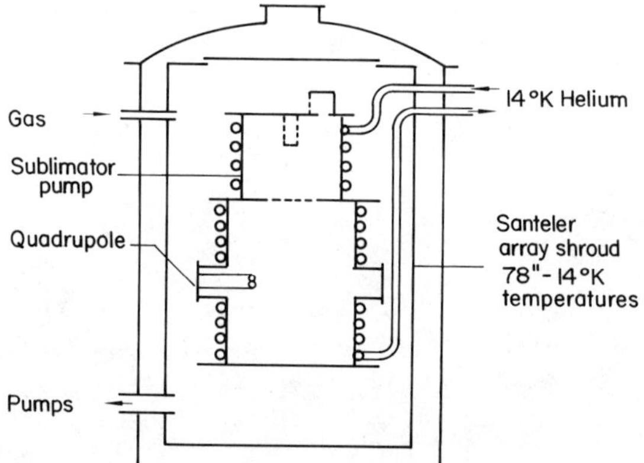

Fig. 5.24 Calibration system with cooled gas and cooled calibration chamber (after Bergquist, 1968).

using the methods discussed in Sections 5.4 and 5.5. It should be noted that for accurate calibration, the background gas density must be at least two decades lower than the lowest pressure required for determinations. Their technique consists, in principle, of forming the calibration gas into a molecular beam that crosses a volume with cryo-pumped walls before entering the calibration chamber. Thus, with reference to Fig. 5.25, the calibration gas flows from the gas source H, (isothermal at a certain temperature) to chamber L through a porous plug W_1. Gas in chamber L at a pressure much lower than in H effuses through the orifice W_2 in a molecular beam which enters the calibration chamber Ch through an aperture W_3. On its way from chambers L to Ch the molecular beam crosses a long tunnel t_n with cryogenically cooled walls. The temperature of the walls is selected such as to insure a unity sticking coefficient of the molecular beam. The distribution of the gas density for these conditions was measured and found to be in good agreement with the predictions of kinetic theory (see Section 8.2.2). The gauge head G_l undergoing calibration is protected from streaming effects with baffle b. Calculations show that pressures in the 10^{-4}–10^{-13} Pa (10^{-6}–10^{-15} Torr) range can be achieved by using this calibration technique.

Grosse and Messer (1981) described calibration equipment using the molecular beam and cryogenic techniques described which covers the range 10^{-5}–10^{-10} Pa (7.5×10^{-8}–7.5×10^{-13} Torr). Their approach for attenuating the pressure of the calibrating gas before forming it in a molecular beam differs from that of Fowler and Brock (1970) in that they used a gas flow dosing device instead of a porous plug. The pressure P_0 built up by the calibration gas in the Knudsen cell (referred to as the reference chamber) is measured with a reference gauge (Fig. 5.26). Gas effused

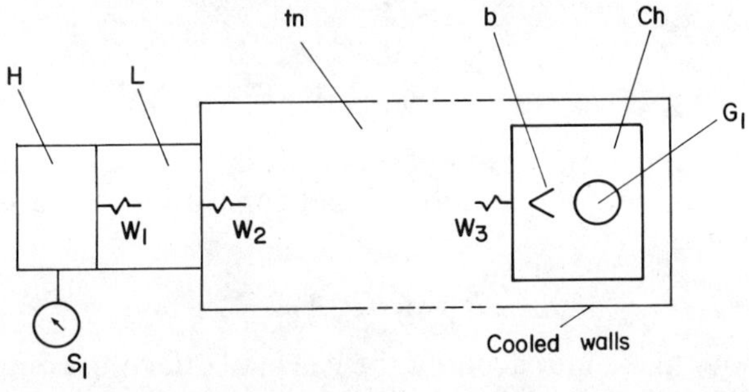

Fig. 5.25 Gauge calibration according to the molecular beam principle. b, baffle; Ch, calibration chamber; G_l, gauge head undergoing calibration; H, chamber accommodating gas at high pressure; L, chamber accommodating gas at low pressure; S_l, pressure standard; tn, tunnel; W_1, constriction (porous plug); W_2 and W_3, constrictions (orifices).

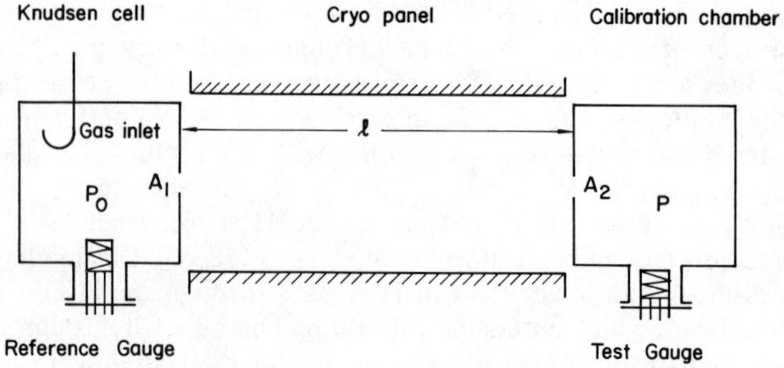

Fig. 5.26 Diagram of the molecular beam calibration method (after Grosse and Messer, 1981).

through the orifice A_1 of the reference chamber travels through a cryopanel to the calibration chamber. There it enters through an orifice A_2. The gauge head to be calibrated is sunk in a recess so that it cannot "see" the beam directly. Either orifice A_1 or A_2 has a very small cross section such that the ratio of the areas of the orifices to the walls of the chamber which they join is smaller than 10^{-3}. This insures an approximately constant impingement rate of the beam on the internal surfaces of the chambers.

Pressure P in the calibration chamber can be calculated upon considering that for stationary equilibrium conditions of molecular flow, the numbers of gas molecules N_i and N_o entering and leaving the orifice A_2, respectively, are equal. Thus, if

$$N_1 = (1/4\pi) n_{m1} \bar{v} (A_1 A_2 / l^2) \tag{5.38}$$

and

$$N_2 = (1/4) n_{m2} \bar{v} A_2 \tag{5.39}$$

where n_m is the number density of molecules, $\bar{v}$ their mean velocity in the reference and calibration chambers, and l the distance between the orifices A_1 and A_2,

$$P = Ct\, P_0 A_1 / l^2 \tag{5.40}$$

Equation (5.39) takes into account the gas load due to the outgassing from the calibration chamber walls, the gas released from the gauge head (ionization) being calibrated, and the gas pumped by the gauge head. The method is not "absolute" in that P is referred to a pressure P_0 measured with another gauge head.

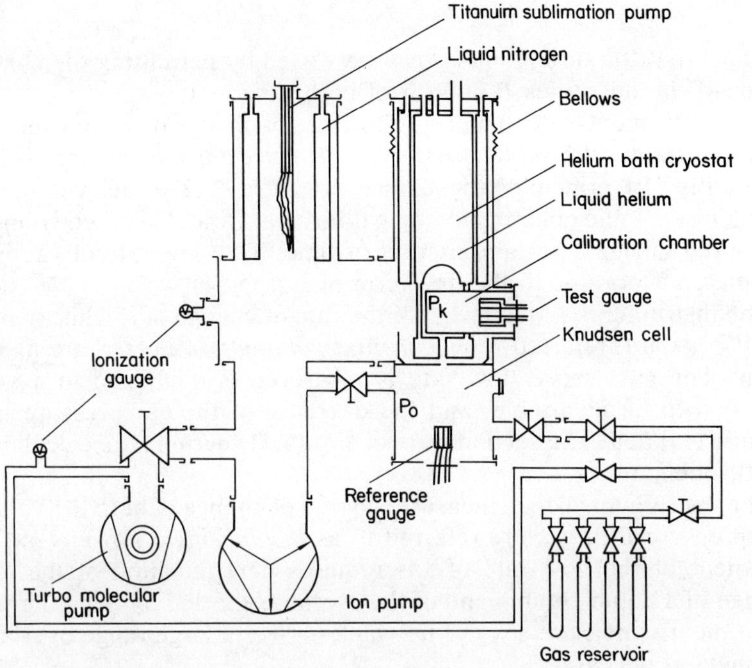

Fig. 5.27 Molecular beam calibration system (after Grosse and Messer, 1981).

The calibration system designed by Grosse and Messer (1981) on this principle is illustrated in Fig. 5.27. They estimated the accuracy of the system as 30% for 10^{-10} Pa (10^{-12} Torr), 15% for 10^{-9} Pa (10^{-11} Torr), and 10% for 10^{-8} Pa (10^{-10} Torr) ranges.

5.7 The Measurement of Gas Throughput

5.7.1 The Physical Basis of Measurement Techniques

The various approaches to the calibration of gauge heads with the dynamic method require the measurement of gas throughput in the range 10^{-1}–10^{-7} Pa L/s (10^{-3}–10^{-9} Torr L/s). This figure is obtained by substituting into Eq. (5.2) the values of pressure P (10^{-2}–10^{-8} Pa, 10^{-4}–10^{-10} Torr) and conductance C (10 L/s) usually encountered in calibration work.

For steady-state conservative temperature and gas flow conditions, the expansion of Eq. (1.17) as

$$Q = P(dV/dt) + V(dP/dt) \tag{5.41}$$

predicts that the throughput Q can be varied by permitting the change of either of the quantities P or V or of both.

The instruments which measure the rate of change of V while maintaining P constant, referred to as *constant pressure flowmeters,* are used most frequently. In principle, the displacement rate of a gas volume by a plunger or a liquid column moving within an accurately known volume are measured during a certain interval of time. The flow rate of gas is then calculated according to the first term of Eq. (5.41).

The instruments which measure the rate of change of P while maintaining V constant, referred to as *constant volume flowmeters,* are used less often. The gas whose flow rate is measured is contained in a volume assumed to be invariable, and the decrease of the gas pressure in that volume is timed. The second term of Eq. (5.41) permits the calculation of the throughput.

The instruments which measure the rate of change of both the volume V and pressure P of gas are referred to as *displacement flowmeters.* Measurement of the flow rate of gas requires determination of the rate of change of a liquid column and of the pressure exerted by that column. To conveniently measure several flowmeters over a large range overlapping in ranges is necessary.

Other techniques that have evolved for the measurement of throughput are based on determinations of the permeation of gas through a solid slab (porous plug) or of the viscosity or thermal capacity of gas.

5.7.2 Constant Pressure Flowmeters

The range spanned by constant pressure flowmeters is 10^{-3}–10^{-7} Pa L/s (10–10^{-4} Torr L/s).

Hayward and Jepsen (1962) developed a flowmeter whose principle is illustrated in Fig. 5.28. The instrument consists of a manifold containing a monitor gauge (G_2), a sensitive differential manometer (D), and a control volume (V_v) as well as a bypass valve (v_b). The control volume V_v is a cylinder in which a sealed plunger can be manually or mechanically moved by a micrometer-driven screw. The sides of the instrument adjacent to the gas supply and calibration chamber are denoted R (reference) and M (monitor).

After evacuating the manifold, a gas flow is established into the calibration chamber with the instrument filled at a pressure of about 100 Pa (1 Torr). The bypass valve v_b is then shut and the time ($t = 0$) noted. The gas flow from section M into the calibration chamber produces a pressure drop which is compensated for by moving the plunger into the control

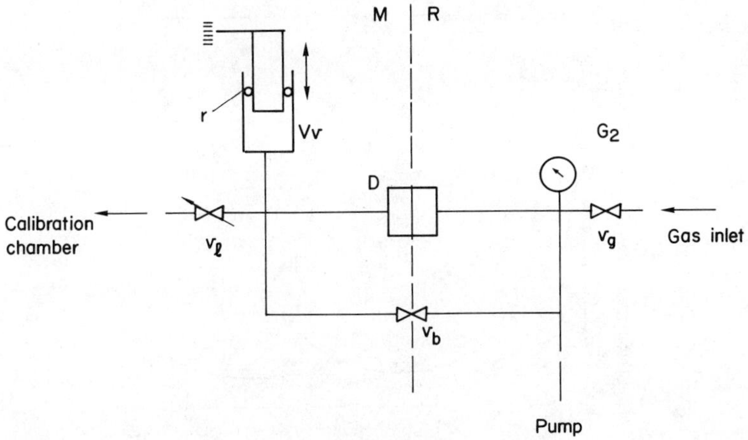

Fig. 5.28 Diagram of the constant-pressure flowmeter designed by Hayward and Jepsen (1962). D, differential pressure manometer; G_2, monitor gauge; r O-ring seals; Vv, control volume; v_b, v_g, and v_l, bypass, gas emission, and leak valve, respectively; M and K, monitor and reference sections.

volume V_v so as to bring the differential manometer D to zero. The time $(t = t_1)$ during which D is maintained at zero differential and the volume displaced by the plunger in V_v in time $t_1 - t_0$ are used then to calculate Q according to the first term of Eq. (5.42). The value calculated this way holds true if the temperature T of the gas flow remains constant in the time interval $t - t_0$.

Equal temperature changes in the sections R and M of the instrument do not affect the accuracy of measurement. Thermal gradients between these sections influence the instrumental drift, which in turn determines the minimum measurable gas throughput. Peggs (1976) showed that a minimum variation of the temperature differential between sections R and M yields an apparent gas throughput of 5.5×10^{-4} Pa L/s (4.12×10^{-6} Torr L/s) in a Hayward–Jepsen-type flowmeter containing gas at 10 kPa (100 Torr) in a 0.1-L volume. The rate of displacement of the plunger to complete a flow measurement as well as the volume displaced by the plunger in V_v are parameters which contribute to the uncertainty of the throughput measurement, as illustrated in Fig. 5.29. The lowest rate of flow, 10^{-4} Pa L/s (10^{-6} Torr L/s), measured by Hayward and Jepsen (1962) during 10 min had an estimated accuracy of ±5%.

All flowmeters developed for use with dynamic calibration systems retain Hayward and Jepsen's idea of controlling the volume rate of flow of gas by means of a vessel with adjustable volume and a differential ma-

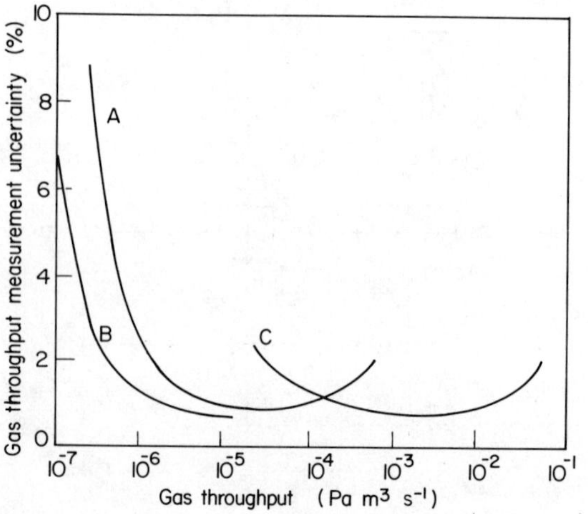

Fig. 5.29 The variation of uncertainty in gas throughput measurement as a function of the cross-sectional area and displacement of the plunger using Hayward and Jepsen's flowmeter. A, 1 mm^2 area plunger displaced 25 mm in 10 min; B, 1 mm^2 area plunger displaced 25 mm in 20 min; C, 100 mm^2 area plunger displaced 25 mm in 10 min. Reprinted with permission from *Vacuum* **26,** 322. G. N. Peggs. Copyright 1976 Pergamon Press.

nometer. The adjustable volume is achieved by the displacement of a plunger (Smetana, 1966; Bannenberg and Tip, 1968; Calcatelli *et al.*, 1974a,b; Hojo *et al.*, 1977) or a liquid column (Bennewitz and Dohmann, 1965; Nakayama and Hojo, 1974).

The differential manometers most used are of the capacitance type. The flowmeter designed by Smetana (1966) uses as a control volume a precision-bore pyrex tube whose volume can be varied by a stainless steel plunger. The plunger, sealed to the tube by a stainless steel bellows, is driven back and forth by a screw operated from a gear box integrated into a servo-controlled circuit. No other means are used to seal the plunger to the tube, so the gas compressed by the plunger can pass it and flow into the volume left by the expanding bellows. This gives rise to an error affecting the volume calculated from the plunger displacement. The author estimates that throughputs in the range $3\text{–}10^{-2}$ Pa L/s (0.3–10 Torr L/s) can be measured with accuracies of 1% or better.

Bannenberg and Tip (1968) designed a flowmeter consisting of a control volume (referred to as a control chamber), a differential manometer, and a servo-controlled motorized circuit (Fig. 5.30). The calibration gas is introduced into the control chamber at pressure P_1 measured with a U-tube manometer. The pressure drop ΔP_1 in the control chamber caused by the flow of gas in the calibration system is compensated for by the movement

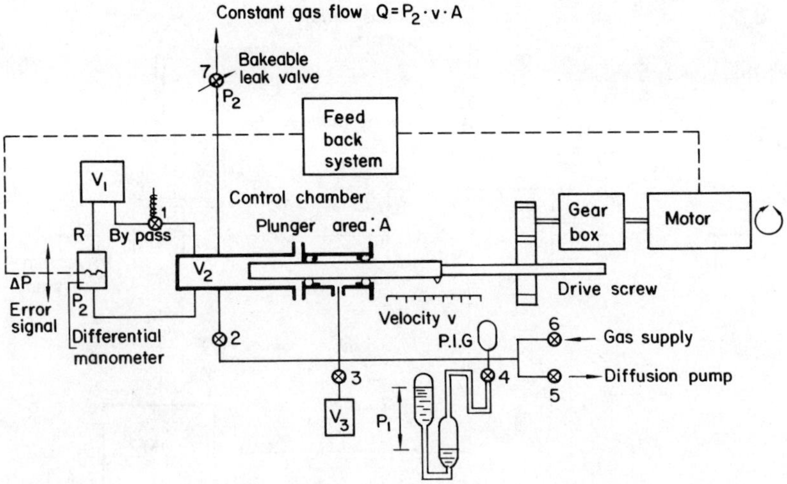

Fig. 5.30 Constant-pressure flowmeter and gas flow system designed by Bannenberg and Tip (1968).

of a plunger actuated by a gear-box–motor system. The motor is controlled by a feedback loop incorporating a differential manometer of special construction. Pressure in the control chamber is maintained constant at values within 10^{-3} to 1.06×10^5 Pa (10–800 Torr). Two plungers with different cross-sectional areas are used for that purpose. Leakage of the calibration gas along the plungers is prevented by double sealing the plungers with O rings and by maintaining the volume balance across the O rings at pressure P_1 produced by gas supplied from a buffer volume V_3. However, the volume changes due to both the O-ring displacement and backlash of the plunger produce inaccuracies in the measurement. These inaccuracies can be minimized by always taking readings in the same direction of plunger movement. Means are provided to maintain the flowmeter at constant temperature.

To prevent the calibration gas from leaking out of the control volume the contact between gas and plunger is avoided by using a bellows as an intermediate element for gas compression. Thus, in the gas flow system used by Calcatelli *et al.* (1974a,b) (Fig. 5.31a) the control volume (Fig. 5.31b) consists of a bellows connected to a piston which compresses the gas. A rotatable magnet, manually or automatically driven, moves the bellows back and forth. A capillary tube filled with oil, in good contact with the bellows, permits determining the height of the piston by measuring the oil height in the capillary when gas is compressed.

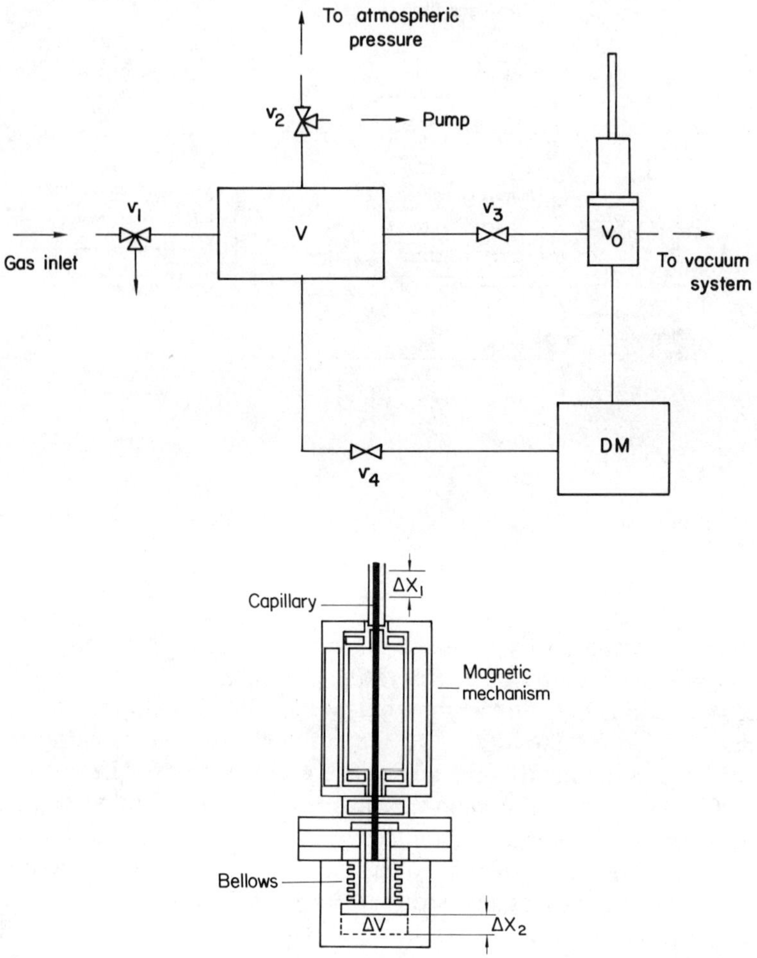

Fig. 5.31 (a) The gas flow system and (b) control volume designed by Calcatelli *et al.* (1974a,b) for metering the gas flow. DM, differential manometer; V, reference volume; V_0, control volume; v_1–v_4, valves.

A flowmeter using basically the same feedback servo control as that used by Bannenberg and Tip (1968) was designed by Hojo *et al.* (1977) to maintain constant the gas flow in the calibration system they used (Fig. 5.32a). The control element consists of a metallic cylinder in which two chambers, separated by bellows, can be moved by a motor-driven plunger (Fig. 5.32b). The plunger, fitted in an oil-filled chamber, actuates the bellows, thus insuring an insignificant leak rate of the calibration gas in

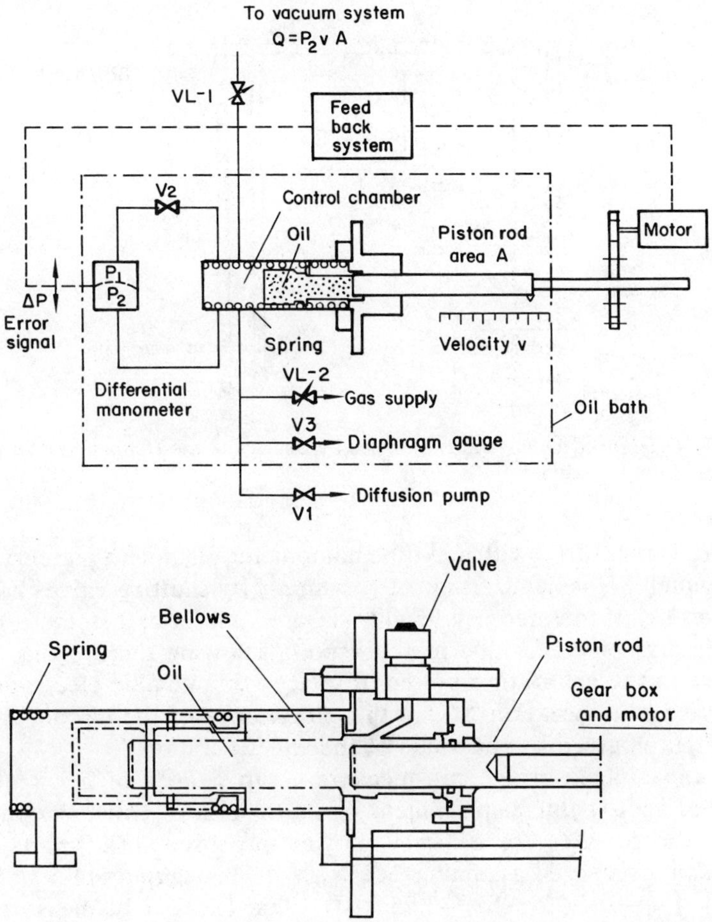

Fig. 5.32 (a) Gas flow system and (b) detailed structure of the control chamber designed by Hojo *et al.* (1977).

the control chamber. Throughputs of 3×10^{-4} to 1.5×10^{-8} Pa L/s (4×10^{-2}–2×10^{-6} Torr L/s) are achieved for plunger movements of 0.01 cm/s.

Bennewitz and Dohmann (1965) designed a flowmeter where the control volume is a U-shaped mercury pool, as illustrated in Fig. 5.33. A steel cylinder, clad with 0.08-cm thick glass, floats in the mercury. The mercury and the insulated metallic cylinder form a cylindric capacitor connected to a capacitance bridge. In operation a certain flow of gas is admitted into the calibration system (valves Z and E open), and the gas

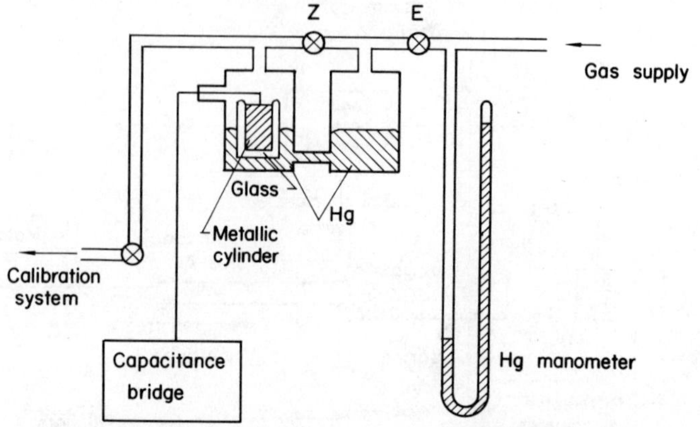

Fig. 5.33 Diagram of the flowmeter designed by Bennewitz and Dohman (1965). Z and E, valves.

pressure is measured with a U-tube manometer filled with mercury. Then the flowmeter is isolated from the gas supply by shutting valves Z and E. The rise Δh_1 of the mercury level (measured with a cathetometer) in the left side arm of the U-tube mercury pool following the drop in the gas pressure in the calibration system displaces the metallic electrode, thus giving rise to a variation ΔC_p in the capacitance of the circuit. The gas throughput is given by the value of the $\Delta h_1/\Delta C$ ratio.

The capacitance bridge can measure a capacitance of 2.5×10^{-3} pF produced by a 1-μm displacement of the merçury. Thus, the variation of the ratio $\Delta h_1/\Delta C_p$ for $\Delta h_1 = 1$ cm was measured with 7 parts in 10^3. $\Delta h_1/\Delta C_p$ is measured assuming $\Delta C_p \propto \Delta h_1$. Throughputs of 5×10^{-1} to 5×10^{-3} Pa L/s (3.75×10^{-3}–3.75×10^{-5} Torr L/s) can be measured with an uncertainty of about 1%.

In the flowmeter designed by Nakayama and Hojo (1974), the conventional piston arrangement is replaced by the movement of mercury in an arm of a U tube, as illustrated in Fig. 5.34. With the mercury at the lowest level in the instrument, the calibration gas is introduced into the right arm and its pressure determined by measuring the height of the liquid in the left arm with a micrometer. Valve c-3 is then shut, thus isolating gas in the glass bulb (reference side R of the system), and valve VL-4 is opened to establish a certain flow to the calibration system. The diaphragm gauge is maintained at zero by raising the mercury with air introduced through valve VL-1. A series of electrical contacts (a, c–j) arranged at known intervals in the right and left arms of the U tube permit measuring the mercury column level. The rate of rise of the mercury column is measured

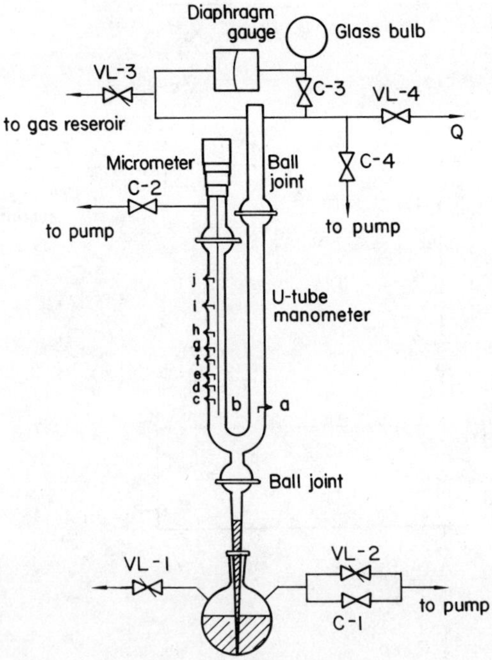

Fig. 5.34 Flowmeter and gas flow system designed by Nakayama and Hojo (1974).

with a timer triggered by the electrical contacts. Throughputs of 6 to 2×10^{-4} Pa L/s (4.5×10^{-2}–1.5×10^{-6} Torr L/s) can be measured with an error of less than 1% by using various pressures in the mercury reservoir as well as U tubes with different bore diameters.

5.7.3 Constant Volume Flowmeter

Bernardet *et al.* (1970b) designed a constant volume flowmeter which they used in conjunction with a dynamic expander with adjustable conductance (see Section 5.5.3). The flowmeter (Fig. 5.35) consists of a reservoir (R) for the calibration gas, a manifold with valves permitting the isolation of the reservoir from volumes v_1 and V_0 and from the calibration system, as well as an interferometric differential manometer (IDM) (see Section 3.2) permitting the measurement of pressure P of the gas flowing into the calibration chamber. The manifold, the reservoir, and the volumes of chambers v_1, V_0, and IDM are evacuated prior to the admission of the calibration gas into R. Gas is then admitted into R at a pressure measured by a Pirani gauge head (G_2) and a certain gas flow established in

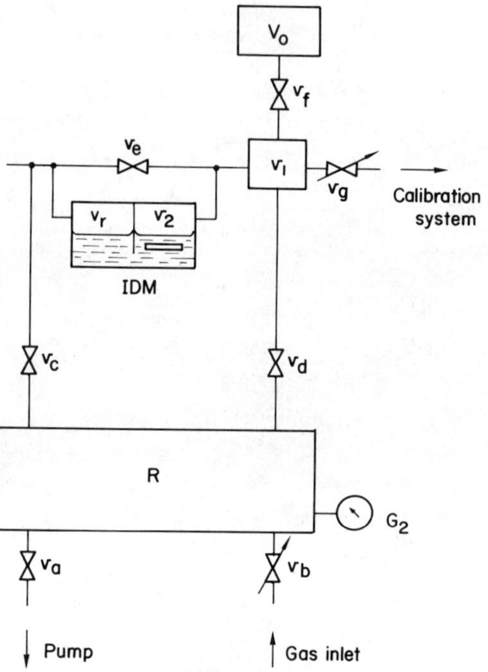

Fig. 5.35 Diagram of a constant-pressure flowmeter. IDM, interferometric differential manometer; G_2, monitor gauge head; v_1, v_2, v_r, and V_0, volumes; $v_a - v_g$, valves; R, gas reservoir.

the calibration system. Measurement starts ($t = 0$) by shutting valves v_e and v_d, thus suppressing the gas supply from R to the calibration system. The rate of pressure decrease resulting from the gas flow in the calibration system is measured by counting the change in fringes in IDM during a time $t = t_1$. The throughput Q is calculated as

$$Q = (\Delta P/\Delta t)|(\Delta V_t/\Delta P)P + V_t| \tag{5.42}$$

where $V_t = V_0 + v_1 + v_2 + v_{\text{manifold}} + v_{\text{valves}}$ and $t = t_1 - t_0$. For the flowmeter designed by Bernardet *et al.* (1970b), $\Delta V_t/\Delta P \leq 10$ L/Pa (1.33×10^{-2} L/Torr) and $P \cong 1$ Pa (10^{-2} Torr). Therefore, $(\Delta V_t/\Delta P)P$ is small compared with V_t, which is usually 10^{-1} L. The inaccuracy resulting from disregarding the term $(\Delta V_t/\Delta P)P$ in calculating the throughput with Eq. (5.43) is less than 1%. The flowmeter, built in a box maintained at constant temperature, can measure throughputs in the range $10–10^{-3}$ Pa L/s ($7.5 \times 10^{-2}–10^{-5}$ Torr L/s).

5.7.4 Displacement-Type Flowmeters

Some of the usual displacement-type flowmeters, sketched in Fig. 5.36a–c, referred to as pipette or burette flowmeters, determine displacement by the gas flowing to a column of fluid within a straight or U-shaped glass tube. The measuring procedure, nearly the same for all types, can be understood best by referring to Fig. 5.36c. The flowmeter is first evacuated (v_1 open; v_2, v closed), and then a certain flow of gas is established (v_2 open) in the calibration system. Measurement starts by closing v_1 (at $t = 0$), causing the fluid to rise in the tube in order to compensate for the gas loss to the calibration system. The rate of rise is measured by timing the passage of the fluid between two diducial marks on the glass tube. If h and A are the distance between the marks and the average cross-sectional area of the glass tube, respectively, and V_a the volume of gas in the instrument at atmospheric pressure at $t = 0$, the volume V displaced by the fluid is

$$V = V_a - hA \tag{5.43}$$

The pressure P of gas contained between the marks is

$$P = P_a - h\rho_{oil}/\rho_{Hg} \tag{5.44}$$

where P_a is the atmospheric pressure and ρ the fluid density. If $t = t_1 - t_0$ is the time it takes for the fluid to rise from the lower ($t = t_0$) to the higher

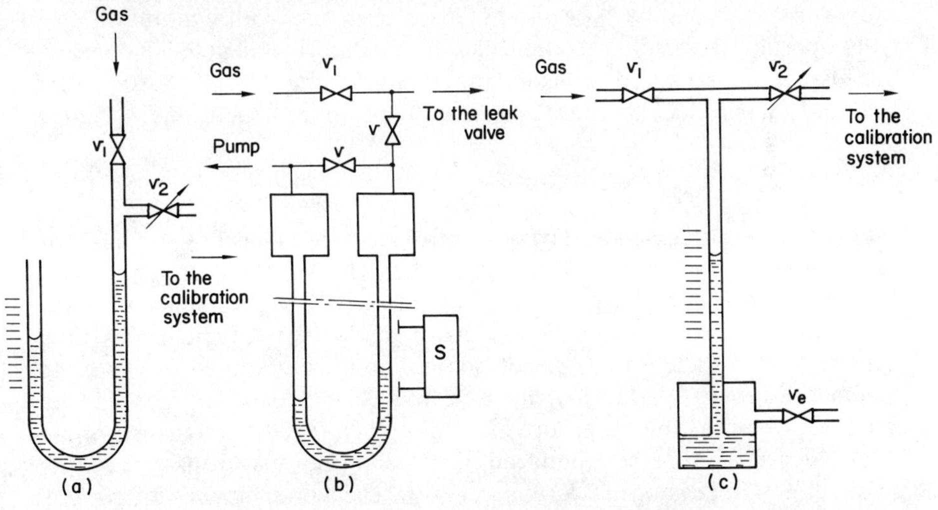

Fig. 5.36 Diagram of some displacement-type flowmeters. (a) and (b) are U types, (c) is a burette type. v, valve; v_1, gas inlet; v_2, leak; v_e, evacuation; S, height sensor.

($t = t_1$) mark, the gas throughput is expressed by

$$Q = (P_a V_a - PV)/t = h/t|P_a A + \rho_{oil}/\rho_{Hg}(V_a - hA)| \qquad (5.45)$$

For $hA \ll P_a$, the term ρ_{oil}/ρ_{Hg} is small compared with P_a, and Eq. (5.45) reduces to

$$Q = h/t(P_a A + \rho_{oil}/\rho_{Hg} V_a) \qquad (5.46)$$

The fluids used are mercury and oil. Mercury can be used with a large variety of calibration gases but requires large hydrostatic head corrections of the second term in Eq. (5.45). Such corrections are not necessary when oil is used, because of its low density. However, the use of oil is limited because some gases can be absorbed and desorbed during the operation of the instrument. The lowest range of flowmeters of the type discussed is 1 Pa L/s (10^{-2} Torr L/s) when $h/t = 0.1$ cm, $V = 0.01$ L, and $a = 0.1$ cm^2.

The main drawbacks of these instruments are

(a) the uncertainty in the volume of the gas in the region of valves v_1 and v_2,
(b) the variable flow rate caused by the changing pressure of the gas confined in the instrument, and
(c) thermal fluctuation of the gas during measurement.

Flowmeters of the type shown in Fig. 5.36a and b have been used by Hayward and Jepsen (1962) and Morrison (1966b) and those of the type shown in Fig. 5.36c by Normand (1962). Their use with calibration systems operated according to the dynamic method is restricted, however, because of their limited pressure range (1–10^{-3} Pa L/s, 10^{-1}–10^{-5} Torr L/s) and their reduced accuracy at the lower limit of the range.

5.7.5 Porous Plugs

Porous plugs are used as a basic control element of gas flow (mainly for inert gases) in the range 10^{-1}–10^{-6} Pa L/s (10^{-3}–10^{-8} Torr L/s). Such elements consist of a large number of tubes with very small cross-sectional areas, permitting molecular flow of gas at relatively high pressures applied at the inlet. Pore sizes of about 1 μm are obtained by sintering nonmetallic (usually glass) or metallic powders. Molecular flow can be obtained when the inlet pressure exceeds 100 Pa (1000 Torr) using porous glass, or a few Pa (a few hundred Torr) using porous bronze, stainless steel, etc. (Hultzman and Krause, 1974). The conductances of porous plugs to gas flow are as low as 10^{-6} L/s. Metallic plates with small holes or orifices of less than 30 μm, more than 300 μm apart, have also been used as control elements for gas flow in the range 10^{-1}–10^{-4} Pa L/s (10^{-3}–10^{-6} Torr L/s) (Weaver, 1971).

The conductance of porous plugs is inversely proportional to the square root of the molecular weight of the gas crossing them. Christian and Leck (1966) experimentally demonstrated this inverse proportionality within ±1% for inert gases (helium, neon, argon, krypton, xenon) and ±3% for the others. The gas flow through a porous plug can be controlled to an accuracy of ±1% for inert gases. This accuracy can be extended to the flow measurement of permanent gases such as nitrogen and oxygen. For other gases, accuracies of the same order of magnitude can be obtained provided that either the pressure upstream from the plug is lower than that used for inert gases or one accepts the fact that the gas flow rate varies nonlinearily with the pressure difference up- and downstream from the plug (Christian and Leck, 1966).

The long-term stability of porous plugs is a function of drifts in their conductances with time caused by the progressive clogging of pores. Christian *et al.* (1966) measured the variation of the conductance of a SiC porous plug over some months. Large changes (about 6%) were observed during the first 50 days, and they were attributed to the fact that the first baking of the porous plug and of the manifold containing it was done only 30 days after starting the experiment. Hultzman and Krause (1974) observed 0.15% drift per day for a platinum porous plug.

5.7.6 Thermal Capacity Flowmeters

The principle underlying thermal capacity flowmeters consists of inferring small mass flow rates from the rise in temperature of a gas subjected to a heat flux. The flowmeter based on this principle proposed by Hawk and Baker (1969), schematically illustrated in Fig. 5.37, is composed of a

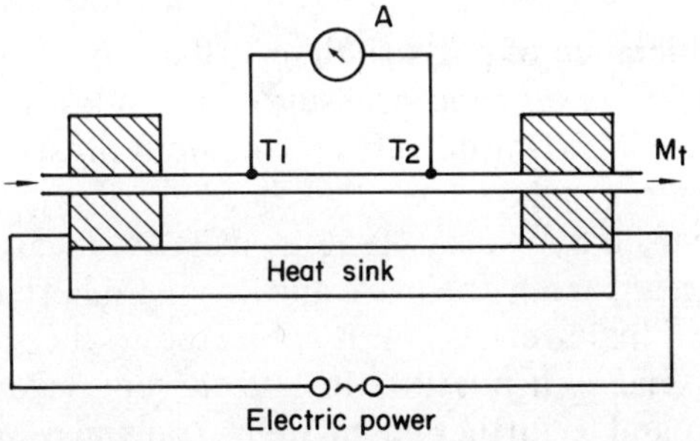

Fig. 5.37 Principle of a thermal capacity flowmeter. A, ammeter; M_t, metallic tube with small diameter; T_1 and T_2, thermoelectric elements.

small-diameter metallic tube (M_t) whose ends are in good thermal contact with a heat sink electrically powered with constant input. Two thermoelectric elements (T_1, T_2) symmetrically spaced from the midpoint of the tube ends are connected to a meter (A), thus making it possible to measure the out-of-balance indication of the elements T_1 and T_2. With the heat sink unpowered and zero flow through M_t, the temperature distribution around the midpoint is symmetrical, resulting in a null meter reading. The instant that gas starts flowing through M_t and the heat sink is powered with constant power input, heat is transferred from the tube to the gas and back to the tube, thus giving rise to a temperature gradient. The out-of-balance indication of the meter is a measure of the mass flow rate. The mass flow of the gas M is expressed for this situation

$$M = H/c_p \, \Delta t \tag{5.47}$$

where H is the heat input, c_p the specific heat of the gas at constant pressure, and Δt the change in temperature due to the gas flow. Rather large variations in temperature and pressure produce small changes in c_p (Hawk and Baker, 1969). Clogging of the metallic tube M_t can be prevented by the use of an upstream filter (Benson *et al.*, 1970). The flowmeter can be operated with chlorine, fluorine, tungsten hexafluoride, and uranium hexafluoride if monel alloys are used for the constructions of M_t.

5.8 The Measurement of Conductance to Free Molecular Flow

5.8.1 Techniques for Conductance Determination

The techniques that have evolved for the measurement of the conductance of constrictions to molecular flow apply to (a) orifices, tubes, or other components of a vacuum system, and (b) porous materials.

(a) The conductance of parts such as orifices or tubes is usually computed by measuring *in situ* the pressures up- and downstream from the part as well as the gas flow rate. This throughput pressure-drop technique is rather inaccurate.

The technique used by Oatley (1954) consists of measuring the pressure up- and downstream from the part and from a short tube of acurately known cross-sectional area A_0, both subjected to the same experimental conditions. This makes it possible to find the equivalent cross-sectional area A of the part and to further calculate its transmission probability with the ratio A/A_0. A slightly modified technique is that used by Levenson *et al.* (1961a,b) and Davis *et al.* (1964) for measuring the conductance to molecular flow of straight cylinders, cylindrical elbows, and other geometries such as cubic and bulged elbows. Instead of a short tube, the comparison elements they used were several orifices of known cross-sectional areas in a thin plate.

Ruthberg (1963, 1964, 1972) determined the conductances of various parts used in vacuum technology by substitution procedures. In his method the accurately determined conductance to the gas flow of a constriction with adjustable geometry served as transfer reference.

(b) Determining the conductance of porous materials made into plugs or slabs involves measuring only the rate of pressure variation of gas in a reservoir. The two approaches that have been developed, the measurement of the pressure drop or of the pressure rise, are used for the determination of either the conductance or the gas leakage of such materials.

5.8.2 Conductance Determination Using Substitution Procedures

A schematic of the experimental setup used by Ruthberg (1972) for conductance determination is illustrated in Fig. 5.38. The experimental flow line (Fig. 5.38a), referred to as a series line, comprises the constriction W_x whose conductance C_x must be measured, a constriction W with adjustable geometry of accurately known conductance C, two monitor gauge heads for pressure measurement at the inlet (G_{2i}) and outlet (G_{2o}) of the constrictions array, a gas supply under controlled conditions and a pumping system. Assuming that

(a) the rate of flow of gas Q in the line, the pumping speed of the monitor gauge heads, and sorption effects over the whole array of the flow line are constant during a determination;
(b) the sensitivity and reproducibility of G_{2i} and G_{2o} does not shift during a determination; and
(c) the setting of constriction W is highly reproducible,

the measurement of conductance is accomplished in the following sequence.

(a) For a certain known set of W_x, gas is admitted into the line, and its flow is monitored and maintained in the molecular range. The indications of G_{2i} and G_{2o} are noted when steady-state conditions of gas flow are reached. At this instant

$$P_{2i} - P_{2o} = Q(1/C + 1/C_1 + 1/C_x) \tag{5.48}$$

where C_1 is the conductance of the remaining flow line between G_{2i} and G_{2o}

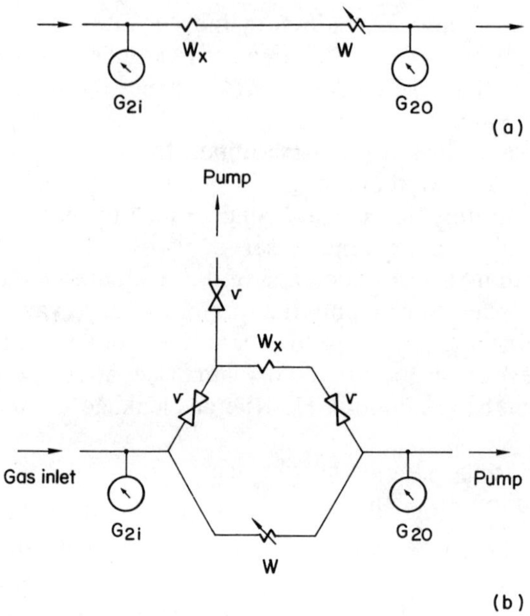

Fig. 5.38 Diagram of the flow line for conductance determination by substitution procedures (after Ruthberg, 1972). G_{2i} and G_{2o}, monitor gauge heads; W_x, constriction with conductance C_x; W, constriction with adjustable geometry having variable known conductance C; v, valves.

(b) The constriction W_x is removed, and with the gas flow rate maintained constant at the former value, the setting of W is adjusted to rebalance the line at the original pressure.

The conductance of constriction W_x is calculated from the difference between the values taken by C_x in sequences (a) and (b).

In the parallel line approach, illustrated in Fig. (5.38b), the gas flow is not disrupted when W_x is removed from the line, a manifold and several valves providing for this shift. For calculations, Eq. (5.49) must be supplemented with the conductances of the arms (a and b), and the changes in the conductance of the line between G_{2i} and G_{2o} is produced as a result of the shift of Q from one branch to the other.

Constriction W consists of two triangular apertures set in opposite directions, cut at the same height into the ground members (rotable and fixed) of a tapered glass seal. The location of the aperture is selected so as to make the flux distribution in the flow line independent of the orifice opening. The conductance of the aperture is calculated and measured by comparison with orifice plates of "zero thickness."

The calibration gas, contained in a reservoir at approximately atmospheric pressure, is supplied into the flow line through an ultrahigh vacuum valve. The gas pressure in the reservoir is controlled with a vacuum gauge. The temperature of the gas supply assembly is thermostatically controlled so that with a gas flow rate of about 10^{-3} Pa L/s (10^{-5} Torr L/s) the change in flow rate after a day's operation is 1 part in 10^3.

The temperature of various parts of the flow line is controlled with an air bath which maintains ambient temperature constant to within 0.1 K.

5.8.3 Conductance Determination Based on the Rate of Pressure Decrease

When determining conductance based on the rate of pressure decrease, gas contained in a known volume V is permitted to flow through the constriction whose conductance is being measured, and the rate of pressure decrease in V is measured. Thus, with reference to Fig. 5.39, if V is the volume of the reservoir R and C the conductance of the constriction W, usually a porous plug, the rate of decrease of pressure P is

$$-V(dP/dT) = PC \tag{5.49}$$

Equation (5.49) holds true if

(a) the gas flow through the constriction is molecular,

(b) the pressure downstream from the constriction is negligibly small compared with that upstream from the conductance, and

(c) both the gas load produced by outgassing from the walls of R and the sinks of gas (except the pumping from the calibration system) are negligible.

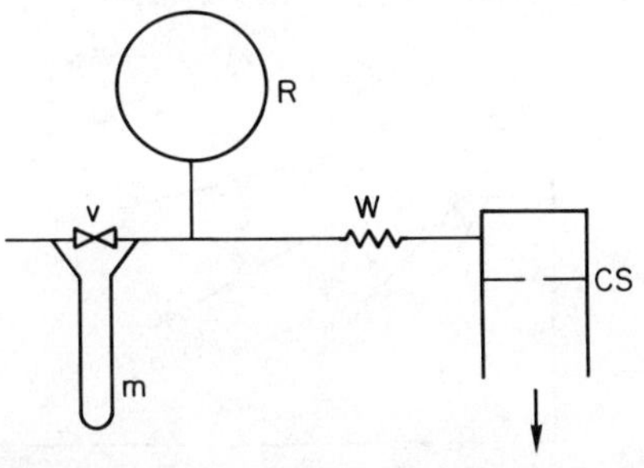

Fig. 5.39 Experimental setup for the determination of the conductance of a porous plug. R, gas reservoir; m, liquid-column manometer; v, valve; W, porous plug; CS, calibration system.

If the pressure in R falls from P at $t = 0$ to P_1 at $t = t_n$

$$-V \int_P^{P_1} dP/P = C \int_0^t dt \tag{5.50}$$

which by integration becomes

$$C = (V/t)\ \ln(P/P_1) \tag{5.51}$$

The plot of $\ln P_1 = f(t)$ yields straight lines (Fig. 5.40); C can be calculated from their slopes.

It is essential that V should be known accurately and that a low pressure (≤ 100 Pa, ≤ 1 Torr) should be maintained downstream from W to insure a continuous gas flow and thus a steady pressure drop.

This technique, described by Leck (1964), was used by other workers to measure the conductance of porous plugs (Mathews, 1964; Owens, 1965; Fowler and Brock, 1970), to measure the conductance of both a porous plug and an orifice (Close *et al.*, 1977), or to evaluate the performance of porous plugs used for gas flow control (see Section 7.5) (Christian and Leck, 1966; Christian *et al.*, 1966).

The accuracy of measurements was improved by Christian and Leck (1966) in that the variable small volume v_m, caused in the manometer m (Fig. 5.39) by the shift of the manometric fluid during determinations, was added to V. Thus, if v_m is expressed in liters,

$$v_m = Ah = A(\rho_{Hg}/\rho_{oil})P \times 10^{-4} \tag{5.52}$$

where A is the mean cross-sectional area (in square centimeters) of the manometer tube bore, h the height (in centimeters) of the manometric fluid (referred to as the head), ρ the density of the manometric fluid, and P the pressure (in millimeters of mercury).

If the volume V is supplemented with v_m, Eq. (5.51) can be written as

$$C = |V + (10^{-4}AP\rho_{Hg}/\rho_{oil})|\ |\ln(P/P_1)|/t \tag{5.53}$$

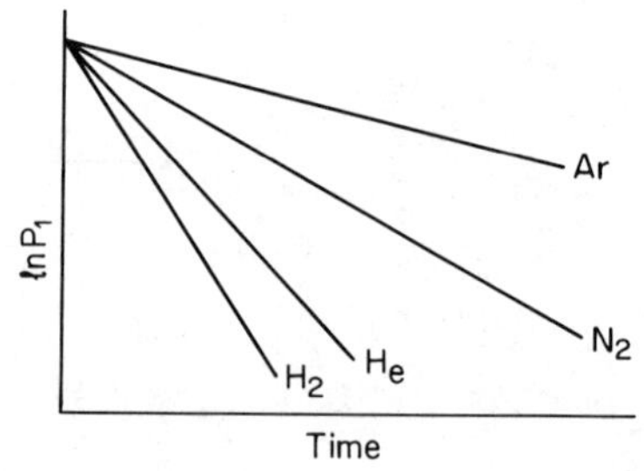

Fig. 5.40 The variation of the pressure P in the reservoir R with time for different gases in the setup illustrated in Fig. 5.39.

Instead of measuring the pressure with a *U*-tube manometer, a piston gauge (Fowler and Brock, 1970) and a hot-cathode ionization gauge head (Close *et al.*, 1977) were used.

The rate of pressure decrease technique was utilized by Close *et al.* (1977) to measure the conductance of the orifice joining two chambers (W_2 in Fig. 5.10) in their calibration system. Twenty determinations with Ar were made over a period of a month. The probable error in measuring C was ±2%. The uncertainty in the linearity of the hot-cathode ionization gauge head used for measurements was less than 1%. The pumping speed of the gauge head during determinations was negligible.

5.8.4 Conductance Determination Based on the Rate of Pressure Rise

Conductance determination using the rate of pressure rise is based on the same assumptions made in Section 5.8.3 and on the same relationship between the quantities V, P, t, and C expressed by Eq. (5.49), where the volume V must be positive.

Hultzman and Krause (1974) used the rate of pressure increase to evaluate the leak through a porous platinum plug, using the setup illustrated in Fig. 5.41. The calibration gas flows through the porous material into a calibration volume V_c previously evacuated, thus raising the pressure in that volume. The gas pressure is measured with a quartz bourdon-tube

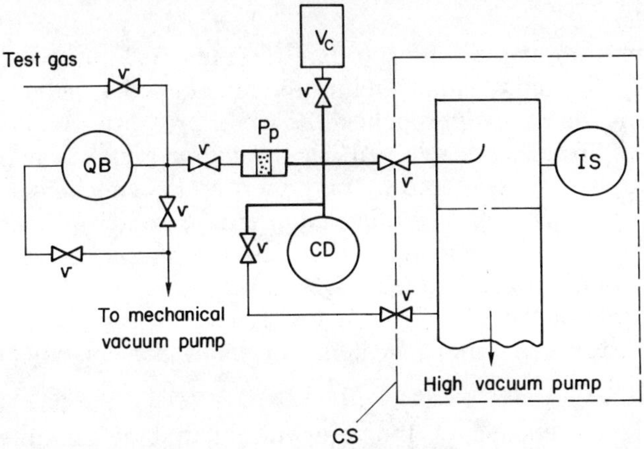

Fig. 5.41 Setup for the determination of a porous leak conductance (after Hultzman and Krause, 1974). CD, capacitance diaphragm manometer; CS, calibration system; IG, ionization gauge head; P_p, porous platinum leak; QB, quartz bourdon-tube manometer; V_c, calibration volume; v, valves.

manometer (QB) and a capacitance diaphragm manometer (CD) up- and downstream from the porous plug, respectively. The volume into which gas enters after crossing the porous plug is composed of the volumes V_c of the calibration volume and that corresponding to the manifold shown in heavy lines in Fig. 5.41. This volume is measured by using Boyle's law of expansion. The timing of the pressure rise in V_c was started 10 s after the calibration gas was admitted into the system so that the initial transient could settle. The rate of pressure rise was measured over a period varying from 5 to 10 min (depending on the gas pressure upstream from the porous plug) until $P_1 \cong 10^2$ Pa (1 Torr).

5.9 Dissociation Equilibrium Pressure Method

5.9.1 Principle

Thermodynamic equilibria of the vapor pressure of a gas with its own solid or of a gas–metal solution are used to set *fixed pressure points* on a low pressure scale. For that purpose, a chemical substance is selected, and its low pressure variation with temperature is experimentally determined. The measured data are then fitted to an analytical expression of the type $P = f(t)$. The pressure points thus generated are used, with suitable extrapolation, to calibrate vacuum gauges. The technique is *absolute* since pressure is derived from the basic thermodynamic quantities of temperature, entropy, and enthalpy. The substance is therefore a *pressure standard.*

For calibration, the relationship $P = f(t)$ must be known and reproducible, i.e., the thermodynamic equilibrium maintained as in the experimental determination. Two approaches, the vapor pressure and the gas–metal solution equilibria, have evolved. In the vapor pressure approach low pressure points are generated by both inert and chemically active gases, whereas in the gas–metal solution equilibria approach, these points are generated only by chemically active gases. The substance selected is suitably accommodated in a leak-tight vessel attached, in turn, to a calibration system of the static or dynamic type.

The main problems met in calibration using dissociation equilibrium techniques are

(a) the maintenance of the temperature stability required for the achievement of thermodynamic equilibrium,
(b) thermal transpiration, and
(c) hysteresis of the pressure–temperature relationship (only for gas–metal equilibrium solution).

5.9.2 The Vapor Pressure Equilibrium Technique

For pressures and temperatures below the triple point, the condensed phase of a substance changes directly to the gas phase without passing through the liquid phase. The gas pressure yielded in this process of sublimation is related to temperature T as

$$\ln P = I_0/R_0T + A \ln T + B \tag{5.54}$$

where I_0 is the heat of sublimation, A and B are coefficients characteristic of the chemical substance, and R_0 has its usual meaning in the kinetic theory of gas. Honig and Kramer (1969) used the expression

$$\log P = A/T + B \log T + CT + DT^2 + E \tag{5.55}$$

to calculate the temperatures corresponding to fixed pressures of substances. Computer techniques aided in establishing the function of temperature for pressures ranging within 10^5 to 10^{-9} Pa (10^3–10^{11} Torr), in decade steps. Handling of chemically active gases (e.g., CO_2, H_2, N_2, O_2) and of inert gases is possible only with the use of cryogenic techniques. For maintaining 1% reproducibility at 10^{-4} Pa (10^{-6} Torr), the cooling temperature (and accuracy required in maintaining it constant) is 25.032 K (7×10^{-3} K) for N_2 and 4.3165 K (1.7×10^{-3}) for H_2 (Poulter, 1973).

Futch (1961) used this technique to calibrate a high-pressure ionization gauge head for water vapor (see Section 6.2). The gas source was water in equilibrium with ice initially maintained at 73 K and allowed to increase to a temperature of 95 K over several hours.

Grishin and Grishina (1972) used H_2 to calibrate a cold magnetron gauge head of the Redhead type. The gauge head (excepting the coils that yield the magnetic field) is accommodated in a small volume chamber in which protrudes a pipe for the gaseous H_2 leak. The chamber is cooled in a cryostat at liquid He temperature. H_2 is then introduced into the chamber, and the liquid level in the cryostat is allowed to vary. This permits the H_2 condensed on the chamber walls to return to the gas phase. The authors calibrated the magnetron gauge head down to 10^{-13} Pa (10^{-15} Torr) with an accuracy of 20%, i.e., 10% accuracy in the temperature measurement and 10% accuracy in extrapolating the pressure–temperature relationship.

Taborek and Goodstein (1979) monitored and controlled the absolute temperature of He on a Grafoil (a form of exfoliated graphite with a specific area of 10 m^2/g) substrate in the range 1–10^{18} Pa (10^{-1}–10^{-20} Torr).

The vapor pressure equilibrium technique requires relatively simple manipulation. Thermal transpiration at 4 K is $P_{300} = P_4 \times 8.66$ and at 70 K is $P_{300} = P_{70} \times 2.07$ (Poulter, 1973). The technique is hysteresis free.

5.9.3 The Gas–Metal Solution Equilibrium Technique

In a two-component system consisting of a gas–solid solution equilibrium in a mixed phase region, the pressure of gas surrounding the solid is a function of temperature as long as the phase region does not change.

Orcutt (1969) studied the generation of nitrogen by thermal dissociation from barium nitride as

$$Ba_3N_2 \rightleftarrows \tfrac{3}{2}\, Ba_3N + \tfrac{1}{4}\, N_2 \tag{5.56}$$

In the temperature range 740–1150 K, the nitrogen pressure is related to the temperature of the reaction as

$$\ln P_{\mathrm{Pa}} = (22.02 \pm 0.14) - (20{,}080 \pm 140)/T \tag{5.57}$$

$$\ln P_{\mathrm{Torr}} = (17.13 \pm 0.14) - (20{,}080 \pm 140)/T \tag{5.58}$$

Equilibrium pressures for N_2 ranging from 4.23×10^{-3} to 96.52 Pa (3.17×10^{-5}–7.24×10^{-1} Torr) were obtained by heating commercial barium nitride powder at temperatures of 738 to 1150 K in an alumina ceramic furnace. A manifold containing gauges for pressure measurement, a residual gas analyzer, and valves were attached to the furnace and a pumping system. The manifold could be isolated from the pumping system, thereby converting the latter to a *static system* for calibration. Pressures over the range mentioned were measured with gauge heads of the Schultz–Phelps and BAG types. Thermal transpiration corrections were made in order to correct for temperature differences between the powder in the furnace and that at which the gauge measuring pressure was operated. Hysteresis was not reported.

Hydrogen was released for calibration purposes from both erbium–hydrogen and niobium–hydrogen systems. Thus, Lundin (1969) investigated the erbium–hydrogen binary system in order to set up a low pressure standard for the calibration of gauge heads used in space flight experiments. He experimentally determined the relationship between the pressure and temperature of a two-phase region of the system where erbium solid solution coexists with erbium dyhidride, which is written as

$$\log P_{\mathrm{Torr}} = -(11{,}490 \pm 18)/T + (10{,}668 \pm 0.019) \tag{5.59}$$

Later, Outlaw and Steel (1971) who also used hydrogen obtained from a binary erbium–hydrogen system to calibrate a nude BAG head in a *dynamic system* found the dissociation pressure related to pressure as

$$\log P_{\mathrm{Torr}} = -(26{,}641 \pm 230)/T + (23.95 \pm 0.32) \tag{5.60}$$

Titcomb and Wallace (1972) suggested the use of the niobium–hydrogen system, which can be brought to equilibrium at relatively low temperatures. The equilibrium pressure of this system in the two-phase region is 2.40×10^{-5} Pa (1.8×10^{-7} Torr) at 295 K and 1.60×10^{-6} Pa (1.2×10^{-8} Torr) at 273 K. The main advantage of using low temperatures for maintaining the two-phase equilibrium is that thermal transpiration effects are much reduced compared with other hydrogen systems. Equilibrium pressures and temperatures are related in the system of Titcomb and Wallace as

$$\log P_{\mathrm{Torr}} = 7.87 - 4313/T \tag{5.61}$$

The authors claimed that the hysteresis effect observed in calibration over the range 10^{-6}–1 Pa (10^{-8}–10^{-1} Torr) is reduced.

The phase equilibrium of the hydrogen–tantalum system was investigated by Poulter and Pryde (1968).

Since inert gases do not dissolve in metals to form solid solutions, only chemically active gases can be yielded by using the gas–metal equilibrium solution technique. The thermal transpiration effect at 1000 K is $P_{300} = P_{1000} \times 0.55$ (Poulter, 1973).

References

ASTM E297/70. *Standard Methods for Calibrating Ionization Vacuum Tubes.*

AVS 6.2 (1969). *Procedure for Calibrating Vacuum Gauges of the Thermal Conductivity Type.*

AVS 6.4 (1969). *Procedure for Calibrating Hot Filament Ionization Gauges Against a Reference Manometer in the Range 10^{-2}–10^{-5} Torr.*

Bannenberg, J. G., and Tip, A. (1968). *Proc. 4th Int. Vac. Congr. Manchester 1968,* p. 609. *Inst. of Phys. and Phys. Soc. Conf. Series No 6.* Addlard and Son, Surrey, England.

Barre, E., Mongodin, G., and Prevot, F. (1956). *Vide* **11,** 25.

Bennewitz, H. G., and Dohmann, H. D. (1965). *Vak. Tech.* **14,** 8.

Benson, J. M., Baker, W. C., and Easter, E. (1970). *Instr. Control Syst.* Feb., 85.

Bergquist, L. E. (1968). *Proc. 4th Int. Vac. Congr. Manchester 1968,* p. 600. *Inst. of Phys. and Phys. Soc. Conf. Series No 6.* Adlard and Son, Surrey, England.

Berman, A. S. (1956). *J. Appl. Phys.* **36,** 3356.

Bernardet, H., Choumoff, P. S., and Iapteff, B. (1970a). *Vide* **25,** 131.

Bernardet, H., Choumoff, P. S., Mativet, J., and Sauneuf, R. (1970b) *Vide* **25,** 84.

Blado, L. D., and Lilienkamp, R. H. (1972). *J. Environ. Sci.* Jan./Feb., 24.

Bureau, A. J., Laslett, L. J., and Keller, J. M. (1952). *Rev. Sci. Instrum.* **23,** 683.

Calcatelli, A., Ferrero, C., Rumiano, G., and Ughi, G. (1974a). *Proc. 6th Int. Vac. Congr. Kyoto 1974,* p. 127. *Jpn. J. Appl. Phys. Suppl. 2* (Part 1).

Calcatelli, A., Ferrero, C., Rumiano, G., and Ughi, G. (1974b). *Proc. 6th Int. Vac. Congr. Kyoto 1974,* p. 131. *Jpn. J. Appl. Phys. Suppl. 2* (Part 1).

Calder, R., and Lewin, G. (1967). *Brit. J. Appl. Phys.* **18,** 1459.

Choumoff, P. S., and Aubry, B. (1965). *Vide* **20,** 177.
Choumoff, P. S., Bernardet, H., Mativet, J., and Saunneuf, R. (1970). *J. Vac. Sci. Technol.* **7,** 270.
Christian, R. G., and Leck, J. H. (1966). *J. Sci. Instrum.* **43,** 229.
Christian, R. G., Leck, J. H., and Werner, J. G. (1966). *Vacuum* **16,** 609.
Clausing, P. (1932). *Ann. Phys.* **12,** 961.
Clausing, P. (1971). *J. Vac. Sci. Technol.* **8,** 636.
Close, K. J., Vaugham-Watkins, R. S., and Yarwood, J. (1977). *Vacuum* **27,** 511.
Das, K. B. (1966). *ISA J.* **13,** 48.
Das, K. B. (1967). *Vacuum* **17,** 587.
Davis, D. W. (1963). *Trans. 10th Natl. Vac. Symp. U.S.A. 1963,* p. 253. McMillan, New York.
Davis, D. W. (1968). *J. Vac. Sci. Technol.* **5,** 23.
Davis, D. H., Levenson, L. L., and Milleron, N. (1964). *J. Appl. Phys.* **35,** 529.
Demarcus, W. C. (1956). *U.S. AEC Report K 1302, AD 124579,* Parts I–IV.
DIN 28416 (1963). *Kalibrieren von Vakuummetern in Bereich von* 10^{-3} *bies* 10^{-7} *mbar.*
Elliott, K. W., Woodman, D. M., and Dadson, R. S. (1967). *Vacuum* **17,** 439.
Feakes, F., and Torney, Jr., F. L. (1963). *Trans. 10th Natl. Vac. Symp. U.S.A. 1963,* p. 257. Mcmillan, New York.
Fletcher, B. (1970). *Vacuum* **20,** 377.
Fletcher, B., and Watts, J. (1967). *Vacuum* **27,** 445.
Fletcher, B., and Watts, J. (1968). *Vak. Tech.* **17,** 145.
Florescu, N. A. (1962). *Trans. 8th Natl. Vac. Symp. U.S.A. 1961,* p. 504. Pergamon Press, New York.
Fowler, P., and Brock, F. J. (1970). *J. Vac. Sci. Technol.* **7,** 507.
Futch, Jr., A. H. (1961). *Rev. Sci. Instrum.* **32,** 1263.
Gosselin, C., and Beitel, G. (1970). *Report NASA CR-1060 N 70 26580.*
Gourjault, J. D., Lefevre, J., and Choumoff, P. S. (1972). *J. Vac. Sci. Technol.* **9,** 206.
Grishin, S. F., and Grishina, E.Ya. (1972). *Sov. Phys.-Tech. Phys.* **16,** 1190.
Grosse, G., and Messer, G. (1970). *Vacuum* **20,** 373.
Grosse, G., and Messer, G. (1981). *Vak. Tech.* **30,** 226.
Hablanian, M. H. (1969). *J. Vac. Sci. Technol.* **6,** 265.
Hawk, Ch.E., and Baker, W. C. (1969). *J. Vac. Sci. Technol.* **6,** 255.
Hayward, W. H., and Jepsen, R. L. (1962). *Trans. 9th Natl. Vac. Symp. U.S.A. 1962,* p. 459. McMillan, New York.
Henry, R. P. (1956). *Vide* **11,** 54.
Hoffman, V. E. (1963). *Res./Dev.* **14,** 62.
Hojo, H., Ono, M., and Nakayama, K. (1977). *Proc. 7th Int. Vac. Congr. Vienna 1977,* p. 117. Berger and Sohne, Horn, Austria.
Holland, L., and Priestland, C. R. D. (1967). *Vacuum* **17,** 461.
Honig, R. E., and Kramer, D. A. (1969). *RCA Rev. 30* June, 285.
Hultzman, W. W., and Krause, L. N. (1974). *J. Vac. Sci. Technol.* **11,** 889.
Iczkowski, R. P., Margrave, J. L., and Robinson, S. M. (1963). *J. Phys. Chem.* **67,** 229.
ISO/DIS 3568. *Ionization Vacuum Gauges—Calibration by Direct Comparison with a Reference Gauge.*
ISO/DIS 3570/1. *Standard Methods for Calibration—Part I. Pressure Reduction by Continuous Flow in the Pressure Range of* 10^{-1} *to* 10^{-5} Pa *(*10^{-3} *to* 10^{-7} *Torr).*
Knudsen, M. (1910). *Ann. Phys.* **31,** 633.
Knudsen, M. (1914). *Ann. Phys.* **44,** 525.

Knudsen, M. (1934). "The Kinetic Theory of Gases," p. 43. Methuen and Co. Ltd., London.

Kreisman, W. S. (1966). *Res./Dev.* **14,** 57.

Kündig Ch, J., and Mennenga, H. (1980). *Proc. 8th Int. Vac. Congr. Cannes 1980,* p. 226. *Suppl. Le Vide les Cuoches Minces* No. 201.

Leck, J. H. (1964). "Pressure Measurement in Vacuum Systems," p. 168. Chapman and Hall, London

Levenson, L. L., Milleron, N. and Davis, D. H. (1961a). *Trans. 7th Natl. Symp. Vac. Technol. U.S.A. 1960,* p. 372. Pergamon Press, New York.

Levenson, L. L., Milleron, N., and Davis, D. H. (1961b). *Report UCRL-6253.*

Lundin, E. Ch. (1969). *NASA Report CR-1271.*

Mathews, A. J. (1964). *AEDC-TR-64-94.*

McCulloch, K. E. (1983). *J. Vac. Sci. Technol.* **A1,** 168.

Messer, G. (1977a). *Proc. 7th Int. Vac. Congr. Vienna 1977,* p. 153. Berger and Sonhne, Horn, Austria.

Messer, G. (1977b). *Phys. Bl.* **33,** 343.

Morrison, Ch.F. (1966a). *Granville–Phillips Technical Rept. 272-5.*

Morrison, Ch.F. (1966b). *J. Vac. Sci. Technol.* **4,** 246.

Morrison, Ch.F. (1969). *Res./Dev.* **20,** 54.

Nakayama, K., and Hojo, H. (1974). *Proc. 6th Int. Vac. Congr. Kyoto 1974,* p. 113. *Jpn. J. Appl. Phys. Suppl. 2* (Part 1).

Nash, P. J., and Thompson, T. J. (1983). *J. Vac. Sci. Technol. A* **1,** 172.

Normand, C. E. (1962). *Trans. 8th Natl. Vac. Symp. U.S.A. 1961,* p. 534. Pergamon Press, New York.

Oatley, C. W. (1954). *Brit. J. Appl. Phys.* **5,** 358.

Orcutt, R. H. (1969). *J. Res. Natl. Bur. Stand. A* **74A,** 45.

Outlaw, R. A., and Steel, R. A. (1971). *J. Vac. Sci. Technol.* **8,** 608.

Owens, Ch. L. (1965). *J. Vac. Sci. Technol.* **2,** 104.

Peggs, G. N. (1976). *Vacuum* **26,** 321.

Peggs, G. N. (1973). *National Physical Laboratory Report No. MOM 6.*

Poulter, K. (1973). *National Physical Laboratory Report No. MOM 3.*

Poulter, K. F. (1974). *J. Phys. E.* **7,** 39.

Poulter, K. F. (1978). *Vacuum* **28,** 135.

Poulter, K. F., and Pryde, J. A. (1968). *Proc. 4th Int. Vac. Congr. Manchester 1968,* p. 111. *Inst. of Phys. and Phys. Soc. Conf. Series No 6.* Adlard and Son, Surrey, England.

Poulter, K. F., Rodgers, M. J., Nash, P. J., Thompson, T. J., and Perkin, M. P. (1983). *Vacuum* **33,** 311.

Power, B. D., and Crawley, D. J. (1954). *Vacuum* **4,** 415.

Redhead, P. A. (1961). *Trans. 7th Natl. Symp. Vac. Technol. U.S.A. 1960,* p. 108. Pergamon Press, New York.

Reich, G. (1980). *Proc. 8th Int. Vac. Congr. Cannes 1980,* p. 222. *Suppl. Le Vide, les Couches Minces No. 201.*

Roehring, J. R., and Simons, Jr., J. C. (1962). *Trans. 8th Natl. Vac. Symp. U.S.A. 1961,* p. 511. Pergamon Press, New York.

Roehring, J. R., and Simons, Jr., J. C. (1963). *Instrum. Control Syst.* **36,** 107.

Ruthberg, S. (1963). *Bull. Am. Phys. Soc. Ser. 11* **8,** 426.

Ruthberg, S. (1964). *J. Vac. Sci. Technol.* **1,** 84.

Ruthberg, S. (1969). *J. Vac. Sci. Technol.* **6,** 401.

Ruthberg, S. (1972). *J. Vac. Sci. Technol.* **9,** 1457.

Santeler, D. J. (1960). *Trans. 6th Natl. Symp. Vac. Technol. U.S.A. 1959,* p. 129. Pergamon Press, New York.
Smetana, F. O. (1966). *J. Vac. Sci. Technol.* **3,** 357.
Steckelmacher, W. (1966). *Vacuum* **16,** 561.
Taborek, P., and Goodstein, D. (1979). *Rev. Sci. Instrum.* **50,** 227.
Titcomb, C. G., and Wallace, W. E. (1972). *J. Vac. Sci. Technol.* **9,** 1253.
Utterbach, N. G., and Griffith, Jr., T. (1966). *Rev. Sci. Instrum.* **37,** 886.
Weaver, D. E. (1971). *J. Vac. Sci. Technol.* **8,** 752.
Willis, D. (1965). *J. Fluids Mech.* **21,** 21.

CHAPTER 6

Calibration of Vacuum Gauges for Different Gases

6.1 Calibration Techniques

In many fields of vacuum application it often becomes necessary to measure low pressures in systems containing gases different from those used for the gauge head calibration (usually nitrogen or argon). This is the situation when the rarefied environment is composed, for instance, of hydrocarbon or halogen gases, isotopes of hydrogen, or water vapor. Vacuum gauges required to measure pressure in such environments are calibrated either by determining their sensitivity for the unknown gas or vapor (ionization gauge heads) or by comparing their response to a pressure standard, in both cases under the same experimental conditions. Thus, Dushman and Young (1945) calibrated hot-cathode ionization gauge heads for H_2O, Hg, and I by comparing the gauge head readings with those of a Knudsen or McLeod gauge. Moesta and Renn (1957) determined the sensitivity of hot-cathode ionization gauge heads for CH_4, C_2H_6, C_3H_8, C_4H_{10}, and ^2H in a calibration system operated according to the static expansion method (see Section 5.4). Nakayama and Hojo (1974) and Tilford (1983) reported determinations of the relative sensitivities of hot-cathode ionization gauge heads for hydrocarbon gases, H_2O, and ^2H performed with dynamic calibration systems (see Section 5.5). In addition to the use of the static or dynamic method, methods based on the dissociation equilibrium pressure (see Section 5.9) were also reported for gauge head calibration with water vapor or uranium hexafluoride.

The type and construction materials of the gauge head to be calibrated for gases other than nitrogen or argon must be selected so as to minimize the effect of the gas on the performance of the gauge. It is known, for instance, that a tungsten cathode is attacked and rapidly destroyed by water vapor even at pressures as low as 10^{-3} Pa (10^{-5} Torr) and that both its electrical resistance and emissivity are changed when operated in the presence of hydrocarbon vapors (see Section 4.4.6).

This chapter will present a brief discussion of vacuum gauge calibration techniques for gases other than nitrogen or argon.

6.2 Calibration with Water Vapor

The method used by Fursov (1970) for calibrating thermal-type gauge heads (see Section 4.3) consists of subjecting the instrument to the pressure exerted by water vapor and measuring this pressure by balancing it against a known pressure. Fursov used the experimental setup illustrated in Fig. 6.1, where a capacitance manometer (CM) differentially measures the pressures in the reference (r) and gauging (g) manifolds of the system. A pressure standard (S_1) of the McLeod type mounted in manifold r permits measuring the pressure of dry air introduced into the system by

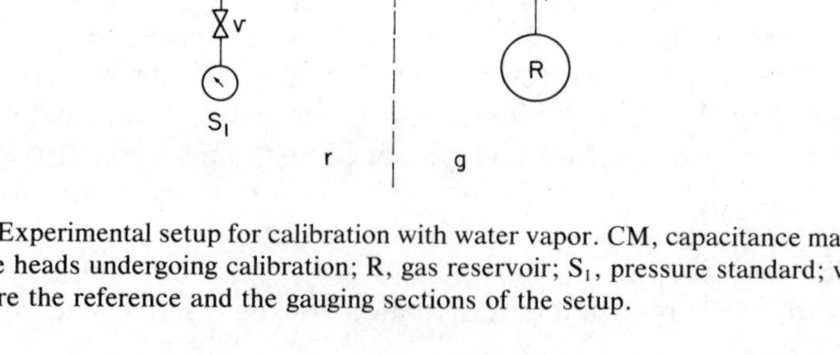

Fig. 6.1 Experimental setup for calibration with water vapor. CM, capacitance manometer; G_1, gauge heads undergoing calibration; R, gas reservoir; S_1, pressure standard; v, valves. r and g are the reference and the gauging sections of the setup.

means of a leak valve. Manifold g contains the gauge heads (G_1) undergoing calibration and a reservoir (R) with water. Both sides of the capacitance manometer (CM) can be evacuated separately through the corresponding manifold by a pumping system. After mounting the gauge heads (G_1), the system is evacuated with the leak valve between reservoir R and manifold g shut, until obtaining a zero pressure differential reading at manometer CM. Then both the manifolds are isolated from the pumps, and water vapor from reservoir R is permitted to enter into the gauge heads. The reading of the capacitance manometer (CM) is restored to the former zero pressure differential by introducing dry air in manifold r. The dry air pressure is measured with the pressure standard (S_1). Calibration of the thermocouple gauge heads for water vapor has been performed in the pressure range 10^{-2}–1600 Pa (10^{-4}–12 Torr).

Futch (1961) used the dissociation equilibrium pressure of water to determine the sensitivity of a Schultz–Phelps gauge head (see Section 4.4.7) to this vapor in the range 2×10^{-2}–20 Pa (1.5×10^{-4}–1.5 Torr). His calibration system was made of a heavy-walled copper box (C in Fig. 6.2) connected to the gauge (G_1) undergoing calibration and to a manifold that can be evacuated with a diffusion pump via a cold trap (Tr). A thermocouple (t) in good thermal contact with the copper walls permits measuring the temperature of the box (C). To operate the system, water was introduced in C and the setup pumped down for about an hour to remove the gas dissolved in water. Then a flask containing liquid nitrogen was fitted around C, and pumping was continued overnight while baking the gauge head with IR lamps. The filament was outgassed and the valve (v) slightly opened so as to have a small conductance for the water vapor compared with the conductance of the tubing connecting G_1 with C. This prevented the accumulation of background gas in the system without affecting the pressure in gauge head G_1 too much. For calibration, the liquid nitrogen in

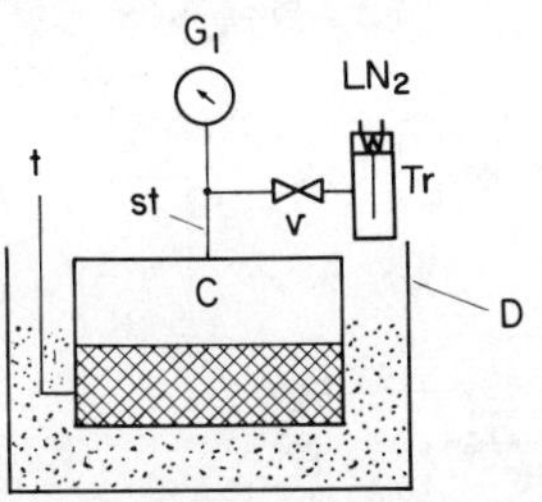

Fig. 6.2 Pressure cell for calibrating gauge heads with water vapor. C, copper box; D, Dewar flask; G_1, gauge head undergoing calibration; N_2, liquid nitrogen; st, stainless steel tube; Tr, cooled trap; v, valve.

D was replaced by a dry ice–acetone mixture and the collected ion current I_c was registered when the temperature of the mixture changed. The emission current I^- was maintained constant throughout the experiment. Readings were taken over a period of several days as the temperature of the mixture increased, thus changing the vapor pressure of water in equilibrium with the ice in box C. The vapor pressure of water versus temperature, extrapolated from higher-pressure data, was used to plot $I_c/I^- = f(P)$. The curve obtained had a slope $s = 4.2 \times 10^{-3}$/Pa (0.56/mm Hg).

Armas *et al.* (1966) used the setup illustrated in Fig. 6.2 to calibrate a radioactive gauge head (^{3}H) for water vapors in the pressure range below 10^{-1}–100 Pa (10^{-3}–1 Torr).

6.3 Calibration with Hydrogen Isotopes

6.3.1 Deuterium

Bergsnov-Hansen *et al.* (1964) determined the sensitivity of a nude BAG head with a thoria-coated tungsten emitter in a calibration system (Fig. 6.3) operated according to the dynamic method (see Section 5.5.2). The two advantages claimed for this system, where the forepump is replaced by a leak detector (L), are

(a) the use of a leak detector as a mass flowmeter reduces calibration to a flow rate measurement; and

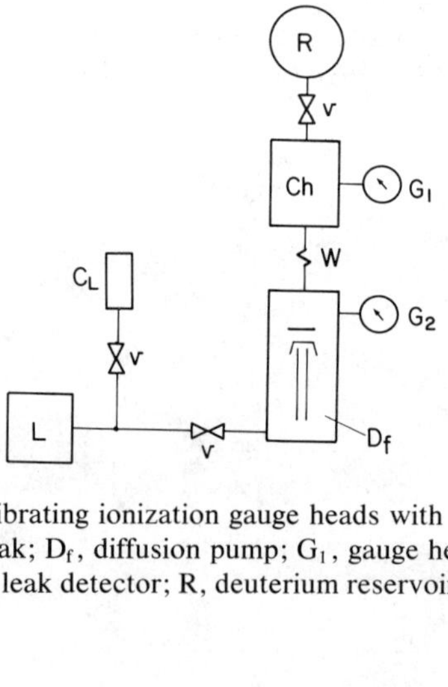

Fig. 6.3 Apparatus for calibrating ionization gauge heads with deuterium. Ch, calibration chamber; C_L, calibration leak; D_f, diffusion pump; G_1, gauge head undergoing calibration; G_2, monitor gauge head; L, leak detector; R, deuterium reservoir; v, valve; W, constriction of the gas flow.

(b) pumping effects produced by the gauge head (see Section 7.5) are ruled out since the calculations for the determination of the tested gauge sensitivity are made by considering the flow of gas *out* of the calibration system.

The sensitivity of the standard leak detector (L) for He was found in a series of measurements in which the dependence of this parameter on the flow rate of gas was also established. Deuterium-standard leaks were also used for sensitivity determinations. It was noticed that a decrease in the flow rate of deuterium does not yield a proportional decrease in pressure measured by the gauge head. This was ascribed to a more complex interaction between deuterium and the gauge head under calibration than simply electronic pumping. The average sensitivity in the pressure range 5×10^{-6}–1.7×10^{-3} Pa (3.7×10^{-8}–1.3×10^{-5} Torr) was $0.0252 \pm 1.35 \times 10^{-3}$/Pa ($3.37 \pm 0.18$ Torr).

6.3.2 Tritium

The technique developed by Barnes and Gibson (1968) for calibration with tritium is based on the fact that the number density of tritium atoms in a known volume can be calculated by measuring the current of He ions generated in that volume in the radioactive decay of ^3H. If a collection chamber of known volume, with a collecting electrode, is attached to a leak-tight volume, the sensitivity of a hot-cathode ionization gauge head mounted in this volume can be determined in an absolute manner. A BAG head and a mass spectrometer (MS) were calibrated with tritium in the setup shown in Fig. 6.4 according to this technique. The system is composed of a stainless steel chamber (Ch) to which are attached the gauge head to be calibrated (G_1) and the collection chamber (Cc). The inner walls of the calibration chamber are electropolished and the joints are

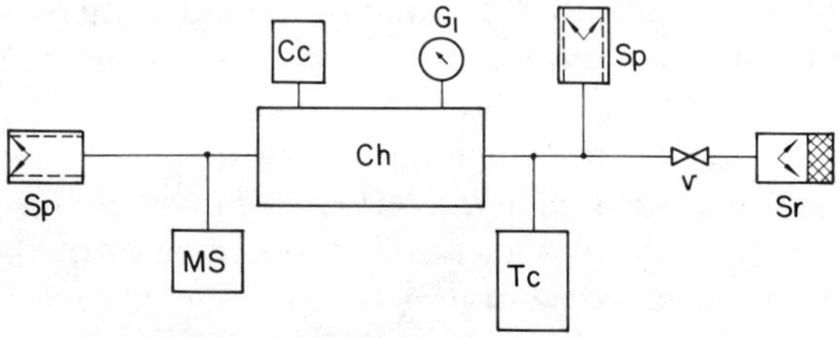

Fig. 6.4 Setup for calibration of ionization gauge heads with tritium. Cc, collection chamber; Ch, calibration chamber; G_1, gauge head undergoing calibration; MS, mass spectrometer; Sr, sorption-roughing pump; Sp, sputter-ion pump; Tc, cell for tritium generation; v, valve.

either gold rings or copper gaskets. The chamber (Ch) is evacuated by a system of sorption-roughing (Sr) and sputter-ion (Sp) pumps. A cell (Tc) for tritium generation and a mass spectrometer (MS) are attached to the manifolds connecting the chamber (Ch) with the pumping system. Tritium gas, obtained from a pyroforic uranium source through a nickel leak, is admitted into the chamber (Ch), and the current from the collection chamber (Cc) is measured simultaneously with the ion current of the gauge head undergoing calibration. The absolute pressure is then determined. The results of pressure determinations cross checked with a McLeod gauge and a ZnS(Ag) scintillator probe show that the systematic error of the technique is less than 10%. Calibration with tritium gas can be performed within the pressure range 10^{-6}–10^{-3} Pa (10^{-8}–10^{-5} Torr).

6.4 Calibration with Reactive Gases

6.4.1 Chlorine

Shaw (1966) calibrated a BAG head with Cl_2 by comparing the readings of the gauge with those of a Knudsen gauge in a dynamic system. It was observed that the reproducibility of the gauge head response to chlorine was poor until the gauge had been operated at about 10^{-4} Pa (10^{-6} Torr) for several minutes with this gas. For chlorine pressures above 10^{-3} Pa (10^{-5} Torr), the response of the gauge was no longer linear with pressure increase. Calibration was performed in the pressure range 10^{-4}–10^{-3} Pa (10^{-6}–10^{-5} Torr).

6.4.2 Tetraethyl Lead, Phenyl Acetylene, Iron Carbonyl

Righby and Wright (1968) determined the sensitivity of a radioactive gauge head to reactive and thermally unstable organic metallic vapors in a metal–glass system operated according to the static method. The output of the gauge head was compared with the output of two capacitance manometers overlapping in the pressure range 10^{-1}–27 Pa (10^{-3}–2×10^{-1} Torr). The sensitivity of the gauge head was linear with pressure in this range. Prolonged use with phenyl acetylene, however, produced a slow shift from linearity. This was ascribed to polymer formation on the foil supplying tritium for gas ionization.

6.4.3 Hydrocarbon Gases

The sensitivity of ionization gauge heads of the BAG type for CH_4, C_2H_4, C_2H_6, and $C_6H_5CH_3$ was determined by Young (1973) in the calibration apparatus illustrated in Fig. 6.5. The setup is based on the princi-

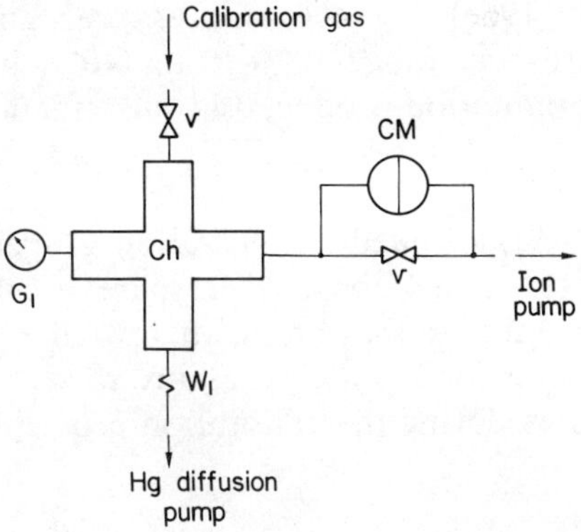

Fig. 6.5 System for calibrating ionization gauge heads with hydrocarbon gases. Ch, calibration chamber; CM, capacitance manometer; G_I, gauge head undergoing calibration; v, valve; W_I, constriction of the gas flow.

ple of single stage of flow division (see Section 5.5.2). A flow rate of gas through the calibration chamber (Ch) of about 10^{-2} Pa L/s (10^{-4} Torr L/s) was selected for nearly all gases in order to reduce dissociation and thermal decomposition at the hot cathode. Before starting calibration, the apparatus and the gauge heads were processed as discussed in Section 5.3. The gas under study was then introduced into the calibration chamber and the normalized ion current (I_c/I^-) compared with the readings of the capacitance manometer (CM). After operating the gauge head for 30 h at nearly 4.26×10^{-2} Pa (3.2×10^{-4} Torr) in CH_4, sensitivity decreased from 0.12/Pa (16.4/Torr) to 1.42×10^{-2}/Pa (1.9/Torr). The emission current was 1 mA. It was noticed that sensitivity decreases more rapidly for higher emission currents. The decrease of sensitivity with time was ascribed to the formation on the collector electrode of a hydrocarbon layer with high resistance, which affects the ion collection. Sensitivity could be restored by connecting the ion collector with the grid and outgassing the assembly thus obtained by electron bombardment (600 V, 75 mA). The black deposit left on the electrodes of the gauge head and on the glass walls after outgassing with electron bombardment can be removed by first operating the gauge head for several hours in O_2 at about 10^{-3} Pa (10^{-1} Torr) and then outgassing it in H_2 or in good vacuum to remove any built-up oxide films.

6.4.4 Uranium Hexafluoride

The dissociation equilibrium pressure of uranium hexafluoride expressed by the reaction

$$Na_2UF_8 \rightleftharpoons 2\,NaF + UF_6 \qquad (6.1)$$

was used by Pappas (1966) to develop a pressure standard for gauge head calibration in the pressure range 10^{-2}–10 Pa (10^{-4}–10^{-1} Torr). The pressure of uranium hexafluoride produced in this reaction is predicted by

$$\log P_{\text{mm Hg}} = 10.88 - 5.09 \times 10^3 T \tag{6.2}$$

Patricia *et al.* (1979) designed an ionization gauge head which can be operated in the presence of UF_6. The pressure standards for calibration were a BAG head for the pressure range below 10^{-2} Pa (10^{-4} Torr) and a capacitance manometer for the high pressure range. The BAG head was replaced several times during the calibration procedures.

References

Armas, B., Blanc, D., and Dagnac, R. (1966). *Vide* **21,** 399.

Barnes, M. R., and Gibson, A. (1968). *Vacuum* **18,** 451.

Bergsnov-Hansen, B., Endow, N., and Pasternak, R. A. (1964). *J. Vac. Sci. Technol.* **1,** 7.

Dushman, S., and Young, A. H. (1945). *Phys. Rev.* **68,** 278.

Fursov, A. I. (1970). *Instrum. Exp. Tech.* **4,** 1773.

Futch, Jr., A. H. (1961). *Rev. Sci. Instrum.* **32,** 1263.

Moesta, H., and Renn, R. (1957). *Vak. Tech.* **6,** 35.

Nakayama, K., and Hojo, H. (1974). *Proc. 6th Int. Vac. Congr. Kyoto 1974,* p. 113. *Jpn. J. Appl. Phys. Suppl. 2* (Part 1).

Pappas, W. S. (1966). *Rev. Sci. Instrum.* **37,** 1217.

Patricia, M. G., Swanson, M. M., and Beauchamp, J. L. (1979). *Rev. Sci. Instrum.* **50,** 1640.

Righby, L. J., and Wright, C. R. (1968). *Vacuum* **18,** 274.

Shaw, M. L. (1966). *Rev. Sci. Instrum.* **37,** 113.

Tilford, Ch. R. (1983). *J. Vac. Sci. Technol. A* **1,** 152.

Young, J. R. (1973). *J. Vac. Sci. Technol.* **10,** 212.

CHAPTER 7

Performance Determination of Hot-Cathode Ionization Gauges

7.1 Characteristics Defining Performance

The majority of gauges which are in use for the measurement of pressure below 10^{-1} Pa (10^{-3} Torr) are hot-cathode ionization gauges. The performance of these gauges is well defined by characteristics such as linearity, sensitivity, residual currents, and pumping and gas release.

Although there is no reason for the linear response of the gauge to be a prerequisite to the accuracy of pressure measurement, knowledge of linearity is critical for calibration. Linearity determination reduces, virtually, to measuring the rate of pressure rise of the test gas in the volume of the gauge head.

The other characteristics, sensitivity, residual currents, and pumping speed and gas release, are functions of several variables each of which must be known and controlled. The determination of these characteristics involves more sophisticated methods than those used for linearity determination (such as calibration or modulation techniques). The results of determinations are then used to correct the gauge readings. As an alternative to the determination, e.g., of the residual current, pumping speed, and gas release, one can resort to several techniques to reduce the influence of this "noise" on the collected ion current measurement.

7.2 Determination of Pressure Linearity with Time

7.2.1 Linear and Quadratic Rise Rate Methods

The principle of the rise rate methods consists of generating within the volume of the tested gauge head a rate of rise of pressure proportional to a function of time and then measuring both the pressure and the time elapsed during the pressure buildup. A linear increase of the collected ion readings with time constitutes a check on the linearity of the gauge head. Thus, with reference to Fig. 7.1, the rate of flow of gas from the reservoir (R) across the constriction (W_1) in the gauge head (G_{11}) of volume V_1 is given by integrating Eq. (5.49) between $t = 0$ and $t = t$, as

$$P_1 = at + P_{10} \tag{7.1}$$

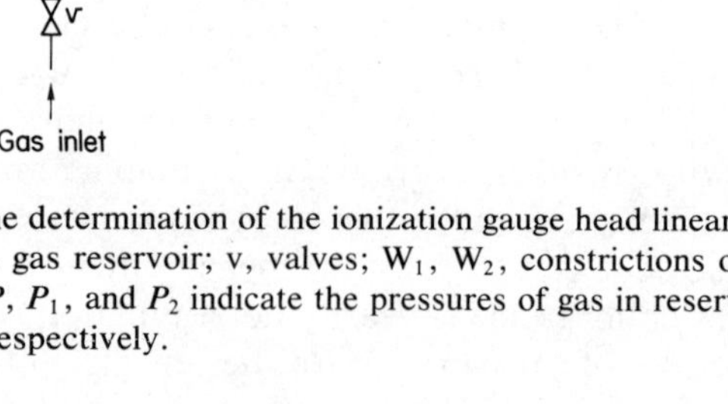

Fig. 7.1 Setup for the determination of the ionization gauge head linearity. G_{11}, G_{12}, gauge heads under test; R, gas reservoir; v, valves; W_1, W_2, constrictions of the gas flow with variable geometry. P, P_1, and P_2 indicate the pressures of gas in reservoir R and in gauge heads G_{11} and G_{12}, respectively.

where $a = (C_1P/V_1)$, (C_1) is the conductance of constriction W_1, P is the pressure in reservoir R, and P_{10} and P_1 are the initial and final pressures in the tested gauge head, respectively. Equation (7.1) holds true if

(a) the pressure of gas P in reservoir R remains constant during determinations,
(b) the gas flow across the constriction of conductance C_1 is molecular,
(c) $P_1 \ll P$, and
(d) there are no sinks or sources of gas.

The method referred to as *linear rise rate* was developed by Dushman and Found (1921) for the determination of gauge head linearity in the pressure range above 10^{-4} Pa (10^{-6} Torr). Since linearity determined by this method depends on maintaining a sufficiently large time constant V/C and pressure differential $P_1 \ll P$, the result is that at pressures in the ultrahigh vacuum range, the time to take meaningful readings would be rather large [about 3 years to cover the range of a BAG head down to 10^{-8} Pa (10^{-10} Torr)]. To circumvent this drawback, Alpert and Buritz (1954) supplemented the original setup shown in Fig. (7.1) with another constriction to the gas flow (W_2) of conductance C_2 and a gauge head (G_{12}). The rate of flow of gas in the volume V_2 of gauge head G_{12} is

$$V_2(dP/dt) = C_2P_1 \tag{7.2}$$

where P_2 is the pressure built up in gauge head G_{12}. On integrating Eq. (7.2) for $t \geq 1$ s and $P_{10} \ll (C_1P_1/V)t$,

$$P_2 = bt^2 + P_{20} \tag{7.3}$$

Here $b = (C_2a/V_2)$ and P_{20} is the pressure in G_{12} at $t = 0$.

The pressure P_2 rises in gauge head G_{11} quadratically with time, which is why the name *quadratic rise rate* is given to the method.

Equation (7.3) holds true if $P_2 \ll P_1 \ll P$. The gas flow through C_2 remains molecular over the whole range of P_2.

If gauge head G_{12} has a linear response, the plot of the square root of its indications against the pressure P_2 on a log–log scale must yield a straight line with *positive slope*. Upon using this method, Alpert and Buritz (1954) completed in about 24 h the determination of linearity of BAG heads in the range 10^{-7}–10^{-1} Pa (10^{-9}–10^{-3} Torr).

The constrictions W_1 and W_2 with variable aperture used by Alpert and Buritz (1954) were standard all-metal vacuum valves (Alpert, 1951).

For determinations, the gauge heads must be located so as to avoid gas beaming effects and must be activated only during pressure read-

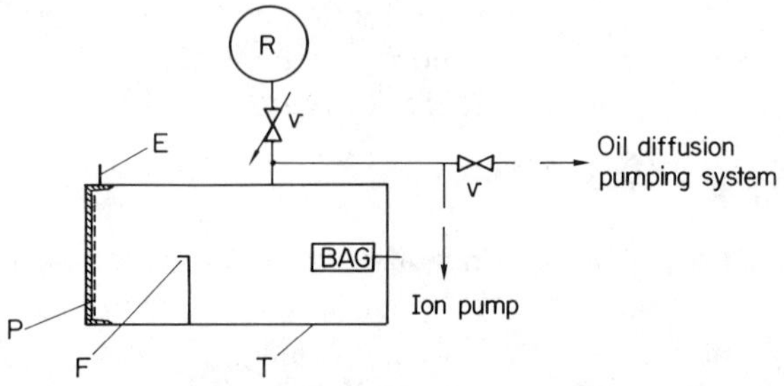

Fig. 7.2 Setup for linearity determination with a field electron microscope. E, electrode; F, tungsten tip; P, phosphor layer on a glass screen; R, gas reservoir; T, leak-tight vessel; v, valve.

ings in order to avoid pumping (see Section 7.5.6). The test gas is usually Ar.

7.2.2 The Field Emission Technique

The procedure described by Little and Whitney (1962) for the linearity determination of a BAG head relies on the fact that the time required for the deposition of a monomolecular layer of gas on a clean surface is proportional to the pressure of gas ambient to that surface. Thus, at any constant pressure, the time elapsed between two successive changes of the pattern of a clean surface on the screen of a field emission microscope is proportional to the pressure.

For linearity determinations, a BAG head was mounted in one end of a field emission microscope opposite the flat screen covered with a thin phosphorus layer (Fig. 7.2). A tungsten wire sharpened to a tip with crystalline structure, at a difference of potential of 10 kV with respect to the screen, was cleaned by heating it to about 3000 K and then allowing it to cool down. For a certain pressure P in the vessel measured with the gauge head tested, the pattern from the tungsten tip was observed, and the time needed for a significant change was noted. The field emission current was recorded and the procedure repeated for different pressures. The plot of the time required to obtain a significant pattern change against the pressure measured with the gauge head tested yields a straight line with negative slope in a log–log plot. Linearity was measured in the pressure range 10^{-8}–6.6×10^{-5} Pa (10^{-10}–5×10^{-7} Torr) for oxygen, nitrogen, and hydrogen. The time required to complete measurements in the pressure range quoted varied from several minutes to about 80 minutes.

7.3 Sensitivity Determination

7.3.1 Absolute Sensitivity

The absolute sensitivity s (see Section 2.5.2) is usually determined by calibration with N_2 or Ar in calibration systems of the static or dynamic type (see Chapter 5). The gauge head is operated from either a specially designed controller or a commercial one. In the latter case, external circuitry or internal rewiring are necessary to supply the interelectrode potentials and stabilize the emission current at the values specified by the gauge head manufacturer. All voltages must be measured with a digital voltmeter and collected ion current with a picoammeter. The filament of the gauge head can be energized with ac or dc current. When dc current is used the ion collector current must be read at least four times, twice with the filament current in one direction, twice with it reversed, and the readings must then be averaged.

The value of the absolute sensitivity can be determined with either Eq. (2.16) or Eq. (2.17). When Eq. (2.16) is used, the plot of the collected ion current I_c versus pressure P for an emission current $I^- =$ const is a straight line terminated in a curve, which is calculated rather than fitted to the measured points. The bending of the straight line into a curve is due to the residual current I_r (see Section 7.4) (Fig. 7.3). If the residual current I_r and the corresponding pressure P_r are predetermined, Eq. (2.17) can be used.

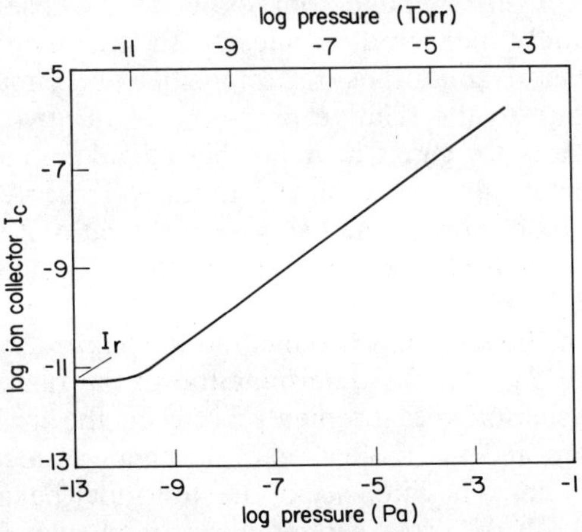

Fig. 7.3 The ion collector current I_c of a BAG head as a function of pressure for $I^- =$ 0.5 mA. The bottom portion of the curve is calculated from the residual current i_R.

Plots of the absolute sensitivity s versus pressure or time reveal the variation of this parameter with pressure or its stability with aging of the gauge head. For tests measuring the stability of sensitivity, the gauge head is left mounted on the calibration system for several hundred hours and intermittently operated at the same pressure P.

7.3.2 Relative Sensitivity

The relative sensitivity s_R is defined as

$$s_R = s_1/s_2 = (I_{c1}/I_{c2})(P_2/P_1) \tag{7.4}$$

Here subscripts (1 and 2) to the absolute sensitivity s, collected ion current I_c, and corresponding pressure P refer to the unknown sensitivity and that for N_2 or Ar. The absolute sensitivities s_1 and s_2 are determined as discussed in Section 7.3.1.

Anderson (1963) measured the relative sensitivity of a BAG head with a mass spectrometer. The principle of his technique is based on the fact that if the collected ion current I_c of a gauge head is the same in the measurement of two different gases, Eq. (7.4) becomes

$$s_R = P_2/P_1 \tag{7.5}$$

In practice the gauge head whose sensitivity is to be determined is mounted in a vacuum system together with a mass spectrometer, and both are calibrated for nitrogen. This establishes a relationship between the mass spectrometer signal and the collected ion current of the gauge head. Then a mixture of nitrogen and some other gas, whose partial pressure has been accurately measured, is allowed to leak into the system. For steady-state pressure conditions, the intensities of the mass spectrometer signals correspond to the relative pressures of the two gases. The collected ion current of the gauge head for the second gas can be determined from the known partial pressure of the mixture. To correct for the difference in the diffusion rates of the two gases, the ratio of the collected ion currents must be multiplied by the ratio of the molecular weights of the two gases.

The molecular beam collision method developed by Labrune and Theóbald (1981, 1983) for the determination of the relative sensitivity is based on cross-section measurements between the molecules of a gas formed in a beam and the molecules of another gas located on a target crossed by the beam. The intensity of the molecular beam is an exponential function of both the cross section of gas molecules in the target and the pressure they exert.

Practically, a gas A is formed in a molecular beam which crosses a buffer containing the gas B whose pressure is measured with an ionization gauge head. The absolute sensitivity of the gauge head for the gas B (s_B) is known. The measurement yields the value of the total collision cross section ($\sigma_{A\to B}$). Another measurement with A and B interchanged yields the value $\sigma_{B\to A}$.

The measured collison cross sections are related as

$$s_A \sigma_{B\to A} = \sigma_{A\to B} \tag{7.6}$$

where s_A is the absolute sensitivity of the gauge head for gas A. The method permits the determination of relative sensitivity in the pressure range 10^{-4}–10^{-1} Pa (10^{-6}–10^{-3} Torr).

7.4 Determination of Residual Currents

7.4.1 Techniques for Estimating the X-Ray Photocurrent

The *graphical technique* described by Alpert and Buritz (1954) is based on the observation that the plot of the ion collector current I_c of a conventional triode (or BAG) head as a function of the grid voltage V_g for P = Const and I^- = Const (solid curve in Fig. 7.4), is the sum of

(a) a curve representing the gas ionization I^+ (dotted curve in Fig. 7.4) having a shape characteristic of the ionization probability curve (see Fig. 4.26), and

(b) a straight line ascribed to the x-ray photocurrent (dotted–dashed line in Fig. 7.4).

Practically, the variation $I_c = f(V_g)$ is measured and plotted. Curves (a) and (b) do not result from this measurement. From previous experience it is known that the plot $I_x = f(V_g)$ is a straight line. For high values of V_g, $I_x = I_c$. Extrapolation of this line from higher to lower V_g, e.g., at the operating voltage, gives the value of the x-ray photocurrent at this voltage.

Accurate determination of the X-ray photocurrent is difficult because

(a) variations of V_g modify the pumping speed and gas release of the gauge head (see Section 7.5) as well as the temperature of the grid; these effects produce a change in the system pressure assumed constant during measurements;

(b) the time necessary for a determination is rather long, thus favoring the change of the experimental conditions (e.g., change of electronically desorbed ion current; see Section 4.4.6).

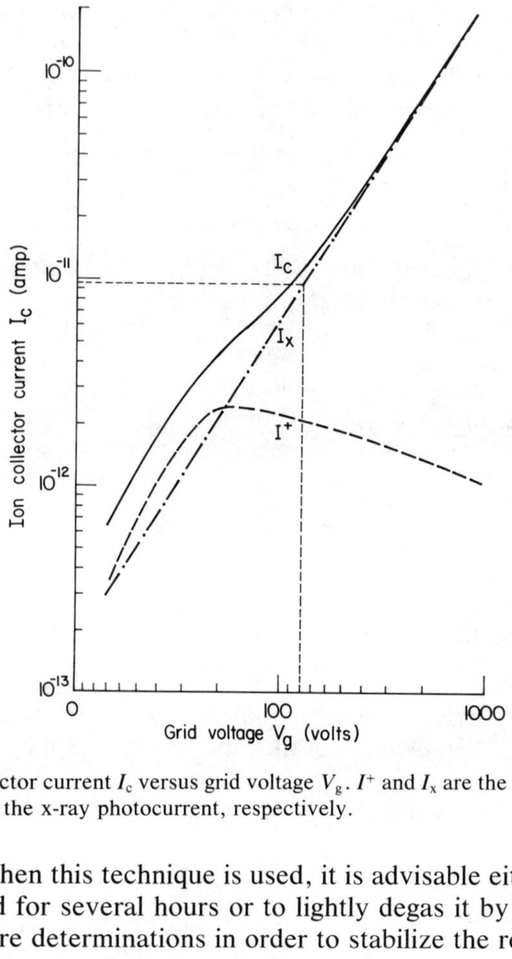

Fig. 7.4 Ion collector current I_c versus grid voltage V_g. I^+ and I_x are the true ion current in the gas phase and the x-ray photocurrent, respectively.

However, when this technique is used, it is advisable either to operate the gauge head for several hours or to lightly degas it by electron bombardment before determinations in order to stabilize the residual current value.

The *cool-down technique* suggested by Hayward *et al.* (1963) consists of chilling the measuring system to the temperature of liquid nitrogen after a pressure near the x-ray limit has been attained. The pressure indicated by the gauge head should be the x-ray limit.

Grosse and Messer (1981) used this method for determining the x-ray limit of a Helmer bent-beam gauge head (see Section 4.4.7). Cooling was achieved in a cryostat operated at the liquid helium temperature (2.3 K). Nearly the same value for the x-ray limit as measured by cooling down the measuring system was obtained by the graphical evaluation previously discussed.

7.4.2 The Modulation Method

The original method proposed by Redhead (1960) for measuring the residual current consists of modulating the total ion current from the gas phase while leaving unchanged the residual current (see Section 4.4.6). This is made possible by using a BAG head of special construction (MBAG, Fig. 4.44).

When the modulator electrode is at grid-to-filament potential V_{gf}, the ion collector current I_{c1} is

$$I_{c1} = \alpha I^{+} + I_{r} \tag{7.7}$$

Here I^{+} is the true ion current from the gas phase (pressure dependent), and I_r and α are the pressure-independent residual current and modulation factor, respectively.

Upon switching the modulator electrode to ground potential, the ion collector current I_{c2} is

$$I_{c2} = I^{+} + I_{r} \tag{7.8}$$

The method permits the determination of both the residual current I_r and the true ion current I^{+} from the gas phase. Thus, if I_c is measured at two different pressures P_a and P_b ($P_a > P_b$) and the corresponding collected currents are designated by subscripts a and b, then

$$I_{r} = I_{c1a}I_{c2b} - I_{c1b}I_{c2a}/(I_{c1a} - I_{c2a}) - (I_{c1b} - I_{c2b}) \tag{7.9}$$

and the true ion current I^{+} is given by

$$I^{+} = (I_{c1} - I_{c2})/(1 - \alpha) \tag{7.10}$$

The modulation factor α can be deduced from measurements at high pressure where $I_c/I_r \gg 1$.

Actually, the residual current is also modulated (Redhead *et al.*, 1968, p. 325) so that Eq. (7.8) has to be written

$$I_{c2} = I^{+} + \nu I_{r} \tag{7.11}$$

where ν is the modulation factor of I_r.

The following phenomena may affect the accuracy of this method referred to as Redhead mode I:

(a) pressure bursts may occur in the system due to the change in the electron current to the modulator electrode, when the modulator potential is changed;

(b) the modulation factor may vary with gas and emission current;

(c) the residual current is modulated.

The variants to this modulation technique have been discussed in Section 4.4.7.

7.5 Pumping Speed and Gas Release Determination

7.5.1 The Physical Basis of the Pumping Mechanism

The kinetic theory of gases states that the time rate of change of pressure P of an ideal gas in a vessel of volume V at constant temperature, evacuated by a pump of constant volumetric speed $S = dV/dt$, is

$$P = P_0 \exp(-t/\tau_k) \tag{7.12}$$

Here P_0 is the pressure of the gas at $t = 0$ and τ_k the time constant of the gauge head (see Section 2.3.6).

Equation (7.12) holds true if

(a) the gas does not interact with the walls of the vessel, but is merely reflected from them,

(b) the gas is removed with constant efficiency,

(c) there is no desorption of the gas within the volume V from the walls of the vessel or from the pump.

The plot of (log P, t) of Eq. (7.12) is a straight line with constant negative slope (I in Fig. 7.5). Such an "ideal" pump is not yet available, since, in practice, the above assumptions are not fulfilled because of operating conditions and pump construction.

The semilogarithmic plot of the pressure reduction with time produced by an ionization gauge head in a limited, closed volume yields a curve of negative slope (II in Fig. 7.5). At the beginning, the pressure decreases rapidly (region a), then at a lower rate (region b), to reach a quasi-equilibrium value P_e. The slope of the curve, which is proportional to the volume rate of flow of gas, varies continuously, thus indicating that the gauge head removes gas at a nonconstant pumping speed. The rapid decrease of the pressure in region a suggests that at the broken line in Fig. 7.5 saturation in pumping is reached, whereas the evolution of the pressure in region b is likely to be the result of a dynamic equilibrium between the gas removed and some other source of gas. The apparent failure of Eq. (7.12) to predict the shape of the plot results from the difference in the concepts of gas pumping by a mechanical pump and an ionization gauge head. Under suitable conditions the gas removed by a mechanical pump *never*

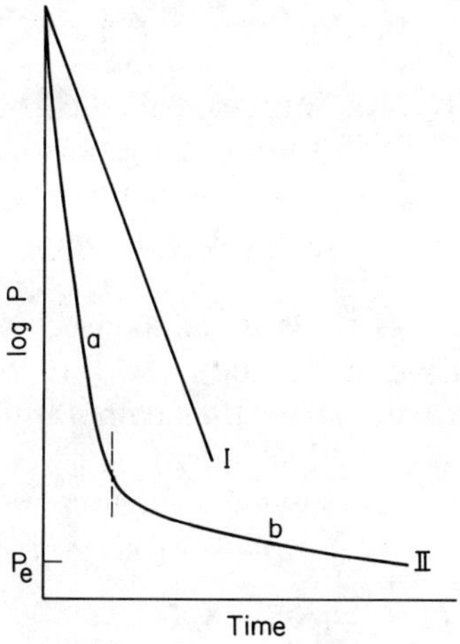

Fig. 7.5 The variation of pressure P with time produced by an ideal pump (I) and an ionization gauge head (II). P_e is the semiequilibrium pressure of the pumping process.

returns, theoretically, to the volume from which it has been evacuated. With an ionization gauge head, gas is *trapped* and *stored* in the very volume where the activating electrical discharge occurs. Since the capacity of the gauge head for these two processes is limited, some of the stored gas is returned to the gas phase. This explains the continuously changing slope of curve II in Fig. 7.5.

The gas kinetics of the sorption–desorption processes in an ionization gauge head are controlled by mechanisms related to each other in a complicated way. In dealing with these processes some workers have interchangeably used the terms "sticking," "entrapment," and "sorption," despite the differences in their physical meaning. Also "re-emission" and "desorption" have alternatively been used to specify the process by which gas or vapor is set free from the solid surfaces where it has temporarily been stored.

7.5.2 The Analytic Expression for the Pumping Process

The time rate of change of the pressure P of a mixture of gases within the volume V of an ionization gauge head operated at room temperature can be written as

$$V(dP_i/dt) = -P_i S_{T,i} + L_{T,i} \tag{7.13}$$

Here S_T is the volume rate of flow of gas produced by either of the pumping mechanisms discussed in Section 4.4.6 and L_T the leak rate contributed by both external sources and the sources discussed in the same section. The subscript i refers to the gas species concerned. The fiducial mark (x in Fig. 4.27) on the time scale for the determination of the duration dt of an experiment to measure S_T or L_T is the instant when the acceleration voltage is switched on ($t < t_x$) or off ($t = t_x$), respectively. If the temperature of the glass walls is reduced below that of the room and maintained at a certain value, S_T and L_T will increase and decrease, respectively, at a rate depending on the temperature equilibrium reached between the gas and the envelope.

The simplest situation occurs when a single inert gas is pumped at room temperature. The solution of Eq. (7.13) for $t < t_x$ is then

$$P(t) = (P_0 - L_T/S_T)\exp(-S_T t/V) + L_T/S_T \tag{7.14}$$

The reduction of the pressure as a result of the gauge head pumping is illustrated by the curve P_d in Fig. 4.27, where $P = P_0$ at $t = 0$. For $t > t_x$, Eq. (7.13) becomes

$$P(t) = (P_0 - L_T/S_T)\exp(-S_T t_x/V) + L_T/S_T + L_T(t - t_x)/V \tag{7.15}$$

and its plot is the curve D in Fig. 4.27. The value of L_T to be substituted in Eqs. (7.14) and (7.15) can be experimentally determined as

$$L_T = SP_{t\to\infty} \qquad \text{or} \qquad L_T = (P_i - P_x)V/(t - t_x) \tag{7.16}$$

for $t < t_x$ and $t > t_x$, respectively. P_i and P_x in Eq. (7.16) are the values of the pressure at the instant of the reading and at $t = t_x$. Provided that the leak contributed by external sources (permeation, leaky valves) is negligible and the gas types formerly and actually pumped are the same,

$$L_T = L_S + L_{IGD} \tag{7.17}$$

The subscripts to the leak rate L refer to the spontaneous (S) and induced gas desorption (IGD). Under carefully controlled experimental conditions L_{IGD} can be held at a reduced level or eliminated. Hence

$$L_T = L_S \tag{7.18}$$

The spontaneous desorption of gas contributing to the term L_S becomes relatively important when small amounts of gas have been pumped, whereas the gas leaks produced by gas desorption (mainly by ionic replacement) become significant only when the pumped quantities are large (Smeaton *et al.*, 1963). Spontaneous gas desorption becomes the limiting factor of the pumping process after ionically pumping more than about four decades of pressure (Redhead *et al.*, 1962, p. 323). Pumping can continue only if the inner surfaces are regenerated. When chemically active gases are pumped, Eqs. (7.14) and (7.15) are not amenable to mathematical analysis, due to complications added by chemical pumping and desorption of chemical species. Thus, only estimates of these equations can be considered. Equation (7.13) can also be expressed in terms of the collected current I_c or of the total number of molecules N according to

$$I_c = KI^- P \tag{7.19}$$

and

$$S = -(V\, dn_m/dt)/n_m = (dN/dt)/n_m \tag{7.20}$$

where n_m is the number density of molecules.

7.5.3 The Experimental Determination of the Pumping Speed

The methods used for measuring pumping speed fall into two classes, static volume or dynamic volume, according to whether the pressure of the gas in the gauge head varies or is maintained constant during determinations.

Static volume method. The static volume method is actually used for inert gases and nitrogen. The gas studied is isolated at a low pressure P in a previously evacuated, constant volume V. This volume may be that of the gauge head itself (*cut-off approach*) or that of a small part of the experimental system containing the gauge head (*seal-off approach*) (Varnerin and Carmichael, 1957; Barnes *et al.*, 1962; Ishikawa, 1965; Völter and Berndt, 1967). After the gas has reached temperature equilibrium with the walls at room temperature, the activated potentials are switched on. As the gas starts to be adsorbed and the pumping process proceeds, the pressure in volume V decreases, and consequently the rate of bombardment of the walls decreases. The rate of gas desorption from the sites where it was trapped starts increasing until it eventually balances the gas adsorption rate. Obviously, constant temperature conditions during determinations are mandatory. The plot (log P, t) assumes a pattern similar to that of curve II in Fig. 7.5. The slope at each point of the curve measures the value of τ_k, and the calculation of the pumping speed according to $\tau_k = V/S$ (see Section 2.3.6) is thus possible. Provided that the chemical pumping speed S_c is insignificant, the pumping speed calculated using the slope of the pump-down curve is that corresponding to the electronic pumping S_E (see Section 4.4.6).

The static volume method is sometimes referred to as the *variable pressure method* because the degree of gas rarefaction in the gauge head is continuously reduced during the measurement. Compared to the cut-off approach, the seal-off approach permits a more accurate determination of the pumping speed.

One of the major drawbacks of the static volume method is that the evolution of the background gas conceals the true pumping speed during determinations. Figure 7.6 schematically illustrates the systems used in the seal-off approach. By combining the adsorption of ions on the surface of a BAG head with the physical adsorption of gas in a liquid-nitrogen-cooled glass finger, Hobson (1961) succeeded in measuring both the electronic and chemical pumping speeds of nitrogen. For this purpose, he used the system illustrated in Fig. 7.7, formerly designed for the measurement of the first adsorbed layer of nitrogen on pyrex (Hobson, 1959).

Dynamic volume method. The dynamic volume method is generally utilized for measurements of the pumping speeds of both inert and chemically active gases which are rapidly removed from the gas phase (e.g.,

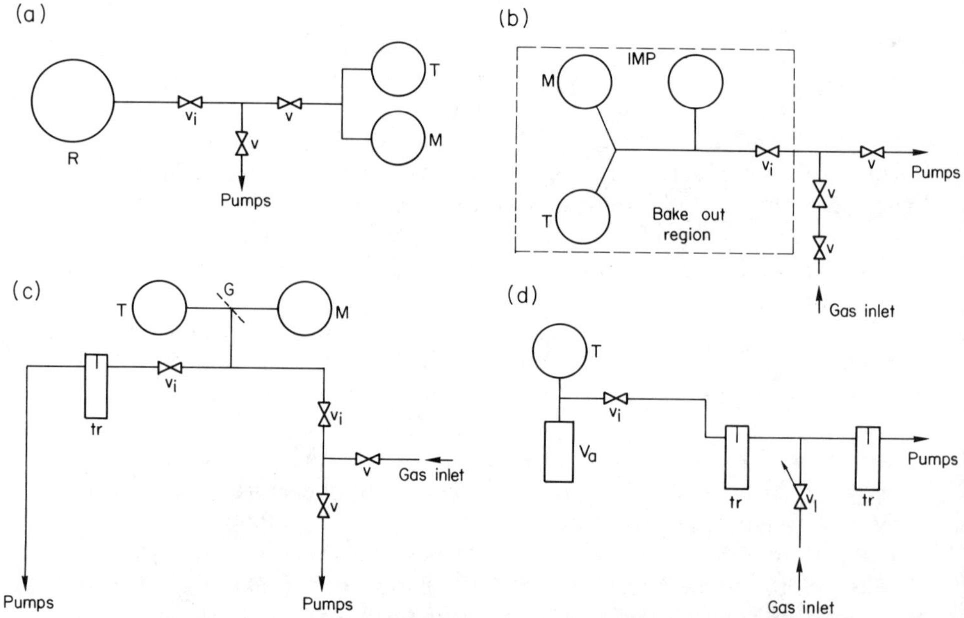

Fig. 7.6 Diagrams of the experimental systems used for the measurement of the pumping speed of an ionization gauge head according to the constant volume technique, used by: (a) Varnerin and Carmichael (1957); (b) Ishikawa (1965); (c) Barnes *et al.* (1962); (d) Völter and Berndt (1967). T, tested gauge head; M, monitor gauge head; R, gas reservoir; IMP, inverted magnetron pump; tr, cold trap; v_i, isolation valve; v_l, leak valve; v, valve; V_a, additional volume; G, grid at negative potential with respect to ground.

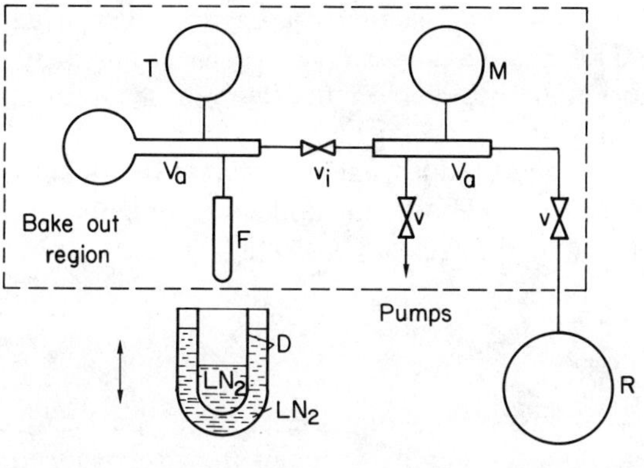

Fig. 7.7 Experimental system for measuring the pumping speed by combining the static volume technique (cut-off approach) with the physical adsorption of the gas. After Hobson (1961). T, tested gauge head; M, monitor gauge head; V_a, additional volume; v_i, isolation valve; R, gas reservoir; F, glass finger; D, Dewar flask; LN_2, liquid nitrogen.

hydrogen). Gas of known chemical composition is leaked under controlled temperature and flow conditions into the gauge head previously evacuated to a low pressure. When the desired gas pressure is reached, the system is left to attain temperature equilibrium with the walls of the gauge head (usually several minutes). Then the activation potentials are switched on. The pressure is selected such that the mean free path of a molecule is large compared to the dimensions of the experimental system. During the pumping of the gauge head, the emission current is maintained at a chosen value, and gas is supplied so as to maintain a constant pressure in the gauge head. Under these conditions the bombardment rate of the walls is constant. Physi- and chemisorption saturation is rapidly reached, and the evolution of the background gas can be maintained at a low level due to the relatively high pressures permitted experimentally. Some of the systems used for the determination of the pumping speed with this technique (sometimes referred to as the *constant pressure technique*) are illustrated in Fig. 7.8. Each system is made of three components: an evacuation setup containing the gauge head, a gas supply, and adequate means (porous plug, setting valve) of monitoring the gas flow into the experimental apparatus.

In another experimental approach, an ultrahigh vacuum valve is substituted for the porous plug to control the gas flow into the manifold containing the gauge head under test (Bills and Carleton, 1958; Bills *et al.*, 1955). The valve, of special construction (Bills and Allen, 1955), can be set manually to maintain constant pressure during measurements. The pressure drop in the reservoir is measured with a Pirani gauge head. The pumping speed of the gauge head tested is deduced by measuring the slope at the point of interest on the plot of the pressure drop in the reservoir with time.

The results obtained by different research workers in the measurement of the pumping speed of both hot and cold ionization gauge heads are summarized in Table 7.1.

7.5.4 The Measurement of the Quantity of Gas Pumped

The quantity of gas sorbed into the surfaces of a gauge head is measured after releasing the gas by induced thermal desorption or ionic replacement. Regardless of the approach taken, the following operations must be performed before starting the measurements.

(a) The gas to be tested is admitted into the gauge head, previously evacuated to background pressure.
(b) The gauge head is operated to pump the gas under test.
(c) The activating potentials are switched off, and the gas is evacuated from the gauge head until the background pressure is reached.

The induced thermal desorption approach. In this approach the gauge head is heated according to a temperature–time schedule by using a furnace closely fitted around it. Rapid and uniform heating is required (e.g., about 30 min with ±10°C variation at 300°C). The pressure transient resulting during heating is monitored by either a mass spectrometer (MS) or a monitor gauge head (M) (Figure 7.8). A typical desorption transient of argon from the glass walls of a BAG head is illustrated in Fig. 7.9. The temperatures reached at various times are approximately indicated. The broken curve, which represents the quantity of gas desorbed (usually expressed in tenths of Pascal liters or micron liters) was obtained by multiplying the effective pumping speed S_e at the manifold by the integrated area under the desorption rate curve.

The ionic replacement approach. In this approach a gas other than that which was pumped in the previous run is leaked into the gauge head, the activated potentials are switched on, and the pressure decay is monitored with a mass spectrometer (Brown and Leck, 1955; Fig. 7.8a). The plot of log P versus time is similar to curve II in Fig. 7.5. The quantity of gas desorbed in a run is not equal to that previously pumped. The main inaccuracies in calculating the quantity of desorbed gas are (Brown and Leck, 1955)

(a) the error (amounting to about ±10%) in estimating the area under the curve (log P, t), and

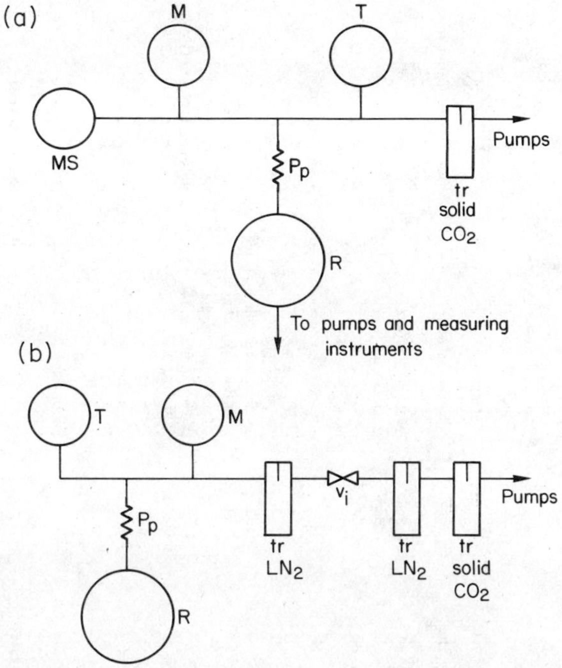

Fig. 7.8 Schematic diagram of the systems for the measurement of the pumping speed of ionization gauge heads according to the dynamic technique. After: (a) Brown and Leck (1955); (b) Cobic *et al.* (1961). T, tested gauge head; M, monitor gauge head; MS, mass spectrometer; R, gas reservoir; tr, cold trap; P_p, porous plug; v_i, isolation valve.

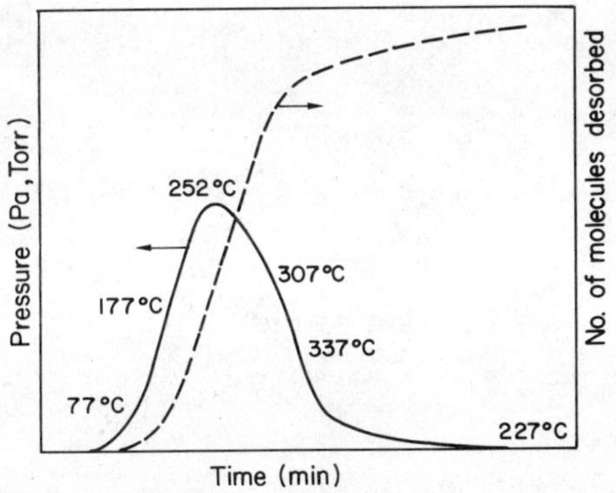

Fig. 7.9 Desorption transient of Ar from the pyrex walls of a BAG head due to thermal energy supplied (where temperatures are approximately indicated). The corresponding quantity of gas desorbed is plotted in the broken curve. After Leck and Carter (1960). Reprinted with permission from Proc. 1st Int. Vac. Congress 1958, J. H. Leck and G. Carter, p. 464. Copyright 1960 Pergamon Press.

Table 7.1

SUMMARY OF THE RESULTS OBTAINED IN THE MEASURING OF THE PUMPING SPEED OF IONIZATION GAUGES[a]

Type of gauge head	Pressure Pa (Torr)	Pumping speeds (L/s)	Anode–cathode voltage (V)	Emission current (A)	Gas	Method	Remarks	Investigator
Conventional triode; spiral ion collector	5×10^{-3} (5×10^{-5})	4×10^{-3}	210	1.25×10^{-3}	Residual	Static	Measured on the upper side of (log P,t) curve	Bloomer and Haine (1953)
BA	1×10^{-5} (1×10^{-7})	8×10^{-2}					Measured on the lower side of (log P,t) curve	
BA	1×10^{-2} (1×10^{-4})	2×10^{-2}		1×10^{-3}	N_2	Static		Alpert (1953)
BA	8.5×10^{-3} (8.5×10^{-5})	1×10^{-4}	180	1×10^{-2}	He		Clean glass wall	Varnerin and Carmichaei (1955)
		5.6×10^{-4}				Not specified	Metallic film on the glass wall	
BA	8×10^{-6} (8×10^{-8})	3×10^{-1}			N_2	Static		Bills and Allen (1955)
BA	2×10^{-3} (2×10^{-5}) to	1×10^{-1}			N_2		Metallic film on the wall	Young (1956)
	1×10^{-2} (1×10^{-4})	3×10^{-2}	145	1×10^{-2}		Static	Walls coated with Aquadag at the grid potential	
		4×10^{-3}					Metallic film on the wall	
BA	5×10^{-2} (5×10^{-4}) to	6.5×10^{-2}					Very thin film on the glass walls	Bills and Carleton (1958)
	1×10^{-2} (1×10^{-4})	5×10^{-1}	118	1×10^{-2}	N_2	Dynamic	Thin film	
		3.7×10^{-1}					Visible film	
	1.25×10^{-4} (1.25×10^{-6})	1.8×10^{-1}			O_2		Metal film	
Inverted magnetron	1×10^{-4} (1×10^{-6}) to	3×10^{-2}	6000		He	Static		Kornelsen (1960)
	1×10^{-8} (1×10^{-10})	2.5×10^{-1}			Ar			
Modified magnetron	1×10^{-3} (1×10^{-5}) to	2×10^{-1}			He			Rhodin and Rowner (1960)
	1×10^{-5} (1×10^{-7})	1.4×10^{-1} 1.5×10^{-1}	5000		N_2 O_2	Static	Baked at 400°C	

BA	1×10^{-2} (1×10^{-4}) to	6.2×10^{-3}			He			Cobic *et al.* (1961)
	1×10^{-3} (1×10^{-5})	7×10^{-3}			Ne			
		1.7×10^{-2}	250	1×10^{-2}	Ar	Static	Clean glass walls	
		5.2×10^{-2}			Kr			
		6.05×10^{-2}			Xe			
BA	5×10^{-1} (5×10^{-3}) to	1.42×10^{-1}					Measured at the beginning of the second pump-down	Cobic *et al.* (1961)
	5×10^{-2} (5×10^{-4})	1.25×10^{-1}	250	1×10^{-2}	N_2	Static	Measured at the beginning of the third pump-down	
BA	1×10^{-4} (1×10^{-6})	1×10^{-2}		1×10^{-3}	Ar	Dynamic		Leck and Carter (1960)
Hot-cathode magnetron	1.2×10^{-5} (1.2×10^{-7})	3×10^{-3}	300	1×10^{-7}	Assumed CO	Static	H = 250 Oe	Lafferty (1961)
BA	1×10^{-5} (1×10^{-7})	2.5×10^{-1}	300	8×10^{-3}	N_2	Dynamic	After saturation of chemical pumping	Hobson (1961)
Redhead's magnetron	6.5×10^{-4} (6.5×10^{-6}) to	1.7×10^{-1}			He			Barnes *et al.* (1962)
	1.75×10^{-4} (1.75×10^{-6})	2.5×10^{-1}			N_2			
		2.0×10^{-1}			H_2			
		3.4×10^{-1}	5000		O_2	Static	B = 0.1 T (1 kG)	
		1.7×10^{-1}			Ar			
		2.4×10^{-1}			CO_2			
BA	1×10^{-7} (1×10^{-9}) to 1×10^{-10} (1×10^{-12})	1.18×10^{-2}	105	8	He	Static	Clean walls at 77 K	Hobson and Edmonds (1963)
BA	1×10^{-5} (1×10^{-7})	1.7×10^{-3}			He			Ishikawa (1955)
		7.2×10^{-3}	105	8×10^{-3}	Ar	Static	Coated bulbs at zero voltage with respect to the collector	
		1.2×10^{-2}			Kr			
BA	1×10^{-3} (1×10^{-5}) to	9.6×10^{-2}			H_2			Byvic and Bradford (1966)
	1×10^{-5} (1×10^{-7})	6.9×10^{-2}		4×10^{-3}	O_2	Static		
		5.4×10^{-2}				N_2		
		5.7×10^{-2}			He			

[a] After Berman (1982).

(b) the difficulty in separating the contribution due to the pressure, the gas desorbed, and the evolution of the background gas.

7.5.5 The Measurement of Spontaneous Gas Desorption

The gas desorption from the glass walls in the period immediately after switching off the accelerating potentials can be determined by using either the static (seal-off) or the dynamic volume method.

Static volume (seal-off) method. After completing a pumping-down cycle of the gas being studied, the accelerating potential is switched off and the subsequent pressure rise measured with the monitor gauge head set to hold its pumping speed to a minimum. Spontaneous desorption can be initiated by switching off the accelerating potential of the gauge had under study after reaching either a selected total gas pressure (Baker and Giorgi, 1960) or any other pressure (Cobic *et al.*, 1961b). In the first case it was observed that the total amount of gas desorbed decreased with successive cycles (Fig. 7.10). Baker and Giorgi (1960) found that the maximum quantity of gas spontaneously desorbed after the initial pump-down was about 10% of the total quantity of gas pumped. They related the maximum quasi-static pressure P_{max} (Fig. 7.10) to the current I_c and the time t of one pumping–desorption cycle and obtained

$$P_{max}^2 \propto (I_c)_{max}^2 \propto a(1/t) - b \tag{7.21}$$

Here a and b are positive constants the value of which vary according to the atomic mass unit of the gas.

Baker and Yarwood (1963) who investigated the spontaneous desorp-

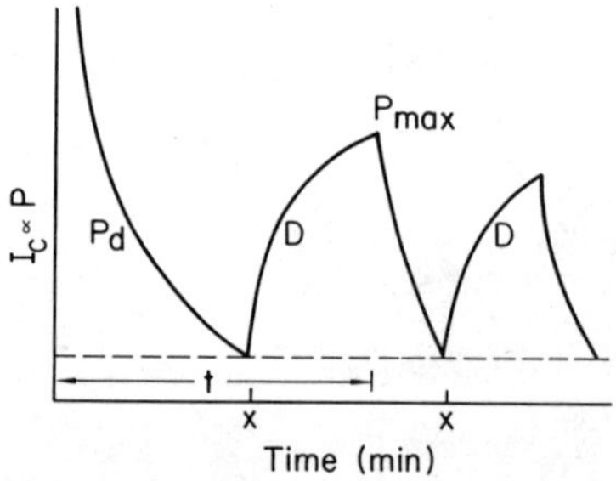

Fig. 7.10 Succesive cycles of pumping (P_d) and spontaneous desorption (D) of a mixture of gases in a hot-cathode ionization gauge head. X indicates that the accelerating voltage is off. Reprinted with permission from Brit. J. Appl. Phys. **11,** 435, F. A. Baker and T. A. Giorgi. Copyright 1960 The Institute of Physics.

tion of Ar, He, and Kr in a modified BAG head found that the desorption rates obtained for He were extremely small. These rates could be confused with the leak rates of He through the all-metal tap of their experimental system.

Dynamic volume method. In this technique the gas studied is pumped down by the gauge being tested, the accelerating potential is switched off, the gas flow into the system is stopped, and the gauge head is evacuated as quickly as possible (usually times of 1 min are reported). The gauge head is then isolated from the pumps and the spontaneous desorption of the gas monitored. On studying the desorption of inert gases from a diode pump of the Comsa–Musa type (Comsa and Musa, 1957), Smeaton *et al.* (1962) concluded that spontaneous desorption critically depends on the ion bombardment rate. During experiments the emission current was maintained constant, and in order to change the bombardment rate, the pressure was varied from one experiment to another. To maintain the temperature of the glass wall constant throughout the pump-down and spontaneous desorption cycles, the filament was held at constant temperature. Background effects due to the pumping action of both the mass spectrometer (monitoring the gas desorption) and the cooling trap between the diode and the pumping system were minimized or suppressed. This was made possible by operating the cathode of the mass spectrometer at low current (10^{-6} A) and separating the cold trap from that part of the system containing the experimental diode. Plotting the pressure rise as a function of time (Fig. 7.11) permits calculating the quantity of gas desorbed. Since the volume of the closed system containing the diode remained unchanged for different determinations, the values read on the ordinate represent the quantity of gas desorbed.

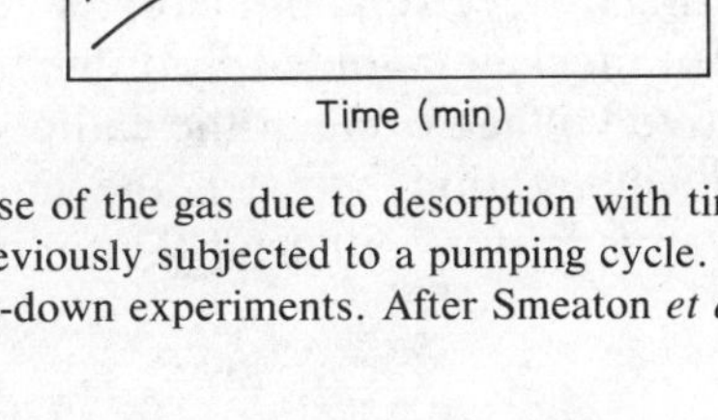

Fig. 7.11 The pressure rise of the gas due to desorption with time upon heating the glass walls of an instrument previously subjected to a pumping cycle. The different curves were obtained in various pump-down experiments. After Smeaton *et al.* (1963).

7.5.6 Experimental Procedures

Criteria for selecting the components of the measuring system include the following.

(a) According to the Liverpool school, the principal components of any measuring system operated by either the static or dynamic volume method are

(i) a gauge head to monitor the pressure in the experimental volume to which the gauge head or instrument under test is connected,
(ii) a mass spectrometer to continuously scan the atomic mass range of the gas in the experimental volume, and
(iii) several cold traps maintained at temperatures which prevent the physical adsorption of the gas tested.

(b) Both the gauge heads (tested and monitoring) and the mass spectrometer must be connected with large-bore tubing (about 40 L/s for N_2 flow). This is necessary to ensure equal pressure within the instruments and thus avoid errors in pressure measurement (Redhead, 1961).
(c) The valves (isolation, leak) used in the high vacuum section of the measuring system must be bakeable. They can be either glass (greaseless) or metallic type (Bills and Allen, 1955; Baker and Yarwood, 1957; Yarwood, 1957).

Control of some possible factors contributing errors to the measurement is maintained as follows.

(a) The pumping process within an ionization gauge head depends on the previous history of the instrument (see Section 4.4.6). Unless rigorous techniques to outgas both the glass walls and the electrode system are followed, the quantity of gas desorbed may exceed that of the gas pumped. Therefore, baking of the envelope and heating of the electrode structure by rf electron bombardment are necessary (see Section 4.4.7). During the outgassing of the electrodes by electron bombardment, the conducting layer on the inner glass walls (if any) must be connected to the cathode. Resistive heating of the grid is usually not recommended, due to the fact that the use of high energies required to avoid EID (see Section 4.4.6) effects will lead to sagging of the grid. The use of a high ac voltage supply for the outgassing by electron bombardment of gauge heads having thoria-coated cathodes must be avoided in order to protect the coating layer during the positive voltage cycle at the cathode (see Section 4.4.6).
(b) The electronic interaction between the monitor and the tested gauge heads can be prevented by either of the following approaches:

(i) inserting a metallic screen, operated at −500 V (Fig. 7.6c) between the two gauge heads; the potentials are measured with reference to the ground potential (Barnes *et al.*, 1962; Hobson and Redhead, 1958);

(ii) connecting the two gauge heads at an angle of 120° with respect to their axes of symmetry and locating the cathodes opposite the opening in the envelope (Fig. 7.6b) Ishikawa, 1965).

(c) When the gauge head (tested or monitor) is in the vicinity of a magnetic field, the influence of this field on the instrument readings can be minimized by locating the instruments at a large distance from the magnet or protecting them with magnetic shields.

(d) The mass spectrometer must be operated at a scanning speed providing adequate resolution and allowing the monitoring of changes in the composition of the gas under study.

(e) The bistable behavior of hot-cathode ionization gauge heads (see Section 4.4.6) can be prevented by covering the inner glass walls with a thin conducting layer (Ishikawa, 1939). A detailed description of the techniques for obtaining metallic stanous chloride layers is given by Espe (1968).

(f) Both the chemical and ionic pumping speeds may be reduced by the following procedures.

(i) The chemical pumping, mainly of N_2, O_2, and H_2, is diminished by dissociative gettering by activating the cathode in the gas tested without switching on the accelerating potential (Langmuir, 1913; Young, 1956; Lafferty, 1961). Saturating the cathode by flushing gas inside the instruments before starting the measurements is also recommended (Ishikawa, 1965). The operation of the instrument at low emission currents has to be avoided, mainly when hydrogen is involved in the measurement, due to the yield of spurious currents bearing no relationship to the pressure (Singleton, 1967).

(ii) The ionic pumping can be reduced by operating the gauge head for sufficiently short intervals of time (Alpert, 1953; Peterman and Baker, 1965; Edelman, 1967) (the drawback to this technique is that pressure cannot be monitored continuously); reducing the value of the emission current (Alpert, 1953, 1958; Peterman and Baker, 1965; Edelman, 1967) [this technique leads to reduced outgassing of the glass walls during measurements and to the increased possibility of the occurrence of Barkhausen–Kurtz oscillations (see Section 4.4.6); under certain circumstances, nonlinearities can appear in the sensitivity of the gauge head (Ackley *et al.*, 1962)], or decreasing the potential of the electrodes [the sensitivities of the gauge head reported using this technique are, however,

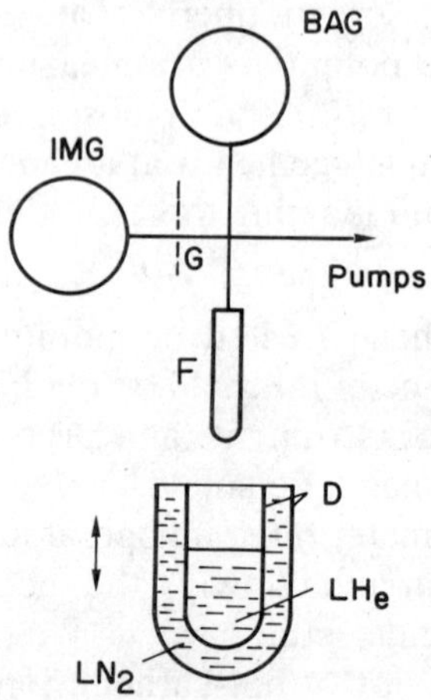

Fig. 7.12 Experimental apparatus combining the removal of gas by both electrical discharge and physical adsorption of the gas. After Hobson and Redhead (1958). IMG, inverted magnetron gauge head; BAG, Bayard–Alpert gauge head; G, metallic screen; F, glass finger; D, Dewar flask.

inconsistent (Ishikawa, 1965; Edelman, 1967)]. Redhead and Hobson (1965) carried out experiments with a magnetron and a BAG head where the grid and cathode potentials were reduced from 145 to 45 V and 45 to 6 V, respectively. The resulting ratio of the pumping speed to the sensitivity reduction was 17/2. When the emission current was diminished so as to obtain the same pumping speed, the sensitivity was reduced by a factor of 17.

(g) Gas leaks due to the diffusion of He, H_2, and ^2H from external sources through the glass envelope of the instruments used in the measur-

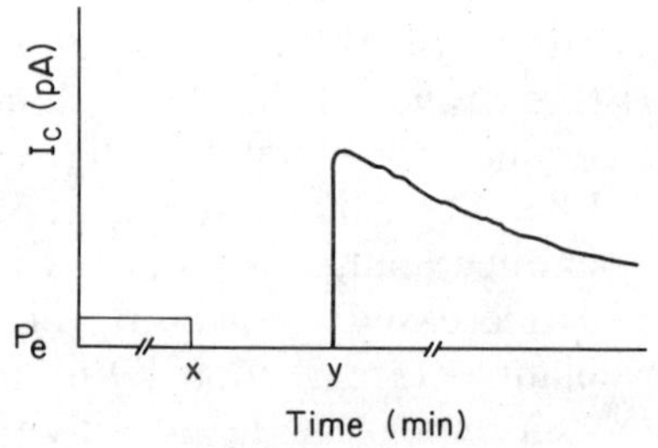

Fig. 7.13 The pressure transient due to a leak in the BAG head in Fig. 7.12. x and y are the times at which the accelerating voltage is switched on and off, respectively. I_c, ion collector current.

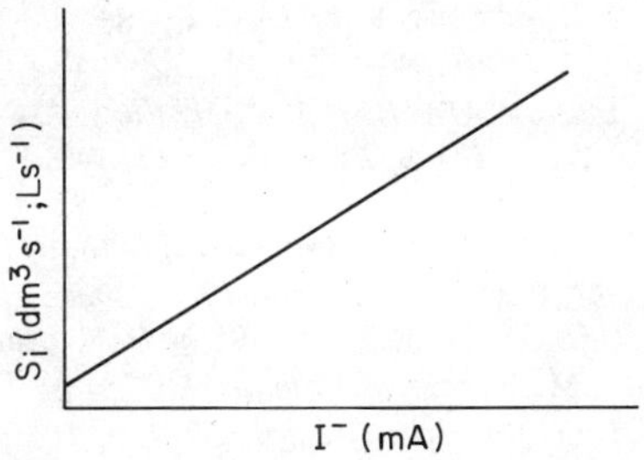

Fig. 7.14 The initial pumping speed S_i as a function of the emission current I^-. After Cobic *et al.* (1961b).

ing system are a function of the gas composition and increase with silica content. ^2H and H_2 permeate through glass at a rate about one tenth that of He (Baker and Yarwood, 1957a). The diffusion of He through glass is measured either by using the mass spectrometer in the system or by the method used by Alpert and Buritz (1954). Hobson and Redhead (1958) determined the He leaked in the monitoring gauge head of their experiment at room temperature (Fig. 7.12) in the following way. Before adding refrigerants to the Dewar flasks, the equilibrium pressure P_e (Fig. 7.13) was recorded. Then the accelerating voltage of the gauge head was switched off for 15 min and restored again (X and Y in Fig. 7.13). The subsequent increase in the collector current I_c represents the pressure rise caused by He leaks into the gauge head.

Cobic *et al.* (1961a,b) determined the leak rate through the bakeable isolation valve v_i of their measuring system (Fig. 7.8b) by plotting the initial pumping speed S_i as a function of the emission current I^- (Fig. 7.14). The intercept of the curve on the pumping speed axis represents the combined action of the leaks through the valve v_i and the gas removed by the pumping action of the monitoring gauge head.

References

Ackley, J. W., Lothrop, C. F., and Wheeler, R. W. (1962). *Trans. 9th Natl. Vac. Symp. U.S.A. 1962,* p. 452. MacMillan, New York.

Alpert, D. (1951). *Rev. Sci. Instrum.* **22,** 536.

Alpert, D. (1953). *J. Appl. Phys.* **24,** 860.

Alpert, D. (1958). *In* "Handbuch der Physik XII Thermodynamic der Gase," chap. 7. Springer Verlag, Berlin.

Alpert, D., and Buritz, R. S. (1954). *J. Appl. Phys.* **25,** 202.

Anderson, H. U. (1963). *Rev. Sci. Instrum.* **34,** 703.

Baker, F. A., and Yarwood, J. (1957a). *Vak. Tech.* **6,** 147.

Baker, F. A., and Yarwood, J. (1957b). *Vak. Tech.* **6,** 186.
Baker, F. A., and Yarwood, J. (1963). *Vide* **19,** 103.
Baker, F. A., and Giorgi, T. A. (1960). *Brit. J. Appl. Phys.* **11,** 433.
Barnes, G., Gaines, J., and Kees, J. (1962). *Vacuum* **12,** 141.
Berman, A. (1982). *Vacuum* **32,** 497.
Bills, D. G., and Allen, F. G. (1955). *Rev. Sci. Instrum.* **26,** 654.
Bills, D. G., and Carleton, N. P. (1958). *J. Appl. Phys.* **29,** 692.
Bills, D. G., Allen, F. G., and Carleton, N. P. (1955). *Phys. Rev.* **99,** 1662.
Bloomer, R. N., and Haine, M. E. (1953). *Vacuum* **3,** 128.
Brown, E., and Leck, J. H. (1955). *Brit. J. Appl. Phys.* **6,** 161.
Cobic, B., Carter, G., and Leck, J. H. (1961a). *Brit. J. Appl. Phys.* **12,** 288.
Cobic, B., Carter, G., and Leck, J. H. (1961b). *Brit. J. Appl. Phys.* **12,** 384.
Comsa, G., and Musa, G. (1957). *J. Sci. Instrum.* **34,** 291.
Dushman, S., and Found, C. G. (1921). *Phys. Rev.* **15,** 7.
Edelman, Chr. (1967). *Trans. 3rd Int. Vac. Cong. Stuttgart 1965,* p. 239. Pergamon Press, New York.
Espe, W. (1968). "Materials of High Vacuum Technology," vol 2, p. 323. Pergamon Press, New York.
Grosse, G., and Messer, G. (1981). *Vak. Tech.* **30,** 226.
Hayward, W. H., Jepsen, R. L., and Redhead, P. A. (1963). *Trans. 10th Natl. Vac. Symp. U.S.A. 1963,* p. 228. MacMillan, New York.
Hobson, J. P. (1959). *Can. J. Phys.* **37,** 1105.
Hobson, J. P. (1961). *Vacuum* **11,** 16.
Hobson, J. P., and Redhead, P. A. (1958). *Can. J. Phys.* **36,** 271.
Ishikawa, K. (1965). *Jpn. J. Appl. Phys.* **4,** 461.
Labrune, J. C., and Théobald, J. G. (1981). *Rev. Sci. Instrum.* **52,** 1774.
Labrune, J. C., and Théobald, J. G. (1983). *Vacuum* **33,** 183.
Lafferty, J. M. (1961). *J. Appl. Phys.* **32,** 424.
Langmuir, I. (1913). *J. Am. Chem. Soc.* **35,** 931.
Leck, J. H., and Carter, G. (1960). *Proc. 1st Int. Cong. Vac. Tech. Belgium 1958,* p. 436. Pergamon Press, New York.
Little, R. P., and Whitney, W. T. (1962). *Trans. 9th Natl. Vac. Symp. U.S.A. 1962,* p. 456. MacMillan, New York.
Peterman, L. A., and Baker, F. A. (1965). *Brit. J. Appl. Phys.* **16,** 487.
Redhead, P. A. (1960). *Rev. Sci. Instrum.* **31,** 343.
Redhead, P. A. (1961). *Trans. 7th Natl. Vac. Symp. U.S.A. 1960,* p. 108. Pergamon Press, New York.
Redhead, P. A., and Hobson, J. P. (1965). *Brit. J. Appl. Phys.* **16,** 1555.
Redhead, P. A., Hobson, J. P., and Kornelsen, E. V. (1962). *In* "Advances in Electronics and Electron Physics" (L. Marton and C. Marton, eds.), p. 323. Academic Press, New York.
Redhead, P. A., Hobson, J. P., and Kornelsen, E. V. (1968). "The Physical Basis of Ultrahigh Vacuum," p. 325. Chapman and Hall, London.
Singleton, J. H. (1967). *J. Vac. Sci. Technol.* **4,** 103.
Smeaton, P. G., Carter, G., and Leck, J. H. (1962). *Trans. 9th Natl. Vac. Symp. U.S.A. 1962,* p. 491. MacMillan, New York.
Smeaton, G. P., Leck, J. H., and Carter, G. (1963). *Nuovo Cimento Suppl.* **1,** 548.
Varnerin, L. J., and Carmichael, J. H. (1957). *J. Appl. Phys.* **28,** 913.
Völter, J., and Berndt, H. (1967). *Nuovo Cimento Suppl.* **5,** 234.
Young, J. R. (1956). *J. Appl. Phys.* **27,** 926.
Yarwood, J. (1957). *J. Sci. Instrum.* **34,** 297.

CHAPTER 8

Pressure Measurement in Confined Environments

8.1 Physical Principles

Low total pressure is the quantity most quoted to specify the degree of rarefaction of the gas environment in the vessel of a vacuum system. The direct determination of pressure, valid only in a gas with a Maxwell–Boltzmann distribution of velocities, is based on the measurement of force. Under certain conditions of molecular flow with a stationary distribution of velocities, pressure is a scalar and is determined by the indirect method. The indirect method measures any of the quantities dependent on gas viscosity, thermal conductivity, or density. Pressure is then determined by using the conventional kinetic theory relationships.

When the pumping process and nonuniform temperature distribution disturb the equilibrium, pressure becomes a vector. The rate of transfer of the normal component of the momentum associated with the passage of molecules in both directions through a small area depends on the orientation of that area. Under such circumstances the molecular flux, its density, and distribution characterize the rarefied environment more meaningfully than pressure. Attempts made to describe analytically the real distribution of gas under nonisotropic conditions are restricted to simple geometries (Moore, 1965). Practically, the rarefied environment can be quantified upon measuring the molecular flow and its characteristics with pressure converters and beam detectors. Both of these instruments can be used for new applications. Thus, the ionization gauge head in the pressure converter can be replaced by a mass spectrometer head, permitting the analysis of the molecular flow (Joo, 1977). The beam detector can also be used in the study of the desorption rate of materials and as a means of establishing the environment in domes used for pumping speed measurement (Holland and Priestland, 1967a).

The techniques used for the measurement or the assessment of pressure in sealed-off products are useful in monitoring the industrial manufacture of some vacuum products.

8.2 Continuously Pumped Vessels

8.2.1 Isotropic Conditions

To measure the pressure of the gas in a vessel, the gauge head is set fairly far from the pump aperture, in a region that is not disturbed by the in-and-out flow of gas from possible local sources of thermal energy. In many industrial systems the temperature of all components including the gauge head usually is uniform, to a good approximation, throughout and equal to the ambient level. Hence temperature corrections are not needed for the gauge head readings. When during the measurements the gauge head assumes a temperature substantially different from that of the rest of the vessel, or from that at which it was calibrated, the gauge readings have to be corrected for temperature. If the gauge head is activated by the injection of some kind of energy, a sufficient time must elapse before taking pressure readings in order to permit the gas to reach thermal equilibrium with the gauge head walls.

According to the degree of rarefaction, pressure is measured by resorting to one of the methods discussed. The main criteria for gauge selection are the range of pressure to be measured, the accuracy desired, the reproducibility of the gauge readings, the gauge head bakeability, and the possibility of remote control of the gauge head. As a general rule, the interference of the gauge head with the environment being measured must be held at a reasonable level. Active gases, such as oxygen or hydrogen, present in the gauge head lead to erroneous results. Thus, Blauth and Scherzer (1963) reported results several orders of magnitude higher than expected in measurements with a thermal gauge head in the presence of hydrogen. For experimental work where the purity of the gas being measured has to be maintained at a very high level, special measurement conditions must be planned. Thus, Gascoigne (1971), who performed scattering experiments where impurity levels in the target gas as low as 1 part in 10^8 or 10^9 are critical, utilized a quartz bourdon and a capacitance manometer to avoid any interference between the gauge head and the gas being measured. He measured pressures within the range 66.6 Pa (0.5 Torr) to 66.6 kPa (500 Torr). Elford and Hergeberger (1979) measured pressure in one of their beam experiments, where gas purity was also a critical parameter, with a capacitance manometer used as a null detector. They balanced the unknown pressure of gas acting on one side of the elastic membrane of the manometer with an accurately known pressure acting on the other side. The balancing pressure was generated in a static expander (Berman, 1979). They succeeded in maintaining the gas purity within the required limits and estimated the absolute error in their determinations to be 0.4% at 0.5 Pa (3.75×10^{-3} Torr).

8.2.2 Nonisotropic Conditions

The problems associated with evaluation of the nonisotropic conditions of gas in vacuum vessels have been the subject of many reviews and investigations, involving rather elaborate calculus (Moore *et al.* 1966; Moore 1962; Steckelmacher, 1966, 1974; Haefer, 1967a,b, 1971a,b; Haefer and Kleber, 1968; Lewin, 1968; Pisani, 1968a,b; Von Paul, 1973). These investigations promoted the development of directional response instruments: the *pressure converter* ("transducer") and the *beam detector*. Adequate techniques for using these instruments to measure the molecular flow as well as its intensity and direction have been elaborated as well.

(a) *Pressure converters.* The operating principle of pressure converters is based on the conversion of the incidence density of molecules on a surface to an isotropic pressure at known temperatures. This pressure is a measure of the molecular impact rate on the surface, regardless of the velocity and the direction of the molecules. The instrument built on this principle consists of a spherical or cylindrical hollow chamber, with an aperture permitting the gas to enter the chamber, and an ionization gauge head. The ionization gauge head can be attached by a tube to the chamber (Fig. 8.1a) or located in a recess of the chamber (Fig. 8.1b–d). The chamber is designed such that an entering molecule reaches thermal equilibrium with the walls of the instrument before reaching the ionization region. All molecules entering the aperture of the instrument directly are randomized by collision with the wall. They thus contribute to the gauge reading before they find their way out, regardless of their direction of entry (Fig. 8.1a). In the case of tubulated construction, where a straight or bent tube is inserted in the aperture of the instrument, measurements in

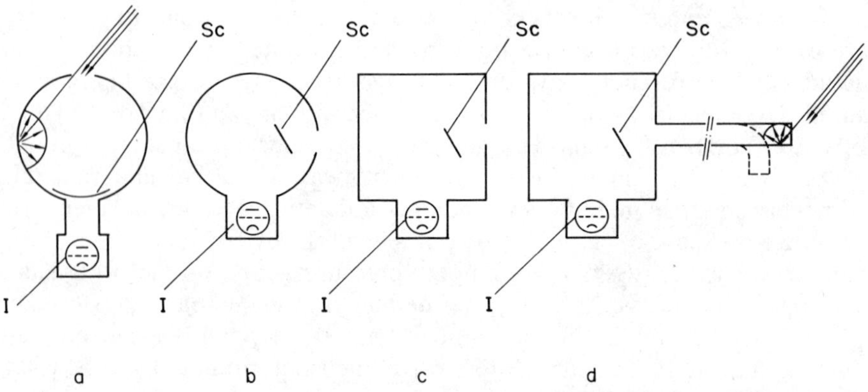

Fig. 8.1 Various types of pressure converters. (a) After Buhl and Trendelenburg (1965); (b–d) after Kleber (1973). Sc, screen; I, hot-cathode ionization gauge.

the presence of a directed beam incident on the tube opening will depend on the angle of incidence relative to the axis of the tube (Steckelmacher and Turner, 1966) (Fig. 8.1d). A screen (S) prevents gas from beaming into the gauge head and shields the thermal radiation emitted from the filament when using the instrument in the vicinity of a cold sink. Buhl (1966) recommended the use of a supplementary screen between the instrument and the sink when the distance is particularly short.

Tubulated instruments are used to measure molecules reflected from cryogenic panels. The tube prevents "shadowing effects" produced by the chamber of the instrument for molecules coming from an emitting source, when the instrument is located between that source and the cryopanel (Kleber, 1975).

The gas in the chamber of a converter can be assumed to have a Maxwell–Boltzmann distribution of velocities under the following conditions:

(i) the walls of the instrument are at constant temperature,
(ii) the inlet aperture is drilled in a "thin wall,"
(iii) the entering molecules collide at least once with the walls of the chamber,
(iv) the ratio of the surface areas of the aperture to that of the chamber is small (usually 1/100 or less),
(v) the amount of gas in the chamber does not change as a result of the effect of sinks or sources inside the instrument.

Stationary equilibrium is reached when the number of molecules entering the instrument from any direction and at any speed equals the number of molecules leaving the instrument through the aperture. Provided that the composition of the gas is known, the ionization gauge head will measure the density of the isotropic gas at the temperature of the instrument, in terms of pressure or ion currents. For a given orientation of the instrument, the pressure readings of the gauge head indicate the pressure P^* which would exist relative to the direction to which the aperture (or the tube) points, if the gas were in a chaotic state.

Hot-cathode ionization gauge heads usually are utilized due to their low pumping speed. Cold-cathode ionization gauge heads are avoided because of their high pumping speed and the considerable time lag in breaking down the discharge at low pressures.

According to the theoretical investigations of Kleber (1973, 1975), one or a combination of the following factors control the response of the pressure converter:

(i) the change in the total amount of gas in the instrument as a result of the pumping speed of the gauge head (S_g),

(ii) the pressure differential which develops between the aperture and the gauge head,

(iii) the gas desorbed from the inner walls of the instrument,

(iv) the ratio of the temperature of the instrument T_c to the temperature at which the ionization gauge head was calibrated T_g,

(v) the gas species which surround the instrument,

(vi) the lowest pressure read by the ionization gauge in the absence of gas in the facility (P_0^*).

The pressure read by the instrument is given by

$$P^* = P_0^*[(1 + S_g/C)(T_c/T_g)^{1/2} + \alpha(2\pi MkT_g)^{1/2}]I_Q/[(1 + S_g/C)(T_c/T_g)^{1/2}] \qquad (8.1)$$

where C is the conductance of the path between the aperture and the gauge head, I_Q is the number of molecules in the gas source, and M and k are the molecular weight of gas and Boltzmann constant, respectively. Kleber (1975) used converters with build-in ionization gauge heads of both the Bayard–Alpert and conventional triode types to verify Eq. (8.1) experimentally. The ratio of the response from "warm" to "cold" of the conventional triode gauge head was found to differ from Eq. (8.1) as a result of the ion-collector temperature.

Buhl and Trendelenburg (1965) investigated the possibility of using a dome with the same diameter as the mouth of the pump to simulate measurements with a large dome, for the determination of the pumping speed of high-speed pumps. A spherical pressure converter, equipped with a Bayard–Alpert gauge head, was used to measure the pressure

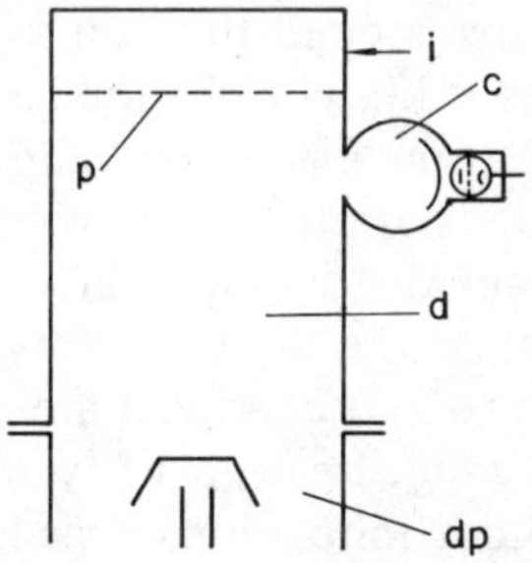

Fig. 8.2 Setup for the measurement of the incidence density of molecules in terms of pressure. c, pressure converter; d, test dome; dp, diffusion pump; i, gas inlet; p, perforated plate. Reprinted with permission from Vacuum **15,** 235, R. Buhl and E. A. Trendelenburg. Copyright 1965 Pergamon Press.

corresponding to the incidence density of molecules of the nonisotropic environment at the wall of the dome (Fig. 8.2). Both the converter and the dome were at room temperature. The nonisotropic gas was produced as a result of nonuniformity of gas distribution due to the high speed diffusion pump.

Kleber (1973, 1975) and Haefer and Kleber (1976) measured the molecular flux distribution under nonisothermal conditions of the 2.5-m diameter chamber of the DFVLR (Deutsche Versuchanstalt fur Luft und Raumfahrt Porz Wahn) space simulator (Fig. 8.3). A homogeneously emitting source E of special construction (Haefer and Kleber, 1970) maintained at constant temperature (300 K) was located on the longitudinal axis of the simulator chamber, the inner walls of which were cooled to 80 K. Two cylindrical pressure converters, one equipped with a bent tube and the other with a nude BAG head, could be moved on a rail along the cold wall. The apertures of the converters pointed to the longitudinal axis of the chamber (Fig. 8.3). The converters, previously calibrated under isotropic conditions, were maintained at constant temperature by electric heaters. The temperature was controlled and monitored by means of thermocouples. The molecular flux distribution measured is in agreement with the theoretical predictions of Haefer (1971b).

(b) *Beam detectors.* In the molecular flow regime, gas molecules in a parallel beam that strike a surface normal to the beam are reemitted with chaotic distribution. For the same temperature of the beam and of the surface, the equilibrium equation of the process, expressed in terms of impingement rate, is

$$n_{\mathrm{mb}} = n_{\mathrm{mc}}/4 \tag{8.2}$$

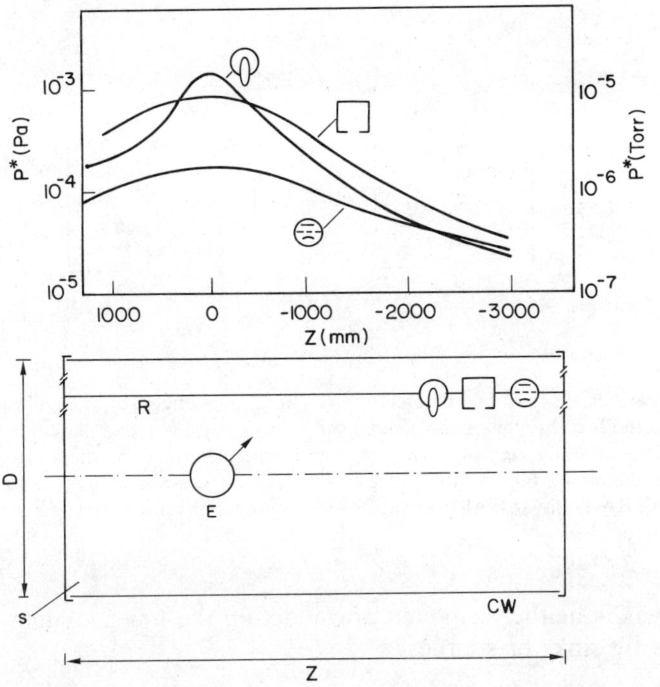

Fig. 8.3 Variation of the pressure P^* read by the converter as a function of the distance from the source E [after Kleber (1973) and Haefer and Kleber (1976)]. CW, cold wall; S, simulator chamber; R, rail; D, diameter of the simulator chamber; Z, length of the simulator chamber.

where n_m is the number density of molecules. Subscripts b and c refer to the directional and chaotic flow of molecules, respectively.

The beam detector is an instrument which is based on the principle expressed by Eq. (8.2). It consists of a sensor located in a tubular envelope and may be of one of the design types shown in Fig. 8.4. The sensor is a hot ionization gauge head of the Bayard–Alpert type which measures an ion current proportional to the gas density in its ionization volume. The gauge head is located in the tubular envelope with the ion collector electrode along the geometrical axis of the tube. The gas density in the detector at equilibrium conditions will be four times larger than in the beam provided that

(i) gas is uniformly distributed inside the volume of the tubular envelope,

(ii) the gas and the sensor are at the same temperature,

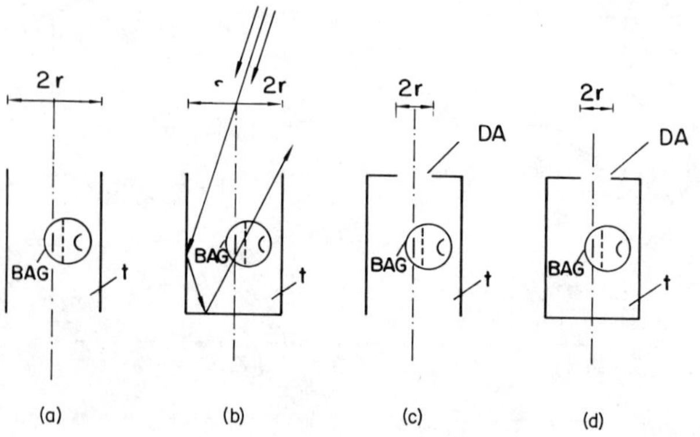

Fig. 8.4 Various types of beam detectors. (a) Both ends open; (b) one end open, the end adjacent to the feedthrough of the gauge head is closed; (c) one end diaphragmed, the end adjacent to the feedthrough of the gauge head is open; (d) one end diaphragmed, the end adjacent to the feedthrough of the gauge head is closed. DA, diaphragm with aperture in a "thin" wall; BAG, Bayard–Alpert gauge head; t, tubular envelope; r, radius of the aperture of the detector.

(iii) gas is neither removed nor added in the inner volume of the instrument by sinks or sources.

Since the calibration of the gauge head in the detector takes place in a chaotic environment, the readings of the instrument in terms of pressure will be

$$P_c = 4P_b^* \tag{8.3}$$

Despite this amplification, the response of the detector in terms of ion current is small in measuring molecular flow of reduced density at different angles of incidence between the normal to the emitting source and the instrument plane (Fig. 8.5). To increase the value of the collected current, the gas density in the ionization volume of the gauge head must be enhanced. Various techniques, or a combination of them, are used for this purpose.

(i) The tubular envelope is closed at the end adjacent to the feedthrough of the gauge head, while the opposite end is left open (Fig. 8.4b). This leads to a re-emission, with cosine law distribution, of the molecules scattered at the wall of the envelope. Most of the reversed molecules will cross the ionization volume of the gauge head once more before leaving the instrument. To a first approximation this approach will double the collected current. Holland and Priestland (1966a) experimentally verified the enhancement obtained by closing one end of the tubular envelope.

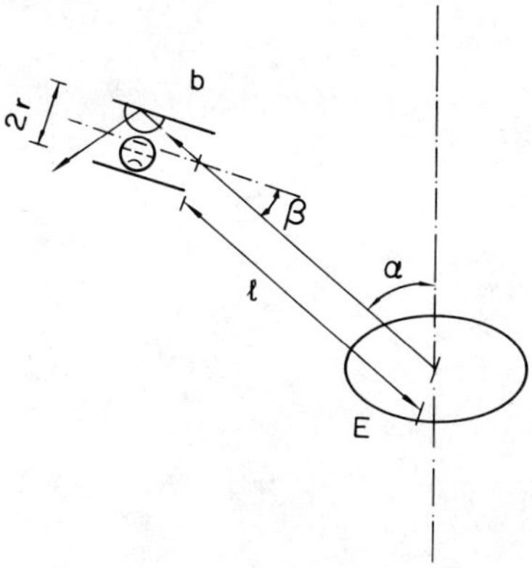

Fig. 8.5 Relative positions of beam detector b with respect to the molecular flow emitted from source E.

They measured the density of a beam emerging from an aperture in a "thin" plate into a rarefied environment with random distribution of the molecular velocities. Beam detectors of types A and B (Fig. 8.4) were rotated at a constant distance from the center of the source with their aperture always maintained normal to the beam direction (Fig. 8.6). The position of the beam detectors with respect to the emitting source is indicated in Fig. 8.6b below each response curve. The curves are given in terms of ion currents as a function of the angle of rotation. Both curves show the cosine dependence of the ion current on the rotation angle.

(ii) Gas is admitted into the instrument via a thin-walled diaphragm located opposite the feedthrough of the ionization gauge head. The center of the aperture and the axis of the ion collector electrode are on the same line (Fig. 8.4c). Molecules will stream unimpeded into the instrument, but to leave it they must find the aperture by a random process.

(iii) The tubular envelope is closed at one end, such as in (i) and diaphragmed at the opposite end as in (ii) (Fig. 8.4d).

In addition to the above, the enhancement factor depends on: (iv) the incidence of the gas beam and (v) the electrode geometry of the gauge head.

(iv) Holland and Priestland (1966b) experimentally verified the influence of the gas beam incidence angle on the enhancement factor. Detectors of the types a–d (Fig. 8.4) were rotated so as to vary the angle of incidence β of the aperture plane with respect to a beam of constant density as shown in Figs. 8.5 and 8.7a. The experimental conditions were

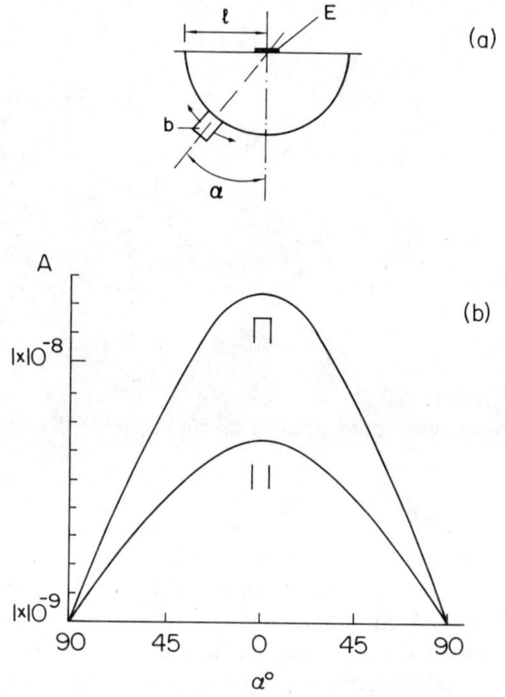

Fig. 8.6 (a) Angle of rotation of detector b with respect to the beam emitted from source E; (b) response curves showing the enhancement of the gas density in a beam detector with one closed end compared with that obtained with both ends open (after Holland and Priestland, 1966a). The orientation of the beam detector with respect to the emitting source is indicated.

like those described in (i). Figure 8.7b illustrates the response of the detector as a function of the variation of β and of the design type. The two maxima on the broken curve are ascribed to molecules scattered at the wall of the instrument for $\beta > 0$. The heights of the plots A and B in Figure 8.7b cannot be compared due to the slight beam divergence measured.

(v) The dependence of the enhancement factor on the electrode geometry of the BAG head was experimentally investigated by Priestland and Holland (1968). The ion collector of a BAG head was measured in a chaotic environment, as well as in a carefully collimated beam. Argon was used to avoid adsorption effects. For a chaotic gas environment, the value of the collected current was 1.55 times higher when the grid was capped than when it was open. When a molecular beam was directed so as to minimize the influence of the asymmetrical electron ionizing stream resulting from the offset position of the filament, the capped grid construc-

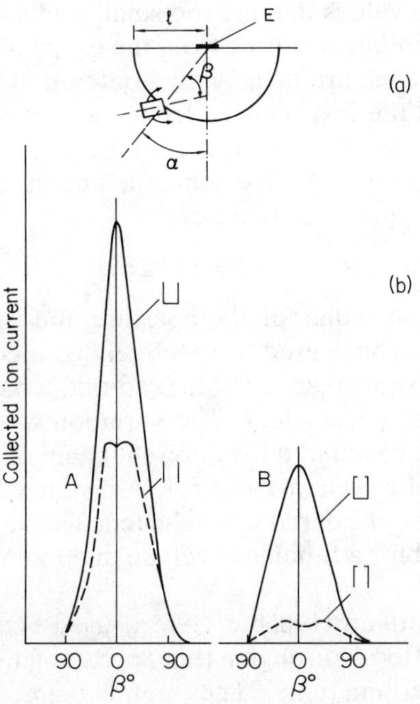

Fig. 8.7 Enhancement of the gas density in the beam detector as a function of the beam incidence, for different beam detector constructions (after Holland and Priestland, 1966c).

tion yielded a 2.87 enhancement factor. A pressure reading enhanced by a factor of 2.2 was obtained by closing the end of the gauge head. The increased gauge head sensitivity in the experiments discussed was explained as due to the efficiency of ion collection obtained when the grid is capped. A computer program for the simulation of ion collection in a BAG head showed, however, that the ion collection efficiency only partially explains the loss of sensitivity of an open gauge head without end caps (Turner and Priestland, 1968).

The enhancement factor of a detector type (Fig. 8.4d) is determined by comparing the collected current it yields with the collected current of a detector of the open type (Fig. 8.4a) for identical experimental conditions (beam of constant density evolved in a chaotic environment) (Holland and Priestland, 1966c).

In addition to the enhancement factor, the response time and the resolution of the beam detector must be known. The response time (time constant T_t) is calculated from the ratio of the volume of the instrument to the conductance of the aperture for the gas considered (see also Section

2.3.6). Conductance values that are too small lead to large response times and pressure differentials resulting from the gauge head pumping. Values in the millisecond range are usually encountered in high vacuum work.

The resolution of the instrument, defined as the percentage of the gas flux arriving at the aperture out of the total gas distributed over a hemisphere, depends on the solid angle subtended by the aperture. The resolution R in percentage points is therefore

$$R = 100\pi r^2/2\pi l^2 \tag{8.4}$$

where r and l are the radius of the aperture and the distance from the aperture to the point considered, respectively. A high resolution conflicts with the need for a short response time, so that a balance must be maintained between these parameters. The variation of the response time T_t and the resolution R calculated for different beam detectors used for high vacuum studies can be seen in Table 8.1. A distance l of 150 mm from the aperture to the point considered was selected. The volume V of the instrument was calculated for a tubular envelope of 60-mm length and 12.7-mm radius.

Holland and Priestland (1966b,c, 1967a) used beam detectors for the study of gas flow patterns through a thin aperture and short tubes, as well as for density determinations. They found agreement between their results and the theory.

To obtain information on both the characteristics of a BAG head, under molecular streaming conditions of the gas distribution in a vacuum gauge calibration vessel, a beam detector of type d (Fig. 8.4) was used (Holland and Priestland, 1967b). The instrument could be made into a

Table 8.1

CALCULATED VALUES OF THE RESPONSE TIME (TIME CONSTANT τ_t) AND RESOLUTION R FOR USUAL BEAM DETECTORS[a]

	Instrument type (according to Fig. 8.4)			
	A	B	C	D
r (mm)	12.7	7.1	3.2	1.6
C (argon)(L/s)	49.90	15.48	3.06	0.88
V (L)	0.03040	0.03040	0.03040	0.03040
$\tau_t = V/C$ (s)	6.0×10^{-4}	2.0×10^{-3}	1.0×10^{-2}	3.4×10^{-2}
l (mm)	150	150	150	150
$\pi r^2/l^2$ (st)	2.3×10^{-2}	7.0×10^{-3}	1.0×10^{-3}	4.0×10^{-4}
$R = 100\pi r^2/2\pi l^2$ (%)	0.4	0.1	0.02	0.06

[a] After Berman (1981).

pressure transducer of the type discussed by Buhl and Trendelenburg (1965) by removing the aperture plate and replacing it with a small spherical chamber and a baffle. The calibration system consisted of two hollow spheres, each with an aperture, tangentially welded together at the openings. Thin-walled diaphragms of different conductance could be inserted in the aperture between the spheres (Fig. 8.8). A rotatable arm permitted changing the beam detector angle (or the pressure transducer) with respect to the center of the sphere. "Pressure" variations were thus measured and the values found were used in explaining the uncertainties that could be met in vacuum calibration systems.

8.2.3 Magnetically Confined Plasma Conditions

Knowledge of the neutral gas density in the vicinity of a plasma is basic to the assessment of various diffusion processes in the confining environment (Jacquinot *et al.*, 1972) and the study of atomic processes in the plasma (Rěpa, 1970; Webb, 1974). Particular features of gas density determination close to the plasma are influenced by the presence of a magnetic field pulsed from zero to several tesla, the noise background produced by

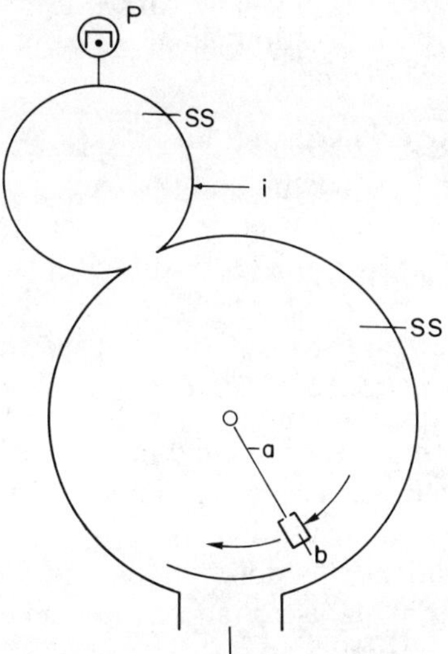

Fig. 8.8 Calibration system for the measurement of molecular impact rate (after Holland and Priestland, 1967b). b, beam detector; i, gas inlet; P, Pirani gauge head; SS, stainless steel spheres; a, magnetic drive rotatable arm.

both charged and uncharged particles and by radiation emitted from the plasma, and a quasistationary discharge.

Following bakeout, the pressure of the residual gas in some controlled fusion reactors is as low as 10^{-7}–10^{-8} Pa (10^{-9}–10^{-10} Torr) (Zubarev, 1970). During plasma injection and firing, the pressure rises suddenly and then remains constant throughout the plasma lifetime (Jacquinot *et al.*, 1971). After the plasma decays, the pressure may rise to 10^{-1} Pa (10^{-3} Torr).

Ionization gauges which measure gas density in terms of pressure are the most adequate for determinations in the range 10^{-1}–10^{-8} Pa (10^{-3}–10^{-10} Torr). Customary gauge heads of the Bayard–Alpert, cold magnetron, or inverted cold magnetron type have been utilized in fusion reactors by making certain provisions (Damm *et al.*, 1965; Butuzov *et al.*, 1966, 1968; Zubarev and Penchko, 1968). Several workers reported the use of specially designed ionization gauges, affording reliable readings under specific experimental conditions.

It is necessary to analyze the problems of neutral gas density determination in plasma machines and to critically discuss the design and operation of instruments reported in literature in the light of the above analysis.

The ionization gauge is the instrument most utilized in vacuum physics for the determination of neutral gas density. For pressure determinations in plasma machines, the ionization gauge head must meet the following requirements:

(a) sensitivity largely unaffected by the presence of a magnetic field,

(b) high resolution in discriminating between the current produced in the gas phase by electrons emitted from the cathode (true current) and spurious and parasitic currents contributed by the plasma or other sources,

(c) short gas kinetic time constant permitting tracking of the pressure variations during plasma buildup and decay.

In addition, the ionization gauge head must be installed so as to permit baking of the gauge head, large signal-to-noise ratio, and remote control in order to prevent radiation health hazards.

Among ionization gauges, the hot-cathode type without a magnetic field meets all these conditions. Hot-cathode ionization gauges with a magnetic field (hot magnetron, virtual cathode) or cross-field cold-cathode ionization gauges (Penning, magnetron conventional, or inverted) have several shortcomings which limit their use. The main shortcomings are

(a) currents not related to pressure, contributed by the plasma, which are difficult to assess,

(b) oscillations of constant amplitude of the ion current, the frequency of which increases with the emission current and pressure (hot magnetron),

(c) delayed break down, instability, and nonlinearity (Penning, magnetron) at low pressures.

Currents not related to the gas density probed usually are produced as a result of background noise contributed by the plasma, as well as the design and operation of the gauge head. Particles (ions, electrons, neutral atoms) and UV radiation emitted from the plasma which enter the gauge head yield currents that are referred to as "spurious" currents. Thus, incoming ions, together with those produced by UV radiation, add to the current. Incoming electrons can reach the modulation electrode, become modulated, and give rise to an erroneous signal. Fast neutral atoms which strike the collector may produce a secondary electron emission exceeding the magnitude of the true current (Zubarev, 1970; Jacquinot *et al.,* 1972) by several decades.

Means of hindering particles and radiation from entering the gauge head are of limited efficiency, as they conflict with conditions necessary for a reduced gas-kinetic time contant.

Unsuppressed spurious currents are discriminated from the true current by differentially monitoring the signals received at the collector electrode. The technique involves the use of elaborate electrode structures, which promote capacitive currents between electrodes. Thus, the addition of a supplementary modulator grid leads to a capacitive coupling between the cathode and the collector electrodes. The capacitive modulated-cathode current is negligible (Voznyuk *et al.,* 1965) whereas the capacitive modulator current may exceed the spurious current by at least one decade (Vinogradov *et al.,* 1973). These currents together with the x-ray currents, referred to as "parasitic" currents, can be minimized or measured.

(a) *Orientation and location of the gauge head.* The following discussion, based on the analysis of Martin (1962), emphasizes the effect of a magnetic field on the average path length of electrons in a hot ionization gauge head located in that field. For cold-cathode ionization gauge heads, the variation of the collected current as a function of the magnetic field will also be mentioned. For convenience, the vector of the magnetic field B is considered as reference. The notation $\mathbf{E}\|$ and $\mathbf{E}\perp$ is used to specify the respective positions of the electric and magnetic field vectors (Fig. 8.9). Three positions are considered: $\mathbf{E}\|$, $\mathbf{E}\perp$, and arbitrary orientation of E and B.

In orientation $\mathbf{E}\|$, electrons practically are emitted from a thermionic

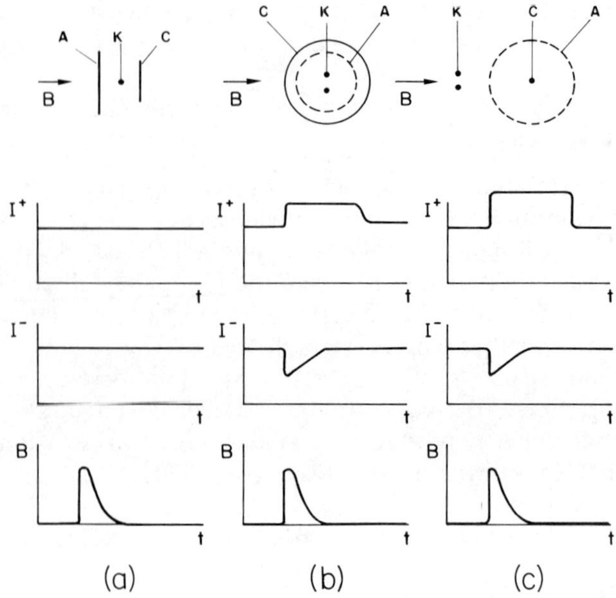

Fig. 8.9 Ideally predicted oscilloscope traces of hot-cathode ionization gauge heads (after Martin, 1962). (a) Schultz–Phelps gauge head; (b) conventional triode gauge head; (c) BAG head. A, anode; B, magnetic field; C, collector; K, cathode; I^+, collected current; I^-, emission current.

cathode at different angles in space. The distribution of the electric potential between the electrodes is not uniform because of the space-charge buildup. The distribution of electron velocities, therefore, is far from uniform. In addition, the lines of the electric field are not straight in the regions close to the edges of the electrodes. Under such circumstances, the motion of each electron can be considered as the resultant of two components: a drift along the cathode due to $\mathbf{E}\perp \times \mathbf{B}$ and a translation parallel to the lines of the electric field, due to $\mathbf{E}\|$.

Electrons starting with a high $\mathbf{E}\perp/\mathbf{E}\|$ ratio move close to the cathode along a curve approximating a cycloid. The kinetic energy acquired by electrons in the field before being forced to move along this curve is reduced. Notwithstanding the long paths the electrons travel in a spiraling, their ionization yield is reduced. A few ions, however, are produced close to the cathode but are lost in recombination processes. The electron drift adjacent to the cathode originates space charges which constrain the emission current at the sites where they build up. Attempts to increase the temperature of the cathode to enhance the emission current may produce favorable conditions for Barkhausen–Kurtz oscillations (see Section

4.4.6). It is known that such oscillations lead to instabilities in gauge head operation and to the generation of parasitic currents. In addition, high emission currents cause an increase in the gauge head pumping speed (see Section 7.4.1). This, in turn, results in an erroneous pressure indication and an apparent change in sensitivity.

Electrons starting with a high $\mathbf{E}\|/\mathbf{E}\perp$ ratio are accelerated along the electric field lines between cathode and anode. They thus gain enough kinetic energy to yield efficient ionization. The average path length of the electrons depends on, among other parameters, the electrode geometry of the gauge head. The present design of ionization gauge heads is of either planar or cylindrical symmetry. Gauge heads with planar symmetry (Schultz–Phelps), where the cathode is located within the collecting space, show practically no change in sensitivity with or without a pulsed magnetic field (Fig. 8.9a). For gauge heads with cylindrical symmetry (conventional triode, Bayard–Alpert), the magnetic field enhances sensitivity (Fig. 8.9b and c). This is due to the additional energy acquired by electrons, deviated from their translational motion when entering the anode space and forced to orbit along complicated paths because of the $\mathbf{E}\perp$ component. Low values of the magnetic flux density occurring when the pulsed field decays have no pronounced effect on sensitivity. In gauge heads with symmetrical electrode geometry, sensitivity is further affected by the mode of electron injection and the mechanical support of the anode grid. Experimental evidence was presented by Hammond and Riviere (1971) who measured the variation of sensitivity as a function of the azimuth of the magnetic field vector with the longitudinal axis of the cathode. Tests were performed on a conventional triode and a modified BAG head (Fig. 8.10). The modification, aimed at conferring symmetry to the injected electrons, consisted of winding the cathode outside the grid. For a conventional triode (Fig. 8.10 curve 1), the grid was supported by two wires, and for the BAG by a single wire (Fig. 8.10 curve 2). The plot of the normalized sensitivity s_B/s_0 (s_B and s_0 are the sensitivities measured with the magnetic field on and off, respectively) as a function of the azimuth shows that

(i) the normalized sensitivity of the conventional triode is a minimum in the regions where the magnetic field lines are interrupted by the support wires of the grid (Fig. 8.10 curve 1); in these regions sensitivity is as if the magnetic field were off;

(ii) the normalized sensitivity of the modified BAG is almost independent of the azimuth of the magnetic field vector (Fig. 8.10 curve 2).

Curves 1 and 2 cannot be compared since they are plotted for different emission currents and magnetic field densities.

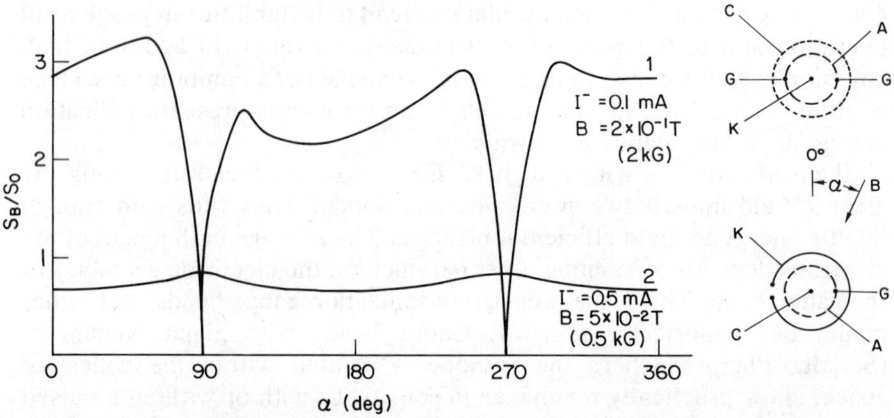

Fig. 8.10 Variation of the normalized sensitivity s_B/s_0 as a function of the magnetic field B for a conventional triode gauge head (curve 1) and a modified BAG head (curve 2) [Interpolated from the polar diagram of Hammond and Riviere (1971)]. A, anode; C, collector; G, grid support wires; K, cathode.

In orientation $\mathbf{E}\perp$, electrons starting with a high $\mathbf{E}\perp/\mathbf{E}\|$ ratio predominate and do not contribute efficient ionizing collisions. Sensitivity decreases with increasing magnetic flux density.

In arbitrary orientation of **E** and **B**, sensitivity is altered in proportion to the magnitude of $\mathbf{E}\perp/\mathbf{E}\|$ or $\mathbf{E}\|/\mathbf{E}\perp$. Hot magnetrons and crossed-field cold-cathode (Penning, cold magnetron) gauge heads practically show little change in sensitivity if oriented with $\mathbf{E}\|$ (Bernardet and Gourdon, 1963; Webb, 1974).

The location of the gauge head is selected so as to avoid thermal transpiration and the effect of dynamic gas flow on measurement.

(b) *Techniques for reducing spurious currents*. Incident particles and radiation are hindered from entering the gauge head by protecting the gauge inlet with one of the following: a metallic mesh (Lewin and Martin, 1962, 1963; Bannenberg and Insinger, 1963), a wire shield (Kotel'nikov, 1967), a plate (Voznyuk *et al.*, 1965), a collimator (Perkins and Barr, 1968) or Venetian blinds (Jacquinot *et al.*, 1972; Webb, 1974), a collimating array of plan parallel metallic plates biased with an electric potential (Kotel'nikov *et al.*, 1967; Gorbunov *et al.*, 1967), a metallic tube (Ellis, 1967). The conductance for gas flow of any of these means is selected such that the gas-kinetic time constant is not significantly affected. The metallic tube is designed so that any incoming particles bounce at least once before entering the ionization volume. Despite these approaches, spurious currents are still generated as a result of particles and radiation entering unobstructed. To discriminate between the spurious currents and the true current, the following techniques are available.

(i) *Off-filament techniques*. The gauge head, suitably connected to an oscilloscope, is operated according to the sequence: cathode cold, plasma discharge on; cathode hot, plasma discharge off; cathode hot, plasma discharge on. Since spurious currents can be detected when operating with the cathode cold, the subtraction of the oscilloscope traces obtained for each sequence permits the determination of the true current (Lewin and Martin, 1962).

Two gauge heads of identical design located in the same box are operated simultaneously. During determinations the cathode of one of the gauges is energized, the other is run cold. Both heads are provided with modulation grids (see below). The modulation grids of both heads are connected in parallel and biased with a suitable frequency. The currents from the cathode and collector electrode of both gauge heads are, respectively, subtracted and amplified in the differential amplifier of a double oscillograph (Voznyuk *et al.*, 1965). This technique, called *double pickup*, is claimed to permit determinations in regimes of plasma instability and nonequilibrium (Gorbunov *et al.*, 1967).

(ii) *Modulation techniques*. The emission current is modulated in the electric field created by a grid (modulator, control) located close to the cathode. This grid is biased with a square wave or sine pulse of adequate amplitude or frequency. During this period the main current to the collector is contributed by the pickup. The shortcoming of this technique is that parasitic x-ray currents are not discriminated from the true current.

The true ion current is modulated by changing the potential distribution in the collection region (Redhead, 1960, 1967; Lange and Singleton, 1966). Simultaneously, the emission current is modulated as shown above. A special design of the electronic circuitry and selected frequencies permits the "tuning-out" of spurious currents in the collector circuit (Kotel'nikov, 1967).

(c) *Main design features*. Spurious and parasitic currents are discriminated from the true current by using the techniques discussed above and compensating techniques (Kotel'nikov, 1967; Vinogradov *et al.*, 1973). A gauge head with a small gas-kinetic time constant is obtained by building rectangularly shaped electrodes and housing them in a flat box. The gas inlet is on the box face with the largest cross-sectional area. In practical determinations of the gas-kinetic time constant, fast acting valves are used to inject known amounts of gas (Hotz and Manaka, 1968; Inutake and Kuriki, 1972; Fisher *et al.*, 1978; Gentry and Giese, 1978; Keyser *et al.*, 1980). Means for degassing the electrode and the box, as well as

Table 8.2

PARAMETERS OF HOT-CATHODE IONIZATION GAUGE HEADS DESIGNED FOR USE IN PLASMA MACHINES[a]

Design shown in Fig. 8.11	Electrode symmetry	Cathode heating		Deflectors	Modulation characteristics		
		ac	dc		Frequency (kHz)	Shape	Amplitude (V)
Without magnetic field	Planar	5×10^4 Hz		Screen	—	∼ Grid 4	48 RMS
	Planar		7 V 22 A	Screen	≈150	⊓ Grid 4	200
	Planar	30 V		Screen	2.5 (grid-4)	⊓	120
					3.0 (grid-9)	∼	100
	Planar			Screen and plates	2 10	⊓ Grid 4[b]	
	Planar		11 V 6 A	Venetian blinds	—	⊓ Grid 4	200 (maximum)
	Cylindrical (conventional triode)			Screen	—	—	—
With magnetic field, hot magnetron	Cylindrical			Venetian blinds	—	—	—

[a] After Berman (1981).
[b] The grid close to the energized cathode.
[c] Magnetic strength.
[d] RT = room temperature.

reducing the gas load of the gauge head resulting from the heating during operation, must be provided (Ellis, 1967; Koteĺnikov, 1967; Jacquinot *et al.*, 1972; Webb, 1974). Adequate shielding and grounding of the gauge head and connecting cables against radio-frequency and magnetic fields is compulsory (Levin and Martin, 1962; Bannenberg and Insinger, 1963; Voznyuk *et al.*, 1965; Koteĺnikov *et al.*, 1967; Jacquinot *et al.*, 1972). The main types of gauge heads reported in use in controlled fusion reactors are schematically shown in Fig. 8.11, and their technical characteristics are given in Table 8.2.

(d) *Calibration and operation of ionization gauges.* The gauge head must supply reliable readings regardless of the effect of the magnetic flux. Calibration is accomplished by the static or dynamic method (see Chapter 5) with or without the magnetic field turned on (Koteĺnikov *et al.*, 1967; Jacquinot *et al.*, 1972; Webb, 1974). Due to the presence of the magnetic

Pressure Pa (Torr)	Magnetic flux density T	(kG)	Gas-kinetic time constant s/gas/temp Calculated	Measured	Machine or experiment where gauge was used	Reference
10^{-1} (10^{-3})	0.9	(9)	$130/D_2/RT^d$	—	B-65 Stellarator	
10^{-4} (10^{-6})	1.5	(15)	$130/H_2/RT$		C Stellarator	Levin and Martin (1963)
10^{-1} (10^{-3})	0.2	(2)	$\simeq 100$	—	2X machine	Ellis (1967)
10^{-5} (10^{-7})						
$\simeq 10^{-10}$ ($\simeq 10^{-12}$)	>600	Oc^c	—	—	Tokamak	Koteĺnikov (1967)
10^{-1} (10^{-3})	2.5–30	kOc^c	$50/H_2$	—	Todamak 2	Koteĺnikov *et al.* (1967)
10^{-5} (10^{-7})					Todamak 3	
10^{-3} (10^{-5})			$<200/H_2/RT$	250	Bet-MB mirror	Jacquinot *et al.* (1972)
10^{-5} (10^{-7})	0.3 (maximum)	(3)				
10 (10^{-2})			$160/H_2$	—	IXION rotating plasma device	Bannenberg and Insinger (1963)
10^{-1} (10^{-3})	0.8 (maximum)	(8)				
7×10^{-6}			$44/N_2/RT$	$300/H_2/RT$	Proto-Clco	Webb (1974)
(7×10^{-8}	0.3 (maximum)	(3)	$12/H_2/RT$			
3×10^{-2}						
3×10^{-4}						

field, vibrations may damage the cathode when energizing with a current alternating at 50 Hz. Hence, direct current or high-frequency alternating current are used. During transients, when the magnetic field is pulsed, the emission current must be maintained constant. If this is not possible, a double oscilloscope tracking of I_c and I^- can be used (Bannenberg and Insinger, 1963). The signal-to-noise ratio is increased by using a differential amplifier and a bandpass filter installed at the gauge head rather than close to the oscilloscope.

8.3 Sealed-Off Vessels

The maintenance and optimization of the characteristics of some mass produced products, such as those manufactured by the lighting or elec-

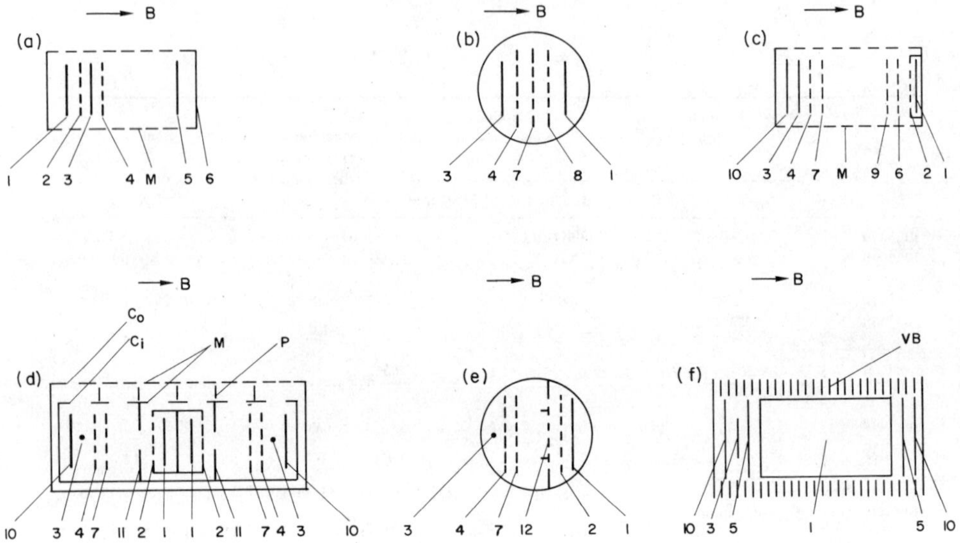

Fig. 8.11 Schematic diagram of six hot-cathode ionization gauge heads, (a)–(f), designed for use in plasma research (details are given in Table 8.2). 1, ion collector; 2, electron-repelling grid; 3, cathode; 4, modulator grid for the emission current; 5, anode; 6, auxiliary electrode; 7, accelerating grid; 8, electron-retarding grid; 9, modulator grid for the ion current; 10, electron reflector; 11, collimator plate; 12, ion-accelerating grid; C_1, inner box; C_o, outer box; M, mesh screen; P, protecting plates; VB, Venetian blinds.

tron tube industries, require knowledge of both the residual pressure and gas species in the sealed-off product. Statistical data concerning these quantities permit quality control of the technological process, thereby preventing a heavy incidence of rejects. The residual pressure determination can be made either destructively or nondestructively.

8.3.1 Destructive Tests

Water or gas at a known pressure slightly in excess of that to be determined is permitted to enter the product, usually an electric light bulb. The required unknown pressure is then calculated by resorting to one of the following approaches usual in the incandescent lamp industry.

The uncapped bulb of the lamp is immersed in water and the evacuation tube broken open below the water level. The weight of the water displaced in the bulb is then measured, and provided the atmospheric pressure and the inner volume of the lamp are known, the residual pressure is calculated by using Boyle's law. Usually the water temperature and the amount of residual gas dissolved in water are disregarded. Another uncer-

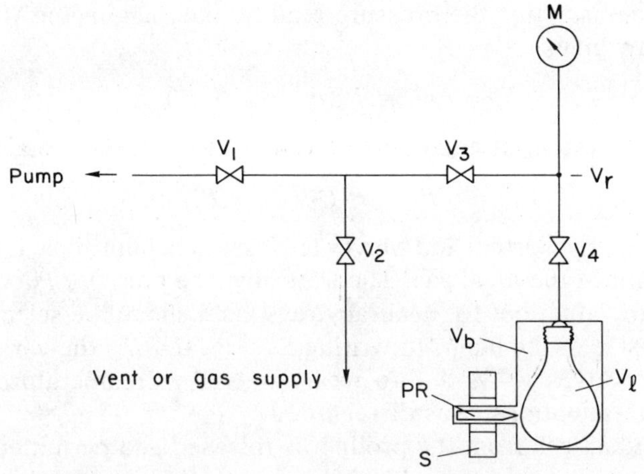

Fig. 8.12 Apparatus for pressure measurement in small lighting lamps (after Fortucci and Meyer, 1979). M, capacitance manometer; PR, puncturing rod; S, solenoid coil; v_r, reference volume; v_b, volume of the box and connecting tubing to valve v_4; v_1–v_4, valves; v_l, inner volume of the lamp.

tainty in the determination of the residual pressure is due to ascribing the same inner volume to all the samples in the batch.

According to the other technique the lamp is accomodated in a leak-tight box, connected to a vacuum system (Fig. 8.12) (Fortuci and Meyer, 1979). The volume v_b includes that of the box and the connecting tubing up to valve v_4. Different valves in the manifold permit removal or supply of the gas in the box. The sequence of operations is as follows:

(a) The reference volume v_r is filled with gas to pressure P_r measured with the capacitance manometer M. Then the gas is expanded into the box. If P_1 is the gas pressure after expansion in v_b, Boyle's law can be written

$$v_r P_r = P_1(v_r + v_b) \tag{8.4}$$

(b) The gas is removed from the apparatus, the glass bulb of the lamp is broken open, and the residual gas expanded in the volume $v_r + v_b + v_l$. If the pressure read by the capacitance manometer is P_2, then

$$v_l P_l = P_2(v_r + v_b + v_l) \tag{8.5}$$

where P_2 is the pressure of the residual gas in the lamp.

(c) The gas is evacuated from the box and the manifold and the apparatus refilled with gas at pressure P_r. Upon expanding the gas in the volume $v_r + v_b + v_l$, the pressure read by the manometer M will be P_3. Boyle's law gives

$$v_r P_r = P_3(v_r + v_b + v_l) \tag{8.6}$$

Solving the system of equations (8.4)–(8.6) for P_1, one obtains

$$P_1 = P_1 P_2/(P_1 - P_3) \tag{8.7}$$

The gas used by Fortuci and Meyer (1979) was helium since it approaches the behavior of the ideal gas. Theoretically, the pressure P_r is not limited to a certain value, but for accuracy reasons it should be selected according to the size of the lamp. In writing Eqs. (8.4)–(8.7) the variation of the volume of valves v_1–v_4 due to wear, as well as temperature variations during determinations, was disregarded.

The residual gas from the product is released and permitted to enter a setup equipped for pressure determination. The gas pressure in this setup, as well as in the ducts which connect it with the product, is much lower than that of the residual gas in the product. Sinks and sources which might change the residual gas both qualitatively and quantitatively are carefully avoided.

Usually the product is accomodated in an enclosure connected by a manifold to the vacuum system. After evacuating the enclosure, the residual gas is released by one of the following techniques.

The sample, fitted in a copper tube, is broken by squeezing the tube in a vice (Norton, 1955). The procedure, although seemingly simple, involves considerable preparation time.

Charlton and Perqin (1959) released the residual gas contained in an electronic valve by breaking the glass jacket of the valve in a magnetically operated hammer and anvil setup (Fig. 8.13). Breaking open the product is facilitated if the jacket is scored with a file and the mark aligned with the edge of the manner (broken line in Fig. 8.13). This apparatus can also be used to release, for analysis, gas bubbles trapped in glass beads melted around refractory metals.

An alternative technique is that utilized by Grady (1973) to open a metal–ceramic magnetron tube. The tube is placed in a leak-tight flanged system attached to the vacuum manifold. A sharp metal rod, operated through a bellows attached to the flange, is used to puncture the magnetron at a previously thinned point. Gas evolution from the magnetron can be recorded on a video tape (Grady, 1972).

Borghi and Rosai (1979) released the gas from the jacket of a high-pressure discharge lamp by breaking the glass with a magnetic hammer (Fig. 8.14). The puncturing rod is magnetically operated in a T-shaped glass tube sealed to the lamp jacket with epoxy. The rod is aimed at an

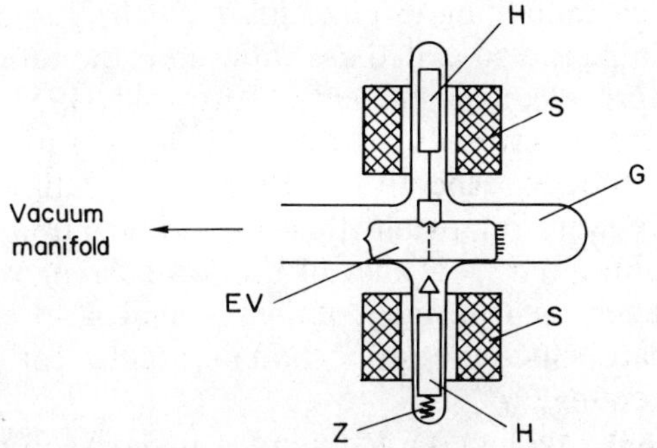

Fig. 8.13 Apparatus for breaking open sealed-off products under vacuum (after Charlton and Perkin, 1959). G, glass tube; H, magnetically operated hammer; S, solenoid; EV, electronic valve; Z, spring.

area where the glass was ultrasonically thinned to facilitate breaking. Since the lamp was on during opening, the epoxy seal had to be air cooled. The authors used this procedure to release the gas for chemical analysis, but the technique can also be used for determining gas pressure.

8.3.2 Nondestructive Tests

In practice, mainly in the manufacture of expensive vacuum valves such as x-ray tubes and glass rectifiers, it is often required that the pressure of the residual gas inside the product be measured nondestructively.

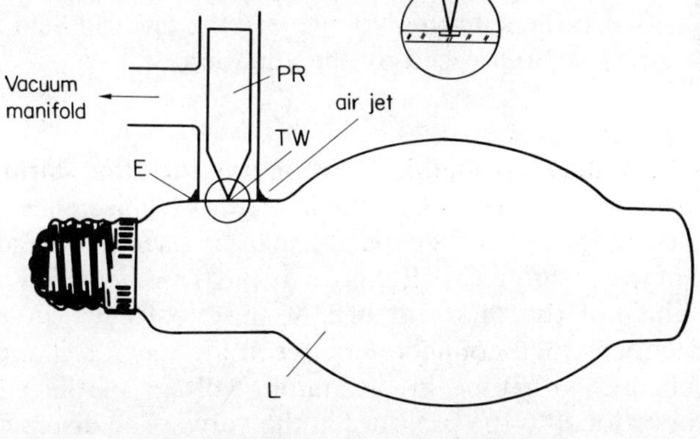

Fig. 8.14 Setup for releasing gas from a high-pressure discharge lamp (after Borghi and Rosai, 1979). TW, thinned wall; L, high-pressure discharge lamp; PR, puncturing rod; E, epoxy seal.

For vacuum valves containing three or more electrodes, pressure is measured by determining the so-called gas ratio, i.e., the ratio of the positive ion to the negative electron current. Ostrowski (1953) and Dimitriev (1959) measured the pressure by determining the gas ratio in picture tubes and cathode ray tubes, respectively. Since in a cathode ray tube the leakage current is more important than in an ionization gauge head, its value must be subtracted from that of the ion current when pressure is calculated. However, measurement of the residual gas pressure by the gas ratio is not accurate, since the proportionality factor for ion collection is not taken into account.

Kobayashi (1963) developed a technique, also based on gas ratio measurement, to determine both the pressure of the residual gas and the potential leaks in an image orticon tube. In such a tube, the fine mesh of the screen electrode hinders the electrons from the cathode from emerging in sufficient amount to produce an ion current unmistakable from the background current. By using the electron-current amplifier of the tube, Kobayashi was able to measure residual gas pressures of about 10^{-4} Pa (10^{-6} Torr).

The measurement of pressure on an approximation of its order of magnitude in a diode is accomplished by one of the following techniques.

(a) The sparking potential between the electrodes of the diode is measured, and pressure is then calculated from Pashen's law. According to this law the sparking potential between a pair of electrodes in gas depends on the gas pressure and the gap between the electrodes. André (1960) used both dc and ac voltage and measured both the potential between the electrode and the current flowing at the instant of breakdown. He experimentally determined a function $\psi(E/P)$ of the reduced field, using the relationship between the sparking voltage V_a, the electric field E, and the residual pressure P as expressed by the equation

$$V_a = \psi(E/P) \tag{8.8}$$

Pressure can be deduced from this equation provided that during determination the space charge does not impede electron emission.

(b) The discharge time of a capacitor can be used as an indication of the pressure of the residual gas (Eckertová, 1967). A metal foil is wrapped around the glass of the diode at one of the electrodes. A previously charged capacitor is then connected between the metal foil and the electrode. The change with time in the output voltage of the capacitor is inversely proportional to the pressure in the valve. The discharge time of the capacitor is a function of several parameters, such as the shape of the external foil electrode, the intensity of the current flowing through the diode during the test, or the voltage across the electrodes of the diode. The reproducibility of measurements is highly dependent upon the potential of the glass wall of the diode during determinations.

References

André, P. (1960). *Proc. 1st Int. Congr. Vac. Technol. Belgium 1958,* p. 290. Pergamon Press, New York.

Bannenberg, J. G., and Insinger, F. G. (1963). *Vak. Tech.* **12,** 210.

Berman, A. (1979). *Vacuum* **29,** 417.

Berman, A. (1981). *J. Vac. Sci. Technol.* **3,** 1017.

Bernardet, H., and Gourdon, C. (1963). *Vide* **18,** 472.

Blauth, E. W., and Scherzer, B. M. U. (1963). *Vak. Tech.* **12,** 110.

Borghi, M., and Rosai, L. (1979). *Vacuum* **29,** 67.

Buhl, R. (1966). *J. Vac. Sci. Technol.* **3,** 83.

Buhl, R., and Trendelenburg, E. A. (1965). *Vacuum* **15,** 231.

Butuzov, S. S., Knoyaev, V. P., Maslenikov, E. A., and Pustovoit, Yu. M. (1966). *Sov. Phys.-Tech. Phys.* **11,** 228.

Butuzov, S. S., Zubarev, V. F., Konyaev, V. P., Maslenikov, E. A., Nekrasov, A. N., Pustovoit, Yu. M., Svischev, V. S., and Chukhin, I. A. (1968). *Sov. Phys.-Tech. Phys.* **13,** 207.

Charlton, M. G., and Perkin, G. M. (1959). *J. Sci. Instrum.* **36,** 49.

Damm, C. C., Foote, J. H., Futch Jr., A. H., Gardner, A. L., Gordon, F. J., Hunt, A. L., and Post, R. F. (1965). *Phys. Fluids* **8,** 1472.

Dimitriev, V. I. (1959). *Instrum. Exp. Tech.* April, 698.

Eckertová, L. (1967). *Trans. 3rd Czech. Conf. Electron. Vac. Phys.,* p. 483. Academia Publishing House of the Czechoslovak Academy of Science, Prague.

Elford, M. T., and Hergeberger, R. (1979). *J. Vac. Sci. Technol.* **16,** 955.

Ellis, R. E. (1967). *UCID-15015.*

Fisher, A., Mako, F., and Shiloh, J. (1978). *Rev. Sci. Instrum.* **49,** 872.

Fortuci, P. L., and Meyer, V. D. (1979). *J. Vac. Sci. Technol.* **16,** 963.

Gentry, W. R., and Giese, C. F. (1978). *Rev. Sci. Instrum.* **49,** 595.

Gascoigne, J. (1971). *Vacuum* **21,** 21.

Gorbunov, E. P., Kotelnikov, Yu. N., Kutukov, G. P., and Simonov, V. A. (1967). *Sov. Phys.-Tech. Phys.* **11,** 1363.

Grady Jr., B. I. (1972). *J. Vac. Sci. Technol.* **9,** 496.

Grady Jr., B. I. (1973). *J. Vac. Sci. Technol.* **10,** 208.

Haefer, R. A. (1967a). *Vak. Tech.* **16,** 149, 185, 193.

Haefer, R. A. (1967b). *Trans. 3rd Int. Vac. Congr. Stuttgart 1965,* p. 581. Pergamon Press, New York.

Haefer, R. A. (1971a). *Vacuum* **21,** 269.

Haefer, R. A. (1971b). *Cryogenics* **11,** 210.

Haefer, R. A., and Kleber, P. (1968). *Proc. 4th Int. Vac. Congr. Manchester 1968,* p. 237. Inst. Phys. and Phys. Soc. Conf. Series No. 6. Adlard and Son, Surrey, England.

Haefer, R. A., and Kleber, P. (1970). *3rd Int. Cryogenic Eng. Conf.,* p. 381. Guilford, England.

Haefer, R. A., and Kleber, P. (1976). *Vide* **31,** 19.

Hammond, D. P., and Riviere, A. C. (1971). *J. Phys. E.* **4,** 650.
Holland, L., and Priestland, C. (1966a). *Nature* **209,** 274.
Holland, L., and Priestland, C. (1966b). *J. Appl. Phys.* **37,** 4594.
Holland, L., and Priestland, C. (1966c). *Vacuum* **16,** 601.
Holland, L., and Priestland, C. (1967a). *Trans. 3rd Int. Vac. Congr. Stuttgart 1965,* p. 141. Pergamon Press, New York.
Holland, L., and Priestland, C. (1967b). *Vacuum* **17,** 461.
Hotz, R. F., and Manaka, C. K. (1968). *Rev. Sci. Instrum.* **39,** 267.
Inutake, M., and Kuriky, K. (1972). *Rev. Sci. Instrum.* **43,** 1670.
Jacquinot, J., Leloup, C., Poffe, J. P., De Pretis, M., and Evrard, P. (1971). "Plasma Physics and Controlled Fusion Research," pp. 595–605. International Atomic Energy Agency, Vienna.
Jacquinot, J., Deck, C., and Godaert, J. (1972). *Rev. Sci. Instrum.* **43,** 905.
Joo, F. (1977). *Proc. 7th Int. Vac. Congr. Vienna 1977,* p. 337. Berger and Söhne, Horn, Austria.
Keyser, C. J., Dembinski, M., and John, P. K. (1980). *Rev. Sci. Instrum.* **51,** 425.
Kleber, P. (1973). *DLR-FB,* p. 73.
Kleber, P. (1975). *Vacuum* **25,** 191.
Kobayashi, H. (1963). *Trans. 10th Natl. Vac. Symp. U.S.A. 1963,* p. 287. MacMillan, New York.
Koteĺnikov, Yu. N. (1967). *Instrum. Exp. Tech.* **5,** 1180.
Koteĺnikov, Yu. N., Kutukov, G. P., and Simonov, V. A. (1967). *Instrum. Exp. Tech.* **5,** 1182.
Lange, W. J., and Singleton, J. H. (1966). *J. Vac. Sci. Technol.* **3,** 319.
Lewin, G. (1968). *J. Vac. Sci. Technol.* **5,** 75.
Lewin, G., and Martin, G. (1962). *Rev. Sci. Instrum.* **33,** 447.
Lewin, G., and Martin, G. (1963). *Rev. Sci. Instrum.* **34,** 942.
Martin Jr., G. D. (1962). *Trans. 8th Natl. Vac. Symp. U.S.A. 1961,* p. 476. Pergamon Press, New York.
Moore, B. C. (1964). *J. Vac. Sci. Technol.* **1,** 10.
Moore, B. C. (1965). *J. Vac. Sci. Technol.* **2,** 211.
Moore, B. C., Bergquist, L. E., Camarillo, R. G., and Larson, R. (1966). *Trans. 3rd Int. Vac. Congr. Stuttgart 1965,* p. 251. Pergamon Press, New York.
Moore Jr, R. W. (1962). *Trans. 8th Natl. Vac. Symp. U.S.A. 1961,* p. 426. Pergamon Press, New York.
Norton, F. Y. (1955). *Rev. Sci. Instrum.* **26,** 238.
Ostrowski, R. E. (1953). *Electronics* **26,** 178.
Paul, G. (1973). *Vak. Tech.* **22,** 210.
Perkins, W. A., and Barr, W. L. (1968). *Phys. Fluids* 11, 398.
Pisani, C. (1968a). *Vacuum* **18,** 327.
Pisani, C. (1968b). *Proc. 4th Int. Vac. Congr. Manchester 1968,* p. 439. Inst. Phys. and Phys. Soc. Conf. Series No. 6. Adlard and Son, Surrey, England.
Priestland, C., and Holland, L. (1968). *Vacuum* **18,** 259.
Redhead, P. A. (1960). *Rev. Sci. Instrum.* **31,** 343.
Redhead, P. A. (1967). *J. Vac. Sci. Technol.* **4,** 57.
Řepa, P. (1970). *Czech. J. Phys. A* **20,** 168.
Steckelmacher, W. (1966). *Vacuum* **16,** 561.
Steckelmacher, W. (1974). *Proc. 6th Int. Vac. Congr. Kyoto 1974,* p. 117. *Jpn. J. Appl. Phys., Suppl. 2, Part 1.*

Steckelmacher, W., and Turner, D. J. (1966). *J. Sci. Instrum.* **43,** 893.

Turner, D. J., and Priestland, C. (1968). *Vacuum* **18,** 319.

Vinogradov, V. I., Grinchenko, V. T., and Pichugin, V. V. (1973). *Instrum. Exp. Tech.* 16 July, 199.

Von Paul, G. (1973). *Vak. Tech.* **22,** 210.

Voznyuk, A. S., Doroshenko, A. N., and Cherepanov, V. N. (1965). *Instrum. Exp. Tech.* 5 Sept–Oct., 1215.

Webb, D. R. A. (1974). *J. Phys. E: Sci. Instrum.* **7,** 453.

Zubarev, V. F. (1970). *Instrum. Exp. Tech.* 1 Jan–Feb., 191.

Zubarev, V. F., and Penchko, E. A. (1968). *Instrum. Exp. Tech.* **5,** 1270.

Index

K

L

M

R

S